12th International Conference on Geotechnical Engineering in Tropical Regions: Fostering Future Through Geotechnical, Sustainable Engineering and Technology & 2nd International Conference on Highway and Transportation Engineering: Empowering Transportation Engineering for Industrial Solutions (GEOTROPIKA & ICHITRA 2021)

IOP Conference Series: Earth and Environmental Science
Volume 971

Semarang, Indonesia
21 – 22 September 2021

ISBN: 978-1-7138-4953-7
ISSN: 1755-1307

Printed from e-media with permission by:

Curran Associates, Inc.
57 Morehouse Lane
Red Hook, NY 12571

proceedings.com

Some format issues inherent in the e-media version may also appear in this print version.

Institute of Physics
Dirac House, Temple Back
Bristol BS1 6BE UK

Phone: 44 1 17 929 7481
Fax: 44 1 17 920 0979

techtracking@iop.org

Additional copies of this publication are available from:

Curran Associates, Inc.
57 Morehouse Lane
Red Hook, NY 12571 USA
Phone: 845-758-0400
Fax: 845-758-2633
Email: curran@proceedings.com
Web: www.proceedings.com

TABLE OF CONTENTS

IOP Conf. Series: Earth and Environmental Science **971** (2022) 011001 doi:10.1088/1755-1315/971/1/011001

PREFACE

International Forum on Integrated Infrastructure Development (InFInID 2021): 12th International Conference on Geotechnical Engineering in Tropical Regions (GEOTROPIKA) & 2nd International Conference on Highway and Transportation Engineering (ICHITRA)

21-22 September 2021, Semarang, Indonesia

The 12th International Conference on Geotechnical Engineering in Tropical Regions (GEOTROPIKA) and 2nd International Conference on Highway and Transportation Engineering (ICHITRA) 2021 was expected to take place in Semarang Indonesia. However, due to the COVID-19 global pandemic, the conference has been changed to a hybrid platform where it involved both physical presentation which was held in Semarang, Indonesia and also virtual presentation using ZOOM application platform. This conference has been organized by Department of Geotechnics and Transportation, School of Civil Engineering, Faculty of Engineering, Universiti Teknologi Malaysia (UTM) and Universitas Islam Sultan Agung (UNISSULA) in collaboration with Rotterdam University of Applied Sciences and Hanze University of Applied Sciences.

The conference had brought together about 100 scholars and engineers in the field of Geotechnics and Transportation from various countries such as Malaysia, Indonesia, Iraq, Nigeria, Saudi Arabia, Japan and Germany. The conference consisted of plenary and parallel sessions. It involved 4 keynote speakers who shared their latest research results and experiences in related research fields. The parallel sessions were conducted in 3 different Zoom conference rooms. Each room was led by a chairperson and co-chairperson. Each participant was given 10 minutes to present the paper followed by a discussion session led by the chairperson. The conference received over 70 abstracts and submission of 60 full papers. All the papers have gone through a rigorous review process to meet the requirements of international publication standards. This conference covered state-of-the-art Civil Engineering research in the field of:

1. Geotechnical Engineering
2. Geological Engineering
3. Tunnel, Subway and Underground Facilities
4. Seismic Engineering
5. Road and Railway Engineering
6. Transportation Engineering
7. Disaster Prevention and Mitigation
8. Cartography and Geographic Information System
9. Surveying Engineering
10. Computer Simulation and CAD/CAE

11. Industrial Revolution 4.0
12. Soft soil engineering

Despite some minor limitations, thanks to the support and advancement of virtual technologies, it was possible for the committees to run the conference on time and excellently.

Among the presented papers, 38 manuscripts have been accepted to be included in this IOP Conference Series: Earth and Environmental Science. Thank you to all participants, reviewers, editors, committees and all who contributed to this event.

EDITORIAL BOARD
GEOTROPIKA – ICHITRA (2021)

ORGANIZING COMMITTEE

IOP Conf. Series: Earth and Environmental Science **971** (2022) 011001 doi:10.1088/1755-1315/971/1/011001

Patrons
Prof. Dr. Khairul Anuar Kassim (UTM)
Ir. H. Rachmat Mudiyono, M.T., PhD (UNISSULA)

Advisors:
Prof. Ir. H. Pratikso
Mrs. Gata Dian Asfari
Prof. Dr. Mohd Rosli Hainin
Prof. Dr. Edy Tonnizam Mohamad
Prof. Ir. Dr. Ramli Nazir
Prof. Dr. Hasanan Md Nor
Prof. Dr. Othman Che Puan
Prof. Ir. Dr. Azman Kassim
Prof. Dr. Nazri Ali
Assoc. Prof Dr. Kamarudin Ahmad
Assoc. Prof. Sr. Dr. Mohd Zulkifli Mohd Yunus
Assoc. Prof. Dr. Mushairry Mustaffar

Chairmen:
Prof. Dr. Ir. S. Imam Wahyudi, DEA (ICCDA 4 2021)
Dr. Nor Zurairahetty Mohd Yunus (GEOTROPIKA 2021)
Dr. Norhidayah Abdul Hassan (ICHITRA 2021)

Co-Chairmen:
Assoc. Prof. Ir. Dr. Rini Asnida Abdullah
Ir. Dr Sitti Asmah Hassan

Secretaries:
Dr. Abdul Rochim
Dr. Dayang Zulaika Abang Hasbollah
Dr. Muhammad Naqiuddin Mohd Warid

Treasurers:
Mrs. Lisa Fitriyana
Mr. Muhammad Azril Hezmi
Dr. Azman Mohamed
Ms. Wulaningrum
Networking & Industrial Advisors:
Prof. Ir. Dr. Azman Kassim
Mr. Che Ros Ismail
Dr. Ir. Rinda Karlinasari, MT
Dr. Hasti Widya Samratri, Ssi, M.Eng

Event, Protocol, Logistic Coordinators:
Dr. Henny Pratiwi Adi (Event)
Dr. Hermin Poedjiastoeti (Event)
Dr. Achmad Sahri, S.Kel., M.Si (Event)
Dr. Khairun Nissa Md Said (Protocol)
Dr. Afikah Rahim (Protocol)
Ir. M. Faiqun Ni'am, MT., Ph.D (Protocol)
Mr. Abdul Ayiz (Protocol)
Ir. Hj. Eppy Yuliani (Logistic)
Mrs. Inda Mardi (Logistic)

IOP Conf. Series: Earth and Environmental Science **971** (2022) 011001 doi:10.1088/1755-1315/971/1/011001

Mr. Yahmin (Logistic)

Web & IT Advisors:
Ir. Suryani Alifah, MT., Ph.D
Dr. Mohd Zul Hanif Mahmud
Dr. Muhammad Irfan Shahrin
Mr. Mohd Nur Asmawisham Alel
Mr. Desri Ardian
Mr. Priyono

Publication and Proceedings:
Dr. Mohd. Khairul Idham Mohd Satar
Dr. Siti Norafida Jusoh
Assoc. Prof. Ir. Dr. Ahmad Safuan Rashid
Dr. Haryati Yaacob
Dr. Nordiana Mashros
Ms. Ardiana Yuli Puspitasari
Mr. Boby Rahman
Mr. M. Rusli Ahyar
Mr. Abied Rizky P.M
Prof. Dr. Ir. Antonius, MT
Ir. Gatot Rusbintardjo, Ph.D
Assoc. Prof. Ir. Dr. Ahmad Safuan A. Rashid
Dr. Nordiana Mashros

IOP Conf. Series: Earth and Environmental Science **971** (2022) 011002 doi:10.1088/1755-1315/971/1/011002

Peer review declaration

All conference organisers/editors are required to declare details about their peer review. Therefore, please provide the following information:

- **Type of peer review: Double Blind Peer Review**

- **Conference submission management system: Confbay System**

- **Number of submissions received: 60**

- **Number of submissions sent for review: 60**

- **Number of submissions accepted: 34**

- **Acceptance Rate (Number of Submissions Accepted / Number of Submissions Received X 100):57%**

- **Average number of reviews per paper: 2**

- **Total number of reviewers involved: 96**

- **Any additional info on review process:-**

  The reviewing process was handled through the confbay system. Reviewers received the manuscript and the manuscripts have been assessed based the following criteria:
  - relevance to the conference
  - originality of work
  - suitability of research methodology
  - novelty of findings
  - contribution to research advancement and policy

  Authors did the corrections and amendments accordingly. They need to submit the corrected version with turnitin less than 20%. Editors went through again the final corrected version of the manuscript in terms of format and language. If corrections are needed, it will be returned back to the author until the manuscript is fully complied to the standard of the publication.

- **Contact person for queries: Haryati Yaacob (haryatiyaacob@utm.my)**

Please submit this form along with the rest of your files on the submission date written in your publishing agreement.

The information you provide will be published as part of your proceedings.

IOP Conf. Series: Earth and Environmental Science **971** (2022) 012001 doi:10.1088/1755-1315/971/1/012001

An Assessment of Green Rural Road Criteria and Elements for My Green Rural Road Index: A Case Study in Pengerang, Johor

N A Mustafa[1], V Munikanan[1*], R Zakaria[2], E Aminudin[2], S Langie[1], J A Adzar[2], M H S Abd Rashid[2], S M Shamsudin[3], S Z Sooria[4] and M M Hassan[4]

[1]Civil Engineering Department, Faculty of Engineering, Universiti Pertahanan Nasional Malaysia, Kuala Lumpur, Malaysia
[2]School of Civil Engineering, Faculty of Engineering, Universiti Teknologi Malaysia, Johor Bahru, Malaysia
[3]Faculty of Architecture Planning and Surveying, Universiti Teknologi MARA, Shah Alam, Malaysia
[4]Public Work Department of Malaysia, Kuala Lumpur, Malaysia.
*Corresponding author: vikneswaran@upnm.edu.my

Abstract. The rural road networks were built to connect the rural people to the town area or to other destination required by the local residents. Normally the rural roads will be developed to connect or increase the socio-economic opportunity in rural area. The green road is one of the key areas that can be look into to create the sustainable concept based on three key aspects namely social, environmental and economic factors. The main focus of the economy aspect is to increase the profits using more efficient resources, especially materials, improving the quality of life by meeting the national needs of social aspects and protecting the environment from the effects of CO^2 emissions and efficient use of natural resources for environmental aspects. Therefore, it is important that stakeholders include sustainability criteria in their projects. The application of a sustainable concept on the road can be assessed by the green road evaluation tool. Therefore, the main aim of this study is to build tools to evaluate and declare green rural roads in Malaysia. The data for this study were gathered through group discussions with road construction experts and distributed questionnaires to identify sustainability factors. Then the data were analysed by factor analysis method using SPSS software. Based on the result, there are five main sustainable criteria: Sustainable Design and Construction, Social and Safety, Energy Efficiency, Environmental and Water Management and Materials and Technology that are included in the development of the green road qualification system. The analysis indicates that the social and safety become the prime criteria among other criteria.

1. Introduction

Rural road is defined as low traffic volume roads located in forested settings that serve residential and recreational purpose [1]. They may have been constructed to relatively low standards with a limited budget and with one or two lanes wide using natural, gravel and other surfacing. Nowadays, green rural road has become one of the favoured methods in road construction as it can improve quality of living and lessen the impact to environment. Green road is the new idea of road construction to provide a sustainable road in terms of social, economic and environmental factors. It is designed and constructed with highest level of sustainability compared to the current ordinary road. It is a collection of sustainability best practices that apply to roadway design and construction throughout the stages of planning, design, construction and service life [2]. This kind of project play an essential role in a nation's social and economic as it uses of low impact of development tools, recycle materials and local resources and mitigate the negative impacts on the surroundings to a minimum level through green concepts and sustainable practices. The criteria of green roads are Sustainable Design and Construction, Material and Resources, Environmental and Water Management, Energy Efficiency and Social and Safety [4]. Green rural road needs more consideration, alertness and application of all the parties like the route concessions, the drivers, and the most vital is that the authority which can enhance the property of the green conception of road style which will be sustained at the life cycle basis [3].

Currently in Malaysia, the guidelines for green rural road are not yet available. Thus, this study is looking at the development of My Green Rural Road Index with the purpose of addressing the sustainability elements specifically for green rural road in Malaysia. A proper guidelines and index will help to reduce the impacts of greenhouse effect, air pollution, noise pollution and water pollution. At the same time, it will increase the road safety and efficiency of road networks thus plays an important role in stimulating socioeconomic development for rural communities. Through better road access, rural development can be achieved which can stimulate rural communities' economy, increase well-being and self-realization and also creating changes among rural community from the normal means of their living to an exceedingly progressive way of living [5]. Despite the benefits of rural access, rural road however associated with several environmental impacts, in California for example, rural road has been identified as a major source of sediment production in watersheds. Thus, in order to ensure the rural road development is achieved and at the same time lessen the impact of environmental through transportation facilities, green rural road should be taken into consideration.

1.1. Green Rural Road Criteria

Green Rural Road criteria are very important for the green rural road index development. The factor analysis is based on the green criteria and elements. The SPSS Software was used to analysed the best criteria for rural road index development. Every criterion has different perspective and characteristics. The main criteria will decide the primary construction work for rural road development. The criteria are listed based on Adzar et al. [6], Abd Rashid et al. [7] and Adzar et al. [8].

The first criteria which is sustainable design and construction activities can be one of the criteria that can minimize the impacts of roads to the environment especially for plants and wildlife. The impact can be reduced by better design, construction, management of roads and other facilities [9]. The concept of sustainable design and construction already present in building construction industry but in road construction industry has just started to adopt initiatives that promote environmentally friendly. Sustainability initiatives in this arena also attempt to consider economic concerns that arrive from utilizing new materials, and construction practices [10] [11].

As for the second criteria the energy efficiency means using less energy to perform anything by eliminating energy waste. Energy efficiency brings many benefits such as reducing greenhouse gas emissions, reducing demand for energy imports and lowering our cost on household. Renewable energy technologies help to improve energy efficiency in every sector of economy in construction, transportation and buildings [12].

The third criteria is environmental and water management. One of the numerous aspects of the management are the safety of the surroundings. The green road is more on the environmentally friendly. Minimizing of cutting slope and also the preservation of vegetative downhill could be a wastage of the precious mass and it will cause a harm to the vegetative cover. Instead, having a correct cut and mill technique of mass levelling that's employed in the green road approach, the excavated materials are then used for fill on the depression facet by constructing a toe wall [13]. Water management is incredibly a lot of necessary within the rural mountain roads. Green road approach provides a good water management to improve the water quality. Besides, the slope is containing more water which can cause landslides or erosion. A better water drainage system must be provided to stop the disaster.

IOP Conf. Series: Earth and Environmental Science **971** (2022) 012001 doi:10.1088/1755-1315/971/1/012001

Based on forth criteria, the construction of the green road is the best way for surroundings because the idea can reduce the environmental impact. Green road concept is the best idea to protect the mountain slope and minimize the danger of landslide and erosion

Besides, the contractors should use best techniques in construction to minimize the carbon emission. The environment approach is like using the dozers, excavators and also the different significant equipment which may consume less quantity of the fuels for construction works and this will reduce the emission of the greenhouse gases. The development strategies are environmentally friendly because it approaches the uses of the labour-intensive methods that leads to virtually zero greenhouse emission throughout the construction of the road.

2. Background

In this study, Pengerang, Johor has been chosen as research area. Figure 1 shows the study area to conduct the questionnaire survey. The study area covered overall Pengerang such as Bandar Penawar, Kota Tinggi, Desaru and JKR Pengerang. Pengerang is well known for Malaysia's mega project named Pengerang Integrated Petroleum Complex (PIPC) which is sited in the area. The project was announced in 2011 and Pengerang was chosen for Malaysia's national project due to its strategic location. It offers access to exist major international shipping lanes. The Pengerang Integrated Petroleum Complex (PIPC) is one big step in creating value to the downstream oil and gas value chain in Johor. In less than 2 years, the area become more accessible via new Senai-Desaru Highway which has shortened the travel time to the area by an hour compared to over two hours previously via small road from Kota Tinggi. Development of rural road network in Pengerang, Johor plays an important role in stimulating development. A major portion of Pengerang, Johor population lives in rural areas, sixty percent of the population board rural Johor. In Pengerang there are several villages situated in several piece of ground condition e.g., plains, hilly and, swamp and lots of additional. The green road index based on several criteria and elements has been assessed.

My green rural road index is a research that aimed to produce valuable benefits to the environment by integrates transportation functionality and ecological requirements. It can be defined as a roadway design which is based on a relatively new concept of roadway design in which environmental approach is used throughout the planning, design, and construction stages. Green road will benefit not only to the transportation infrastructure, but towards ecosystem, urban growth, public health, and surrounding communities as well [14]. The main objective of this study is to identify the criteria and elements for the development of My Green Rural Road Index and to categorized the identified element into specified criteria. These were achieved through the method explained below.

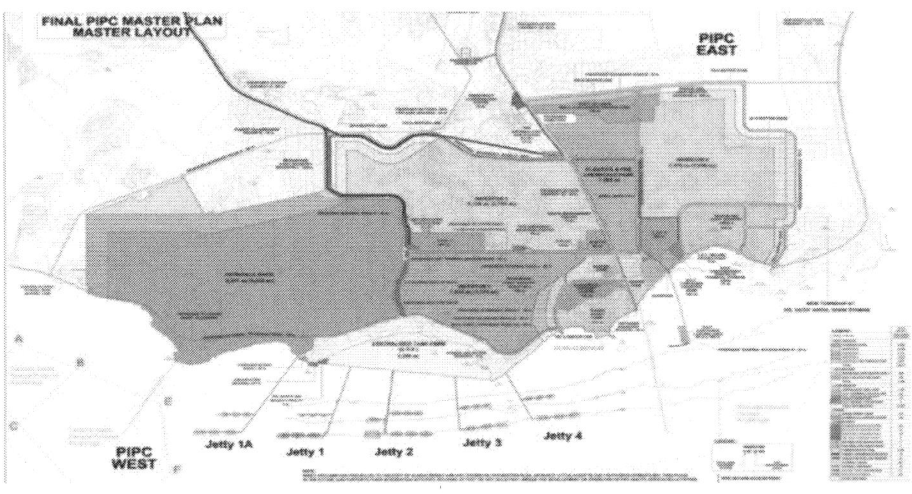

Figure 1. The Pengerang Map (Source: Johor Petroleum Development Corporation).

IOP Conf. Series: Earth and Environmental Science **971** (2022) 012001 doi:10.1088/1755-1315/971/1/012001

3. Methodology

The criterias and elements were identified and analysed through extensive literature review. The questionnaire was designed into 2 categories. The first category based on the rural road criteria selection and the second category based on rural road elements. Both the category were then analysed based on questionnaire survey.

The data collected from the questionnaire survey were analyzed by using the Statistical Package for the Social Science (SPSS) version 22.0. Then, the descriptive statistics for each question was conducted to give a better understanding of the overall results. Finally, factor analysis was performed to identify the criteria of green rural road index of the road users and the residents of Pengerang, Johor. Kaiser-Meyer-Olkin (KMO) test was used to interpretate the statistical result. The extraction communalities are estimates of the variance in each variable accounted for by the components. The communalities in this table are all high, which indicates that the extracted components represent the variables well. In rotated Factor Matrix – The table contains the rotated factor loadings which represent both how the variables are weighted for each factor but also the correlation between the variables/elements and the factor.

4. Data Analysis

The results obtained need to be fully interpreted to answer the objectives of the study. The descriptive statistics for each question was conducted to give a better understanding of the overall results. In the end, factor analysis was also conducted to identify the criteria of green rural road index of the road users and the residents of Pengerang, Johor. The data were collected among the residents of one city/municipal council as stated below. There were 75 responses received during the questionnaire survey of the studied area as stated from total 100 questionnaire distributed. From the Table 1, KMO Test for Green Rural Road Factor was achieved at 0.861 which is adequate because KMO test range is between 0.8 to 1.0. The data considered positive to analyse the green factors and criteria.

Table 1. KMO test for Green Rural Road Factor.

Kaiser-Meyer-Olkin Measure of Sampling Adequacy.		0.861
Bartlett's Test of Sphericity	Approx. Chi-Square	213.461
	df	10
	Sig.	.000

From the analyse of communalities of Green Rural Road Factor in Table 2 below, the highest criteria shown is the Social and Safety with value of 0.729. Based on the analysis, it indicates that the social and safety with environmental and water management are the two important criteria that are crucial for rural road development. Through the literature review many researchers emphasized the rural infrastructure development gives much important to these criteria. Table 2 shows the communalities of Green Rural Road criteria that analysed through SPSS Software.

Table 2. The communalities of Green Rural Road Factor.

Criteria	Extraction
Social and safety	0.729
Environmental and water management	0.702
Material and Technology	0.676
Sustainable Design and construction activities	0.639
Energy efficiency	0.433

IOP Conf. Series: Earth and Environmental Science **971** (2022) 012001 doi:10.1088/1755-1315/971/1/012001

The correlation between criteria and the element were tabulated accordingly in Table 3. The highest variance of the criteria that fall on each factor considered the correlation between the element and the criteria is higher. For an instance, in Table 3, for elements noise mitigation control is more correlated on environmental and water management criteria with highest variance of 0.657 as compared to another variance. Highest value for each element representing the highest criteria were correlated with one another.

Table 3. Rotated Factor matrix of Green Criteria and Element.

Element	Factor Loadings				
	Social and Safety	Environmental And Water management	Material and Technology	Sustainable Design and Construction Activities	Energy Efficiency
Noise mitigation control	0.358	0.657	0.154	0.501	0.154
Good water flow	0.620	0.142	0.304	0.080	0.075
Shady area with tall trees	0.151	0.443	0.148	0.209	0.274
Building orientation	0.077	0.226	0.317	0.722	0.243
Road safety audit	0.510	0.205	0.378	0.313	-.0.309
Erosion control	0.512	0.294	0.357	0.555	-0.014
Dust Control	0.297	0.633	0.335	0.197	0.073
Water Quality Control	0.535	0.598	0.157	0.283	0.033
Pavement material	0.445	0.251	0.585	0.418	-0.013
Use LED lights	0.129	0.236	0.650	0.317	-0.047
Application of reuse, reduce, recycle	0.129	0.357	0.420	0.272	0.123
Use less street lights	-0.121	0.384	0.117	0.213	0.504
Less the usage of carbon emission equipment	0.132	0.171	0.301	0.157	0.732
Natural heritage should be preserved during road construction	0.391	0.513	0.191	0.248	0.144
Habitats should be preserved during road construction	0.546	0.487	-0.050	-0.073	0.099
Water runoff drainage system	0.576	0.202	0.316	0.181	0.138
Porous pavement for water flow	0.269	0.371	0.497	0.404	0.084
Overhead pedestrian bridge	0.567	0.429	0.240	0.188	-0.084
Exclusive motorcycle lane	0.552	0.473	0.262	0.216	-0.028

The analysed rotated factor matrix showing that the criteria and elements were well distributed accordingly. All the elements were highly correlated with the main criteria. The two elements which are building orientation and less the usage of carbon emission equipment has shown the highest correlation with sustainable design and construction activities and energy efficiency criteria with 0.722 and 0.732 respectively.

5. Result and Discussion

The results were the mean for every criterion, the factor score and the Green Rural Road Index. All the criteria have been categorized into specified criteria in Table 4. Besides, the factor scores for all criteria have been calculated. Table 4 clearly shows the arrangement of criteria according to the suitable criteria.

Table 4. The Arrangement of Green Criteria and Element.

Main Criteria	Element	Factor Loading	Mean	Factor Score	Criteria Factor Score	Total Factor Score
Social and Safety	Good water flow	0.620	4.21	2.61	18.13	48.05
	Road safety audit	0.510	4.34	2.21		
	Overhead pedestrian bridge.	0.567	4.03	2.29		
	Exclusive motorcycle	0.552	4.20	2.32		
	Erosion control	0.512	4.23	2.17		
	Water Quality control	0.434	4.29	1.86		
	Habitat should preserve during road construction	0.546	4.19	2.29		
	Water runoff drainage system	0.576	4.14	2.38		
Environmental and Water management	Natural heritage should be preserved during road construction	0.513	4.13	2.12	9.15	
	Shady are with tall trees	0.443	4.06	1.80		
	Dust control	0.633	4.20	2.66		
	Water quality control	0.598	4.29	2.57		
Material and Technology	Porous pavement,	0.497	4.20	2.09	7.98	
	Good pavement material	0.485	4.11	1.99		
	Use LED lights for roads	0.521	4.13	2.15		
	Application of 3R	0.420	4.16	1.75		
Sustainable design and construction activities	Sustainable road facilities	0.722	4.10	2.96	7.93	
	Erosion control	0.555	4.23	2.35		
	Noise mitigation control	0.657	3.99	2.62		
Energy efficiency	Less the usage of carbon emission equipment	0.732	4.00	2.93	4.85	
	Use less street lights	0.504	3.81	1.92		

The arrangement was done from the rotated factor matrix data. Besides, the mean was also analysed from every criterion. The mean was calculated from the range of the score (1-5) for every criterion of the questionnaire. The mean is very important data to calculate the factor score by multiplying the factor loadings. Every criterion had different factor score.

The social and safety criteria had the most elements according to the analysis with the highest score 18.13 compared to all other criteria. This huge different may happen due to the importance given to this criterion in all design and development of road networks. Another main reason is to make sure the development of rural road index to consider the social and safety criteria be the main reason in reducing the impact to the human and environment.

The environmental and water management, material and technology and sustainable design and construction fairly distributed between 9.15 to 7.13. The energy efficiency had the lowest score due to infrastructure works that may consume less energy. Beside the index may quantify the energy in term of natural energy usage which may reduce the usage of equipment base energy.

Figure 2 shows the percentage of criteria that have been calculated. From the figure, Social and Safety become the main criteria to be focus in Pengerang with percentage of 38%, followed by Environmental and Water Management (19%), Material and Technology (16%), Sustainable Design and Construction Activities (16 %) and the least criteria is Energy Efficiency with 10%. The gap between highest and lowest percentage of the criteria are larger as compared to My Green Highway Index, (MyGHI). MyGHI is a performance baseline standard which measure the level of greenness for current highways in Malaysia which have small gap of percentage among criteria measured with; Sustainable Design and Construction Activities (22%), Energy Efficiency (21%), Environmental and Water Management (19%), Material and Technology (16%) and Social and Safety (22%). From here, it can be concluded that rural road and urban road had different favoured criterion for the sustainable development.

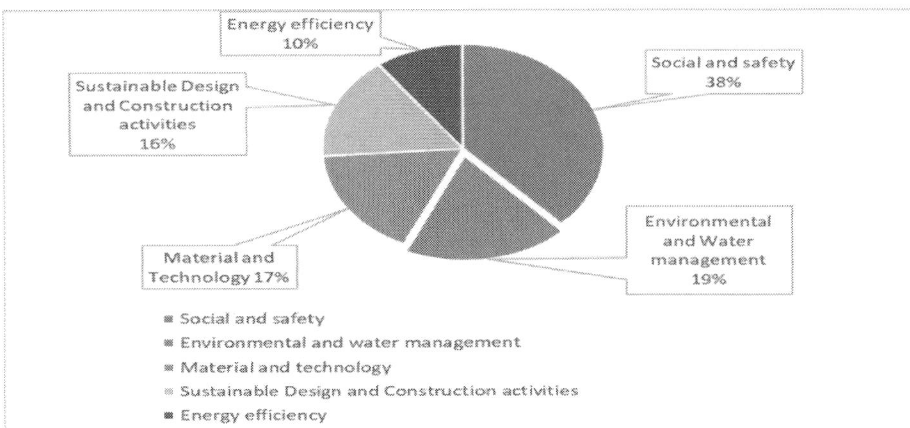

Figure 2. My Green Rural Road Index.

Based on Figure 2, social and safety criteria become the most important criteria with contributing elements of good water flow, road safety audit, overhead pedestrian bridge, exclusive motorcycle, erosion control, water quality control, habitat should preserve during road construction and water runoff drainage system. All these eight elements play an important role in defining social and safety criteria. It shows that the safety of the community within the road network is the main priority and its urgency in conserving the identity and originality of the local areas. Social and safety are very important in developing a better road networks. This is to avoid from harmful incidents and accidents in the future. For an instance, exclusive motorcycle lane is the best way to decrease the accidents of motorist in the future.

Then, second most important criteria are Environmental and Water Management, with four contributing elements, namely; natural heritage should be preserved during road construction, shady are with tall trees, dust control, and water quality control. This criterion is also important due to plantation and farming activity that need a good watering system and fertile land for their plantation.

Third criterion is material and technology with four contributing elements of porous pavement, good pavement material, use led lights for roads, application of 3R (reduce, reuse, recycle). Next for sustainable design and construction activities criteria, it consists three elements of sustainable road facilities, erosion control and noise mitigation control. All these three elements play important roles in defining these criteria since it involves the development of rural road network. The entire road will become a failure if the criteria did not well define and addressed. Finally, the least percentage criteria are energy efficiency with two elements namely; less the usage of carbon emission equipment and use less street lights which have been discussed above.

6. Conclusion

In conclusion, all the element has been categorized into specified criteria which give a very good indication to the development of My Green Rural Road Index. Based on the results, it can be concluded that the people in Pengerang were preferred social and safety as the main priority for the development of green rural road. The priority given may due to possibility of reducing the human and environmental impacts such as accidents and pollution. Besides, this assessment tool can also benefit the rural residential area particularly in terms of socio-economic development and environmental safety.

In addition, the analysis gives the indication that the rural residents show much interest in shifting from traditional roads to green rural road because it is environmental friendly, reduce carbon emission and can conserve the natural environment and resources. It also helps to minimizes and reduces the negative impact to human and environment. Respondent in Pengerang more prefer Social and Safety criteria as the number one priority for the development of green rural road network. The criteria that categorized on social and safety are water quality control, road safety audit, exclusive motorcycle, overhead pedestrian bridge, erosion control, water runoff drainage system, habitat should preserve during road construction and water runoff drainage system. In a nutshell, developing a green rural network can give huge impact to the future generation by saving the environment with a green infrastructure networks. This also may improve the socio-economic benefit to the local community with more green facility and environmentally sound road technologies for the future generations.

7. References

[1] Woodhead P 2003 *Sustainable Transport, Environmental Management* ed R Tolley (US: CRC Press)

[2] Abdullah R G 2017 Accessibility and Development. A Case Study from Rural Sarawak, Malaysia, *Inter. J. of Business and Society* **18**(4) 791-799

[3] Huang R Y and Yeh C H 2008 Development of an Assessment Framework for Green Highway Construction, *J. of the Chinese Institute of Engineers* **31**(4) 573-585

[4] Rafidah R, Muhammad R, Ab N, Umar N Z, Zaimi M, Majid A, and Ismail F 2014 An evaluation of sustainable design and construction criteria for green highway *Procedia Environmental Sciences* **20** 180–18

[5] Najwa F, Nusa M, Endut I, Takim R, and Ishak S Z 2015 Green Highway for Malaysia: A Literature Review, *J. of Civil Eng. and Arch.* **9**(1) 64-71

[6] Adzar J A, Zakaria R, Aminudin E, Rashid M H S A, Munikanan V, Shamsudin S M, Rahman M F A, and Wah C K 2019 Development of operation and maintenance sustainability index for "penarafan hijau jabatan kerja raya" (pHJKR) green road rating system, *IOP Conf. Ser.: Mater. Sci. Eng* **527** 012058

[7] Abd Rashid M H S, Zakaria R, Aminudin E, Adzar J A, Shamsuddin S M, Munikanan V, Alias N E, Sooria S Z and Saha K M 2020 Critical green road criteria for Malaysia green rural road Index, *IOP Conf. Ser.: Mater. Sci. Eng* **849** 012039

[8] Mulmi A D 2009 Green Road Approach in Rural Road Construction for the Sustainable Development of Nepal, *J. of Sustain. Develop.* **2**(3) 149-165

[9] Rafidah R, Muhammad R, Ab N, Umar N Z, Zaimi M, Majid A, and Ismail F 2014 An evaluation of sustainable design and construction criteria for green highway *Procedia Environmental Sciences* **20** 180–186

[10] Eisenman A A P 2012 *Sustainable Streets and Highways: An Analysis of Green Roads Rating Systems* Master Thesis (Georgia Institute of Technology)

[11] John F, Abdul D, Rick W and Michelle A 2012 *Liveable and Sustainable Transportation Rating System and Guide* (Illinois: Department of Transportation)

[12] Energy C 2019 *Suruhanjaya Tenaga (Energy Commission)* Annual Report 2019 (Malaysia: Energy Commission)

[13] Mulmi A D 2009 Green Road Approach in Rural Road Construction for the Sustainable Development of Nepal, *J. of Sustain. Develop.* **2**(3) 149-165

[14] Adzar J A, Zakaria R, Aminudin E, Rashid M H S A, Munikanan V, Shamsudin S M, Rahman M F A, Sooria S Z and Hassan M M 2019 Sustainable operation and maintenance criteria for non-toll road green rating system *IOP Conf. Ser.: Mater. Sci. Eng.* **615** 012128

Acknowledgments
This work was financially supported by collaborative research grant PY/208/03564 from research team members of CReATE JKR, UiTM, UPNM and UTM.

IOP Conf. Series: Earth and Environmental Science **971** (2022) 012002 doi:10.1088/1755-1315/971/1/012002

Investigating Mechanical Performance of Ravelling in Hot Mix Asphalt Containing Plastic Waste

H A Susanto[1*], K Merdiana[1] and E W Indriyati[1]

[1] Department of Civil Engineering, Faculty of Engineering, Universitas Jenderal Soedirman, Purbalingga 53371, Indonesia
*Corresponding author: hery.susanto@unsoed.ac.id

Abstract. One of the road damages on the flexible pavement is ravelling. Ravelling caused by the release of aggregate grains from the asphalt mixture. The interaction between the wheels of the vehicle and the surface of the pavement causes friction that pushes the aggregate out. The plastic waste on road pavements has been widely used to improve the performance of road pavement structures. Most of previous study focused on Hot Mix Asphalt (HMA) ravelling using Cantabro test. The purpose of this study was to investigate the ravelling performance of plastic waste material in hot mix asphalt mixtures. In addition, the use of plastic waste can reduce the environmental impact due to the high volume of plastic waste. The test is carried out in the laboratory using a rotation abrasion device based on the principle of rotating wheel load with loading, speed, and testing time parameters. Furthermore, the ravelling resistance value of HMA was obtained in dry and wet conditions. The addition of plastic waste into asphalt can increase the ravelling resistance of HMA mixes. The results of the wet test have greater ravelling resistance than dry test.

1. Introduction

Ravelling damage is one of three commons damage in hot mix asphalt (HMA) that often encountered in the field. Ravelling damage occurred when some of the constituent materials of the HMA was separated from the mixture [1]. Some of the causes of ravelling damage are the weak bond between asphalt and aggregate, dust in the voids of the mixture, and the aggregate is not well embedded in the mixture [2-4]. In addition, the quality of the material is also very influential on ravelling damage [5]. Therefore, the selection of material quality is one of the factors that influence the occurrence of ravelling. One of the efforts to improve the quality of the material is by modifying the asphalt. Several researchers investigated performance of HMA by adding polypropylene (PP) plastic waste into asphalt. One of the reasons is the PP can improve the binding capacity of asphalt.

Thus far the study about incorporating PP plastic waste into HMA only focused on Marshall characteristic [6, 7, 8]. In addition, several researchers used Cantabro test to investigated ravelling in hot mix asphalt [9,10]. Nevertheless, Cantabro test cannot simulate the interaction between vehicle wheels and pavement surface. Meanwhile, the study of PP plastic waste to ravelling resistance is very limited. Thus, it is necessary to investigate the effect of HMA containing PP plastic waste on ravelling resistance. This research is expected to improve the quality of HMA against ravelling resistance and solution in dealing with plastic waste in Indonesia.

2. Research Methodology

This research was carried out through three stages including material property testing for material physical criteria, Marshall testing for determine optimum asphalt content (OAC) and abrasion testing for revelling resistance. In addition, the HMA samples tested in dry and wet condition. Lastly, the analysis includes a ravelling comparison between HMA without PP plastic waste and HMA containing PP plastic waste. Furthermore, the effect of dry and wet test to ravelling will discuss for both HMA without PP plastic waste and HMA containing PP plastic waste.

2.1. Mix design dan samples

The samples of HMA without PP plastic waste and HMA containing PP plastic waste were prepared before testing. The HMA containing PP plastic waste is the HMA made by asphalt mixed with PP plastic waste. The PP plastic waste (Figure 1) cleaned and cut into small size, then 6% of the PP plastic waste [4] added into asphalt. The mixing plastic waste into penetration asphalt is processed by heating. Asphalt and plastic are heated separately, after the plastic is melted then mixed into hot asphalt, and stirred until homogeneous. Furthermore, the samples made based on two testing scenarios, OAC and ravelling. Table 1 below described the number of samples.

Table 1. Number of samples.

Testing	Sample			
OAC	15			
Ravelling	HMA standard		HMA modified	
	dry	wet	dry	Wet
	3	3	3	3

Figure 1. PP Plastic waste.

2.2. Samples testing

The HMA test divided into two stages, first stage is Marshall test according to SNI 06-2489-1991 [11, 12] and second stage is ravelling test use a rotation abrasion device (RAD). The RAD is a simple modified ravelling device used to determine the abrasion value (Figure 2). The device developed based on the wet track abrasion loss in the ISSA A143 standard [13]. The simple ravelling modification device provides an overview of the real interaction between vehicle wheel rotation and the surface layer of road pavement [14]. The RAD considers several condition variables in the field including load, vehicle speed, testing duration, and submerged sample [15, 16, 17, 18]. In term of the dry and wet test, the samples immersed in a water bath with a temperature of 60 °C for 30 minutes, then an abrasion test was carried out for 5 minutes. For the wet test, the sample submerged in water at room temperature during the abrasion test.

a) b)

Figure 2. Device for a) Marshall, and b) Rotation abrasion device.

2.3. Data analysis

The data analysis includes OAC and ravelling. The OAC was obtained from the graph of relationship between asphalt content, Marshall volumetric and Marshall characteristics. Meanwhile the abrasion value depicted in a graph of relationship between the two HMAs and abrasion value. Furthermore, the graph will present the comparison of abrasion value for both dry and wet samples.

3. Results and Discussion

3.1. Physical property

Table 2-Table 7 present the physical properties for aggregates, asphalt and PP plastic waste. The asphalt used bitumen penetration grade (PG) 60/70. All the materials properties were found to be complied with the required specifications. Thus, the material can be used in the mix design of HMA.

Table 2. Coarse aggregate.

Testing		Unit	Value	Requirement	Specification
Los Angeles Abrasion	100 rotation	%	4	Max. 8	SNI 2417:2008
	500 rotation	%	16	Max. 40	
Bulk specific gravity		gr/cc	2.7	>2,5	SNI 20-1969-2008
SSD specific gravity		gr/cc	2.72	>2,5	SNI 20-1969-2008
Apparent specific gravity		gr/cc	2.78	>2,5	SNI 20-1969-2008
Absorption		%	1.05	<3	SNI 20-1969-2008
Aggregate adhesiveness to bitumen		%	100	Min. 95	SNI 7619:2012

Table 3. Fine aggregate.

Testing	Unit	Value	Requirement	Specification
Bulk specific gravity	gr/cc	2.68	-	SNI 20-1970-2008
SSD specific gravity	gr/cc	2.75	-	SNI 20-1970-2008
Apparent specific gravity	gr/cc	2.88	-	SNI 20-1970-2008
Absorption	%	2.57	-	SNI 20-1970-2008
Passing No.200	%	0.78	Max. 1	SNI ASTM
Filler specific gravity	gr/cc	2.128	-	C117:2012

Table 4. Filler.

Sieve number	Passing (gr)	Retained (gr)	Passing (%)	Requirement (%)
No. 300	120	0	100	100
No. 50	114.6	5.4	95.5	90-100
No. 200	94.2	20.4	78.5	75-100
Pan	-	94.2	-	-

Table 5. Polypropylene (PP) plastic waste.

Testing	Value	Requirement
Sepecific gravity (gr/cm^3)	0.904	0.895-0.920
Melting point (°C)	160	-

Table 6. Asphalt PG 60/70 and modification asphalt.

Testing	Value		Requirement		Specification
	0% of PP	6% of PP	PG 60/70	Modification asphalt	
Penetration at 25° (0,1 mm)	68.65	60.05	60-70	60-70	SNI 2456:2011
Softening point (°C)	58.25	71.25	≥ 48	≥ 54	SNI 2434:2011
Ductility at 25°C (cm)	110	98	≥ 100	≥ 100	SNI 2432:2011
Flash point (°C)	258	241	≥ 232	≥ 230	SNI 2433:2011
Specific gravity	1.016	1.060	≥ 1,0	≥ 1,0	SNI 2441:2011
Kinematic viscosity (cSt)	494	1733	≥ 300	≥3000	SNI 7729: 2011

3.2. Optimum asphalt content (OAC)

The OAC is the optimum asphalt content required to produce a good performance of mixture according to Marshall specification [12]. The OAC obtained by finding the average asphalt content value that meets the Marshall characteristics. Table 7 present the calculation of OAC. Table 7 showed that OAC value is 5.5%. Therefore, the samples for ravelling testing have been prepared by using 5.5% OAC.

Table 7. Determination of OAC.

Marshall parameters	Requirement	Asphalt content				
		4,5%	5%	5,5%	6%	6,5%
Stability (Kg)	> 800	2393.5	2796.9	2934.4	2328.8	2462.6
Flow (mm)	> 3	3.3	4.2	4.3	4.5	4.5
VIM (%)	3-5	13.43	7.20	4.95	9.25	7.19
VMA (%)	> 15	20.69	16.29	16.13	21.15	21.18
VFA (%)	> 65	46.27	69.30	77.68	66.61	72.50
MQ (Kg/mm)	> 250	749.85	688.53	676.11	522.19	551.36
OAC (%)				5.5		

3.3. Ravelling

The abrasion value indicates the ability of the HMA mixture to resist ravelling [19]. The abrasion value describes the amount of material lost in the HMA mixture. The greater the abrasion value, the smaller the ravelling resistance of the HMA, and vice versa. In this study, samples were tested in dry and wet conditions. The result of the ravelling test presents in Figure 3. Figure 3 shows that HMA containing 6% plastic waste can reduce the abrasion value of the HMA 69.56% in dry conditions and 78.65% in wet condition. The addition of PP plastic waste to the HMA can increase the adhesion between asphalt and aggregate, thus the amount of material loss due to abrasion is getting smaller. In addition, HMA containing PP plastic waste had low air voids, thus making it more impermeable compared to control mixture. These results similar to the previous study which found that polymer-modified binder and low air voids content increased the ravelling resistance [20].

Another phenomenon presented from the comparison of the abrasion between dry and wet samples. It was found that in wet condition both of HMA without PP plastic waste and HMA containing plastic waste can increase the abrasion value up to 48.32% and 26.32% compared to the dry samples, respectively. In the wet test, the cavity will be filled with water which can decreased the bonding between asphalt and aggregate, thus lead to the dislodgement of some of the aggregates from the HMA mixture. This will have an impact on increasing the abrasion process.

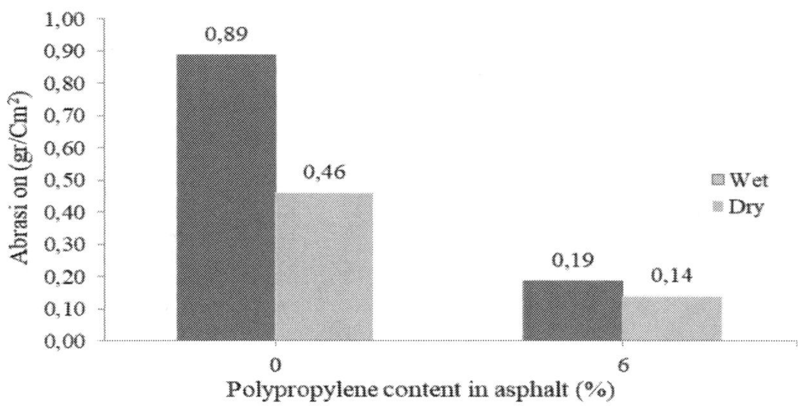

Figure 3. Abrasion value of HMA without PP and HMA containing PP plastic waste.

4. Conclusions

The conclusions of this study are:

- From the Marshall testing obtained 5.5% OAC.
- The addition of polypropylene (PP) plastic waste into asphalt increased the ravelling resistance of HMA mixes.
- The dry test results have greater ravelling resistance than wet. The water affects the increasing of abrasion, thereby reducing the ravelling resistance.

5. References

[1] Hardiyatmo, Hary C 2015 *Pemeliharaan Jalan Raya* (Yogyakarta: UGM Press)

[2] Arifin, Bustanul, and Wijanarko D 2001 Pengaruh Penggunaan Abu Sekam Padi sebagai Filler terhadap Karakteristik Campuran HRS.

[3] Angelo G D, Ojum C, Widyatmoko I, Wayman M, and Khojinian A 2016 Ravelling Resistance of UK Thin Asphalt Surface Course Systems Under A Large-Scale Accelerated Test Method. doi 10.1515ijpeat-2016-0035.

[4] Huurman M, Mo LT, Molenaar AAA, Wu SP. 2006. Raveling in porous asphalt concrete *In: Road constr. conf. in Ede (Netherlands).*

[5] You Q, Zheng N, and Ma J 2018 Study of ravelling failure on dense graded asphalt pavement *In Proc. of the Inst. of Civil Eng.–Trans.* **171(3)** 146-155

[6] Rahmawati A 2017 Perbandingan Penggunaan Polypropilene (PP) dan High Density Polyethylene (HDPE) pada campuran Laston_WC *Media Teknik Sipil Universitas Muhammadiyah Malang* **15(1)** 11-19

[7] Arif S 2018 Alternatif Penggunaan Plastik Polypropylene pada Campuran Aspal *J. CIVILA* **3(1)** 140-145

[8] Winayati, Rahmat A, and Saleh A 2018 Analisis Penggunaan Abu Tandan Kelapa Sawit sebagai Filler ditinjau dari Nilai Keausan Perkerasan (Cantabro Test) Retrieved on 15 July 2021 from https://osf.io/preprints/inarxiv/y7su9/

[9] Mabui D S, Tjaronge M W, Adisasmita S A, and Pasra M 2020 Resistance to cohesion loss in cantabro test on specimens of porous asphalt containing modificated asbuton *In IOP Conf. Series: Earth and Env. Sci.* **419(1)** 012100

[10] Estakhri C K, Kassem E, Arambula E, and Scullion T 2017 *Use of Cantabro Test to Reduce Raveling in Patching Mixes* (No. 17-05967) (Washington DC: United States)

[11] Badan Standarisasi Nasional 1991 *Metode Pengujian Campuran Aspal dengan Alat Marshall* SNI 2489-1991 (Jakarta: Badan Stadarisasi Nasional)

[12] Badan Standarisasi Nasional 2003 *Metode Pengujian Campuran Beraspal Panas dengan Alat Marshall* RSNI-M-01-2003 (Jakarta:Badan Standardisasi Nasional)

[13] International Slurry Surfacing Association (ISSA) 1991 A143 *Recommended Performance Guidelines for Micro-Surfacing* (USA)

[14] Susanto H A, Yang S H, Chou H H 2019 Mechanical Properties of Thin Surface Treatment for Pavement Maintenance, *The Baltic J. of Road and Bridge Eng.* 14(2) 136-157

[15] Khedoe R, de Bondt A, Villani M, and Scarpas T 2016 The development and use of the skid resistance and smart ravelling interface testing device *In Proc 6th Eurasphalt & Eurobitume Congress*

[16] Kelie C J P 2009 Influence of the Type of Road Surface on Tire Wear through Rotating Surface Abrasion Test *In Department of Traffic and Navigation* (The Netherlands)

[17] Mo L T, Huurman M, Woldekidan M F, Wu S P, and Molenaar A A 2010 Investigation into material optimization and development for improved ravelling resistant porous asphalt concrete, *Mater. & Des.* **31(7)** 3194-3206

[18] Nicholls J C, De Visscher J, Hammoum F, and Jacobs M 2015 Review of parameters influencing the propensity of asphalt to ravel *CEDR Transnational Road Research Programme Call-Asset Management and Maintenance*

[19] Arambula-Mercado E, Hill R A, Caro S, Manrique L, Park E S, and Fernando E G 2016 Understanding mechanisms of raveling to extend open graded friction course (OGFC) service life *Texas A&M Transportation Institute and Universidad de los Andes, Colombia*

[20] Gayfutdinov R, Bajmukhametov G, and Hafizov E 2021 Pavement wear process and abrasive wear resistance of asphalt concrete *In E3S Web of Conf.* **274** 02008

IOP Conf. Series: Earth and Environmental Science **971** (2022) 012003 doi:10.1088/1755-1315/971/1/012003

Building Information Modelling (BIM) Implementation for Highway Project from Consultant's Perspectives in Malaysia

E Halim[1], A Mohamed[2] and M S Fathi[1*]

[1] Razak Faculty of Technology and Informatics, Universiti Teknologi Malaysia, 54100 Kuala Lumpur, Malaysia
[2] School of Civil Engineering, Faculty of Engineering, Universiti Teknologi Malaysia, 81310 Skudai, Johor, Malaysia
*Corresponding author: syazli@utm.my

Abstract: In today's world, Building Information Modelling (BIM) is not a new term in the construction industry. Although BIM has been adopted in the AEC industry for a few decades now, the BIM implementation cultivation pace between developed countries and developing countries such as many European which considered as BIM leaders and some Asian countries has still shown significant differences. In Malaysia, there are several remarkable building projects with BIM utilization but very limited in road and highways project. Despite BIM capabilities in facilitating designers during design stage, BIM implementation are still considered unfavourable within the highway design consultants in Malaysia. Many firms seem reluctant to migrate from their traditional 2D drafting centric design process to BIM design process due to several reasons such as high cost in software investment, inadequate knowledge and expertise on BIM and lack of promotion from the government or relevant authority. Although BIM can benefit the project throughout its life-cycle from preliminary to operation and maintenance stage, the coordination and integration during design stage can contributes to positive impacts in progressing towards efficient construction documentation and construction stage. BIM 3D integrated modelling during design stage (conceptual and detailed design) are deemed to facilitate a highway project in terms of optimization through early visualization, design for constructability, road safety enhancement, reduce error and time consumption during construction and early clash detection. Therefore, this paper is intended to gather preliminary outlooks and views from the present practitioners in the highway industry on challenges encounters by consultants using the current traditional design practices, BIM capabilities in transforming the current design approach and BIM adoption barriers in Malaysian highway industry. Series of interviews were conducted with the professionals involved in highway projects especially during the design stage and based on the findings from the interviews conducted, all highway design consultants from various disciplines are aware on how BIM can tackle the challenges of current traditional design practice with 3D modelling approach and collaborative working environment. Although the results showed that highway consultants are willing to migrate to BIM workflow, cost and lack of expertise, urges from government and understanding on BIM are among barriers which hinder the BIM adoption in highway industry.

1. Background

Construction industry has grown rapidly throughout the decades particularly in developing countries. Productivity, quality and value of a project can be maximized by promoting technological adoption in the construction industry [1]. BIM can be simply defined as a process or workflow with collaborations tools to design, analyze, construct and manage a project [2].

While developed countries has shown active inclination in adopting BIM, developing countries are still at the infancy phase due to several drawbacks. For instance, many European countries can be considered as BIM leaders such as Finland, Denmark, Sweden and even Singapore and Hong Kong where their BIM implementation pace has been moving rapidly. In developing countries such as Africa and Nigeria, level of BIM awareness and adoption is relatively low [3]. In Asian countries on the other hand, has also show either almost less or none in BIM implementation in some of the regions [4].

There is no doubt that the application of BIM in a building industry is expanding every day, however, the adoption in the infrastructure and transportation industry is still limited and lagging [5]. Although the construction industry and academia has consistently making effort in promoting BIM application other non-building civil infrastructure field, yet, there is lacking of comprehensive review and publication specifically focus on transportation industry. A research conducted also shows the significant increase of reviewed publication on BIM potential in infrastructure from year 2002 to 2017 [5]. Hence, it fairly can say that, nonetheless, the application of BIM in infrastructure has gathered the attention of the players in the construction industry until today.

The first highway project in Malaysia using BIM is the Pan Borneo Highway in the Malaysian State of Sarawak in Borneo Island which adopt Highway Information Modeling (HIM) which combines the Geographical Information Systems (GIS) with BIM as their main Integrated Project Management System. One of the issues encountered is the project visibility which includes physically monitoring the project progress for the whole project which transverse along highway stretch is time consuming and requires high resources. The incorporation of three-dimensional (3D) parametric model with geospatial datasets has managed to develop the Pan Borneo Highway project from the preliminary-design-construction and operation and maintenance of the highway including the highway asset management [6]. Despite the fact that there a distinctive difference in vertical (hospital, schools etc.) and horizontal (road, highway, tunnel etc.) construction in terms of the process from planning to construction, legal and contractual issues, and even cultural management, it is fair to say that by implementing BIM in the right directions at each project stage can unlock the full potential of BIM [5].

2. Literature Review

2.1. The concept of BIM

Every construction project has its own unique nature which relatively subjected to risks and uncertainties. These uncertainties have to be tackled in order to achieve the standard quality level within allocated budget and time. Poor project performance has higher tendency in leading to cost overruns and project delay. Hence, throughout the research for past years, researchers have attempted to discover various technologies with the intention of controlling these uncertainties, of these where BIM has been proven as one of the best practices.

As part of emerging technology which has been globally employed in the construction industry to date, there are still many misperceptions in understanding the accurate definition of BIM. According to National Building Information Modelling Standard (NBIMS), BIM can be defined as

"An improved planning, design, construction, operation, and maintenance process using a standardized machine-readable information model for each facility, new or old, which contains all appropriate information created or gathered about that facility in a format useable by all throughout its lifecycle". [7]

Generally, it is not a piece of software but more of a way of working throughout the whole process. According to Azam Haron N et al. (2017) also outlined diverse definitions of BIM which generally concluded BIM as a software model which can be used throughout project life-cycle and can benefit project stakeholders to ensure project success [8].

2.2. Causes of Delays in Highway Project

There are several studies and researches worldwide that has been conducted in indicating the factors that leads to a successful highway project throughout its project life-cycle from planning to construction stage. A project completion without time and cost overruns as well as satisfying the required quality standard can be considered as a successful project [9]. However, the defiance of these successful factors will cause negative impacts on the project progress and performance e.g. delays, cost overruns, poor quality, variation orders which are shown in Table 1 based on several studies conducted.

Table 1. Summary of causes of delays in highway project based on countries.

Country	Summary of Causes of Delays in Highway Project	Sources
Saudi Arabia	Change orders by owner during construction, consultant's and contractors' performance, delay of progress payment by the owner	[9]
Thailand	Incomplete drawings, contractor's incompetency in terms of resources and financial, and delay in original infrastructure structures relocation	[10]
Pakistan	Competency of project management team, effective site management, involvement from all parties within the project and also the proficiency of the design team	[11]
Malaysia	Workflow for the design process, competencies of project managers, availability of workforce, availability of materials	[12]

Both academia and practitioners has been investigating on topics related to infrastructure and highway transportation field. Based on these researches, it is apparent that designers or consultants play one of the major roles during design stage towards a successful highway project. According to Cooperative Research Centre (CRC) for Construction Innovation (2007), BIM for road and highway projects are capable to foresee the project performance prior to construction, faster communication on design changes, design optimization, visualization and simulation of the whole project, and also high quality and efficient construction documentation [13]. Hence, it is essential to explore the challenges and disputes faced by the consultants during design stage which influenced to poor performance in a highway project in Malaysia and how BIM can facilitate the challenges.

2.3. BIM Adoption in Transportation Infrastructure (Roads & Highways)

As BIM implementation is not restricted only to structural design, the road and highways industries shall also be benefited from the advantages of BIM implementation. Each level of stakeholders shall be able to experience the full potential of BIM from preliminary design to operation and maintenance stages of a roads or highways project. Although BIM has widely been utilized in building industry, the technology has shown huge potential in advancing transportation infrastructure [5].

One of the significant examples of successful infrastructure project in adopting BIM is by the Wisconsin Department of Transportation (WisDOT) is now has led the movement towards implementing 3D modeling as their core approach in roadway construction. Among their successful projects involved BIM adoption are the Zoo Interchange, west of Milwaukee (1.7 billion project) and Mitchell Interchange. Zoo Interchange which was completed in year 2018, the use of parametric 3D models has provided a firm foundation for the project to move forward in using 4D modelling (staging and scheduling), 5D (cost information) and also 6D (as-built, operation & maintenance). Another significant benefit gained through 3D modeling is the clash detection which helps in reducing the future change orders and design issues [14].

On the other hand, originally WisDOT have developed the design of Mitchell Interchange through traditional 2D CAD design. However, near completion of the project, they have generated the 3D models in order to investigate the potential impact on design modelling versus the cost saving. They have concluded that the project could have saved up to $9.5 million if 3D modeling has been implemented during the preliminary stage. [14]

2.4. Benefits of BIM in Highway Construction Project: Design Stage

In line with the aim of this paper, further evaluation will be given to assess the benefits of BIM adoption during design stage. Based on previous researches and studies, highway construction industry can gain similar benefits from BIM adoption in building construction.

IOP Conf. Series: Earth and Environmental Science **971** (2022) 012003 doi:10.1088/1755-1315/971/1/012003

Referring to subsection 2.2, it is quite clear that design phase in a highway construction can be considered as vital stages which can contributes to smooth and effective project progress. According to Anderson. R (2010), he has outlined the comparisons between traditional design approaches versus designing using technological advance such as BIM workflow through "MacLeamy Curve" as shown in Figure 1 [15]. This curve indicates that the by shifting to BIM workflow, the highest level of effort, effect and time shall be given during the detailed stages. This approach will impact the project performance at higher level and the cost of overcoming design changes is low. Therefore, it is costly and time-consuming to make design changes after the construction documentation started. Figure 1 also clearly shows why it is important to emphasis on efficient and coordinated design stage and how this approach can affect the overall project progress.

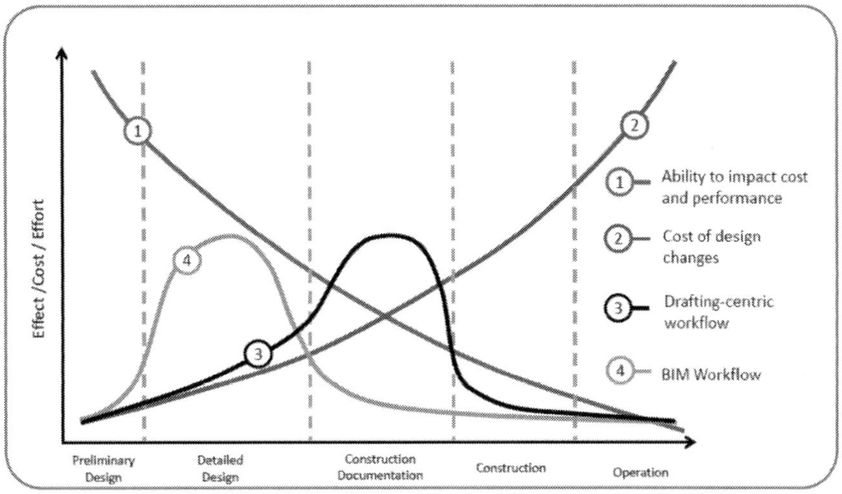

Figure 1. 'MacLeamy Curve' on BIM workflow [15].

2.4.1. Optimization through Early Visualization. During preliminary or planning stage, different alternatives will be evaluated and assessed in order to achieve the optimum alignment. Shetwi et al. (2015) has demonstrated from their research on the Champlain Bridge in Montreal, Canada which uses three levels of BIM: 3D visual, 4D schedule, and 5D cost estimations [16]. The editable realistic rendered 3D model has been used as the basis for visualization which helps the owners, engineers and even the contractors to decide on the form and aesthetics of the project. As most of the proposed highway in the city or urban area nowadays are elevated, visualization through 3D model is useful to envision the alignment as well as helps with early judgement on the pier's location and land acquisition evaluation.

2.4.2. Design for Constructability. BIM allows designers to design for constructability which helps to reduce the mistake before it become a problem [17]. An example of design for constructability is determined by [18] where they have integrated the Bridge Information Modeling (BrIM) with a hybrid model in order to locate the suitable location for mobile cranes placement during bridge construction. Additionally, the constructability and mobility plan of interfacing between two highways can also be tackled by BIM implementation.

2.4.3. Road Safety Enhancement. Another important aspect in highway designs is the road safety features. Passing and stopping sight distances is one of the crucial elements in highway designs. The visual obstruction and horizontal layout can be assessed through interactive visualization and sight distance simulation. Thus, further safety measures can be taken into considerations as well as ensuring the road geometry satisfying the critical safety parameters [17].

2.4.4. Reduce Error and Time Consumption during Construction. As 3d modelling in BIM results with a model with rich attributes and functional datasets associated, any changes can be made at earlier stage for instance halfway through the project, the alignments need an adjustment in vertical profile, and hence by adjusting the profile in the 3D model, all related elements will be updated automatically. Designers will be able to see the impact of the adjustment for example to the cut-and fill as well as the right of way [17].

2.4.5. Earlier Design Clashes Detection. Another key value of BIM adoption is the ability to detect clashes from all the integrated disciplines through model coordination. Clash detection process helps to resolve design issues prior to construction stage and contributes to cost-saving value to the project [19]. Despite the significant and noteworthy capabilities on how BIM can change the game of transportation and infrastructure industry based on the previous studies researches, the number of available literatures which specify on BIM adoption in highway industry are still lacking. Hence, it is crucial to gain the insights on the current situation on BIM implementation in Malaysia particularly from the practitioners in the actual highway industry perspective.

3. Research Frameworks and Methodology

The objectives of this paper are to gather preliminary outlooks and views from the present practitioners in the highway industry on;

- Challenges encounters by consultants using the current traditional design practices
- BIM capabilities in transforming the current design approach
- BIM adoption barriers and future in Malaysian highway industry

3.1. Methodology

Extensive literature review through current publications and sources will be executed in emphasizing the benefits of BIM adoption in highway project. Most of the data were collected from various sources such as books, journal articles, international conference papers, and materials available on the Internet.

However, semi structured interviews also were conducted in order to gain deeper information and perspectives from the highway industry practitioners from various disciplines in achieving the research objectives. A series of questions were outlined in order to get their insights based on the research objectives. All data gained were recorded, transcribed, and analyzed using content analysis and representation through text, images, as well as expressions.

3.2. Research Frameworks

The first steps taken was by outlining the structure of this paper. Figure 2 shows the basis of this paper on achieving the research objectives.

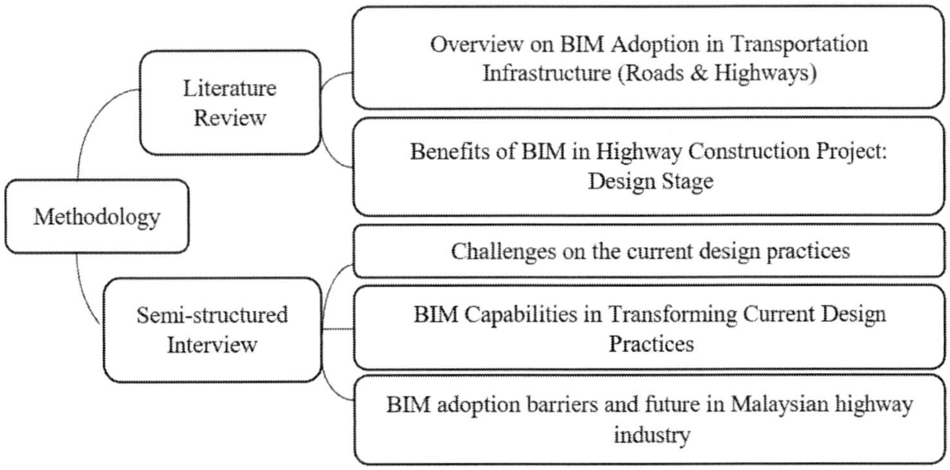

Figure 2. Proposed research frameworks.

As part of the one of the main intentions is to evaluate how BIM can change the paradigm of the current design practices in a highway project, it is crucial to understand how each discipline involves in a highway project deliver their design practices as well as ensuring precise, integrated and satisfactory information can be established in the final design as their approach and tools or software may vary.

In the United States of America (USA), Findley et al. (2015) indicates that their current design practices allow for flexibility in applying the design principal which varies by region, agency and individual. Additionally, Findley et al. (2015) also indicated that the existing highway design practices are not formally documented and typically inherited from earlier generations accumulated with the knowledge from the industry [20].

However, in the United Kingdom (UK), the design manuals and standards does not indicate how to design highways but the information highlighted are rather explained on the general quality assurance (QA) principles related to highway design, roles and responsibility and the procurement of quality plans and records. The manual [21] also indicated on the requirement on pavement design, geometric design, geotechnical design, landscape and road furniture.

Figure 3 illustrated below has been developed in presenting the typical workflow involving the general multi disciplines involved in a highway project from conceptual stage to production of tender or construction documents. This typical workflow was developed based on the information provided from [22-23] and also expert reviews from some of practitioners in highway industry in Malaysia. In order to gain deeper insights from the respondents, Figure 3 will be used part of the interview as basis in obtaining respondent's views on identifying the gaps or challenges on the current design approach by consultants in highway project.

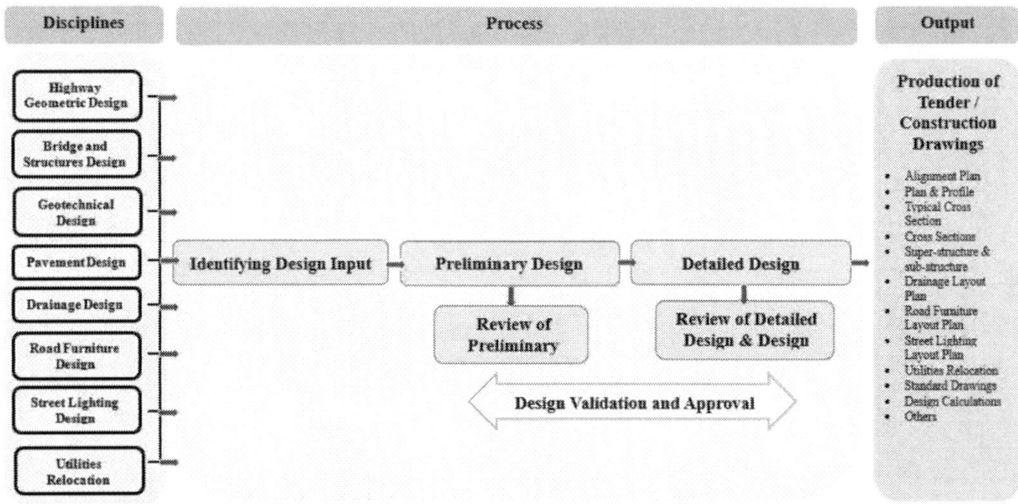

Figure 3. Summary of typical design practices in highway project in Malaysia.

4. Results and Analysis

The results and findings from the semi-structured interview conducted will be further discussed in this section and it shall provide an overview on the current BIM situation in Malaysian highway industry from consultant's perspective. Results and findings from the interview were categorized into four (4) subsections as follows:

- Respondent's profile and background
- Challenges on the current highway design practices
- BIM capabilities in transforming the current highway design practices
- Limitations and barriers on BIM implementation for highway design companies

The results from this exercise are deemed to be valuable in understanding the current BIM situation especially in highway industry in Malaysia and encouraging highway consultants in transitioning to BIM-based design workflow.

4.1. Respondent's Profile and Background

The respondent groups for the interview were divided in two (2) categories: A) design consultants who are yet to utilize BIM and B) design consultants with experiences utilizing BIM in any road or highway projects. Align with the objectives of this study, the respondents were selected based on their involvement within the general disciplines involved in a highway project during design stage. For each group, three (3) respondents were selected and interviewed for their insights and respondent's demographic are summarized in Table 2.

Table 2. Respondent's demographic.

Category	Respondent	Position in Organization	Experience in Highway Industry (years)	Experiences with BIM (years)
A	1	Engineer	7	-
	2	Senior Engineer	20	-
	3	Director	> 30	-
B	1	BIM Modeller	6	< 1
	2	BIM Coordinator	8	3
	3	BIM Consultants	6	9

Based on Table 2, all respondents from category A and B possessed experiences and involvement in a highway project from various disciplines with experiences in highway industry ranging from 7 years to more than 30 years. As level of BIM implementation for highway project in Malaysia are very low, respondents' experiences in using BIM specifically for highway project from category B is considered still very limited where it only ranges from less than 1 year to 9 years. The experiences possessed on BIM by respondents from category B generally involved in building projects or other infrastructure project such as mass rapid transit project (MRT) and only limited experiences on BIM for highway project. It is significant to perceive the respondent's input from different highway background as it will shows different level on BIM awareness, understanding and adoption.

4.2. Challenges on the current highway design practices

As Figure 3 was used as a basis in representing the typical workflow between multi-disciplinary involved in a highway project during design stage, respondents were asked if they agreed on the figure illustrated. All respondents agreed that Figure 3 exemplified the typical approach used in their workflow during design stage of a highway project based on the nature of their organizations and experiences.

Based on the interview, the main challenges faced during the design stage of a highway project is the non-collaborative working environment between disciplines as highway projects generally involved input from various disciplines as shown in Figure 4. The process which described by Strafaci (2010) as "siloed" where the steps from preliminary design to detailed design to construction documentation only starts once each step completed with minimal collaboration [17]. This followed by the likelihood of design changes in highway project are higher.

Design changes are also one of the most common issues which caused conflicts and disputes between disciplines which are then led to further delays in overall project progress [9-12]. Respondent A3 also mentioned that as horizontal construction such as highway project typically transverse across different site conditions such as various types of terrain, requirement for land acquisitions, etc. thus, multiple design changes typically due to client's or relevant authority request may affects most of the disciplines design progress. Other challenges are summarized in Figure 4. Based on the findings from the Figure 4, it fairly apparent that there is the need to improve and transform the traditional highway design process towards efficient and sustainable process

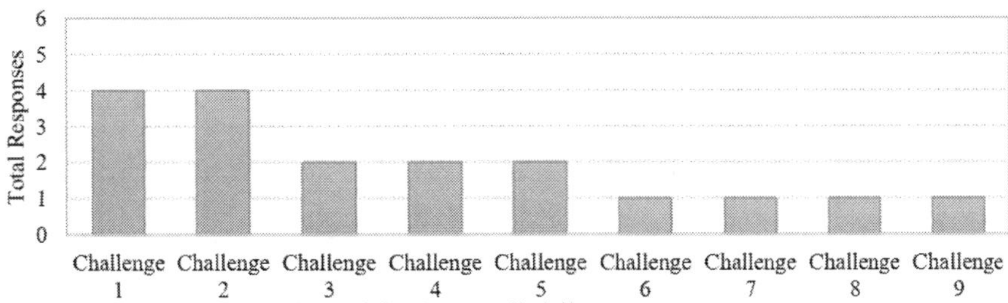

Challenge 1 - Non-collaborative workflow between disciplines
Challenge 2 - Likelihood of design changes during construction are higher
Challenge 3 - Risks of design conflicts in between disciplines are higher
Challenge 4 - Information on any design changes and updates from other disciplines are easily missed out
Challenge 5 - Lack of design integration and constructability between disciplines
Challenge 6 - Ineffective and non-integrated information sharing process between disciplines
Challenge 7 - Too many amount of paper printing
Challenge 8 - Manual drafting is more error-prone
Challenge 9 - 2D drawings provide insufficient information

Figure 4. Summary of challenges on the current highway design practices.

4.3. BIM capabilities in transforming the current highway design practices

As BIM is widely known as technological advancement in construction nowadays with several successful implementation in road and transportation projects across the globe, BIM it deemed to be capable in transforming the current highway design practices.

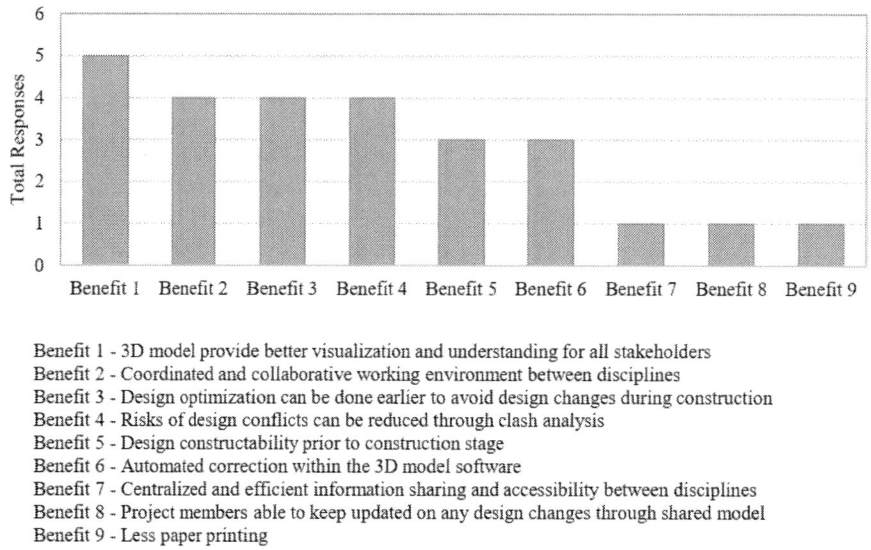

Figure 5. Summary of BIM capabilities in transforming the current highway design practices.

Figure 5 shows the summary of BIM capabilities in transforming the current highway design practices. Based on the interview, the first BIM capabilities in highway design corresponded by almost all of the respondents are the 3D modelling application in providing better visualization and understanding of the projects to all project's stakeholders followed by the collaborative working environment between disciplines and preliminary design optimization to avoid design changes. Generally, the responses from the interview are streamlined with the BIM capabilities highlighted in the previous study [5,17].

Respondent A1 stated that the application of 3D model will helps project teams to obtain clearer picture on the project and also for better comprehension on the highway impact especially to the relevant authorities and public during preliminary stage as also mentioned by respondent A3 and B3. As a highway project especially, urban highway has higher tendency to transverse across developed and populated area, it is crucial for project team members to be able to tackle the public resistance and convince all the affected authorities. In comparison to the current approach of presenting the alignment in 2D CAD drawings, 3D models of the project are easier to be understand and comprehended including for those with limited proficiency in interpreting 2D drawings [24]. Visualization will also help designers to improve road safety features such as road geometric, sight distance and road furniture positioning which sometime may be overlooked through 2D drawings [24]

Secondly, most of the respondents also agreed that another valuable BIM attributes for highway project are the ability of all disciplines to work in coordinated and and collaborative working environment. Respondent B3 and B2 specified that by implementing BIM in highway project, all disciplines can work in a single streamline and communicate on single platform. Hence, any required changes or information can be accessed and delivered at rapider pace.

This attribute also relatively corresponded to the BIM ability in transforming one of the most crucial issues in highway design which is design changes as mentioned in previous section. Application of BIM technology can reduce design changes during construction stage can be reduced by executing design optimization earlier during design stage. Tackling design issues in earlier stage will facilitate the project team to firm up the conforming design which shall lead to lesser design changes during construction. This is also in line with the MacLeamy curve shown in Figure 1 which signifies the significance of focusing on design stage can results in time and cost-effective construction delivery [17].

Another prominent BIM attribute corresponded by majority of the respondents is the ability to carry out clash analysis which can lessened the risks of design conflicts especially between disciplines. Respondent B2 also mentioned that the 3D model integration can also provide better presentation of the clashes. Similarly, in order to undertake amount of design changes, BIM capabilities in conducting clash analysis and design optimization during the design stage as agreed by the respondents are seen as one of the effective ways to reduce design changes especially once the construction commenced. This somehow will reduce additional cost incurred and contributes to smoother and on-time project delivery.

Moreover, BIM also is deemed to facilitate highway design stage by accelerating time consumption for design changes by automated correction within 3D software model and also through design constructability prior to construction stage as it enables consultants to construct the highway virtually with data rich and intelligent 3D model [25].

4.4. Limitations and barriers on BIM implementation for highway design companies

Parallel with the findings from study conducted by several researches on factors contributing to impediments of BIM adoption in construction industry Malaysia [1, 26-27], the results from interviews conducted likewise reveals quite similar results. Results are illustrated in Figure 6 below which summarized responses on the limitations or barriers in implementing BIM for highway project from highway design consultant's perspective.

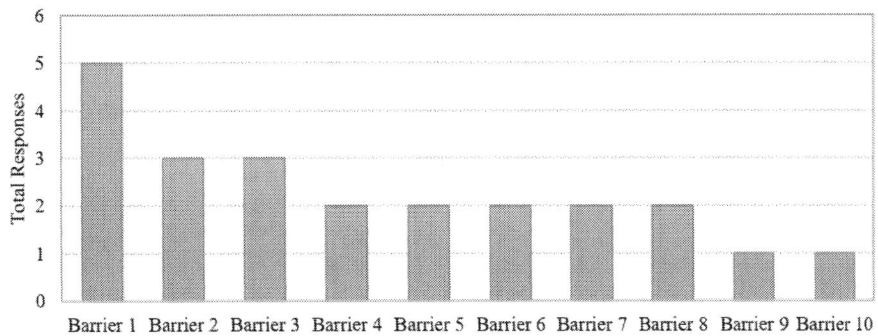

Barrier 1 - BIM software and hardware are too costly
Barrier 2 - High investment on resources and manpower
Barrier 3 - Limited urges and initiatives from the government and relevant agency
Barrier 4 - Lack of expertise and skills in highway industry
Barrier 5 - Lack of understanding on BIM concept and capabilities
Barrier 6 - Minimal involvement from all project members
Barrier 7 - Organizations reluctant to provide trainings due to high tendency of trained team members leaving the organisation
Barrier 8 - Transitioning to BIM technology may takes longer times
Barrier 9 - Lack of commitment for extensive trainings and seminars
Barrier 10 - Unwillingness to shift from current workflow

Figure 6. Limitations or barriers in implementing BIM for highway project

Figure 6 shows that higher cost for BIM technology, resources and manpower, limited initiatives from government and relevant agency, lack of BIM expertise and lack of understanding on BIM concept are the top five challenges agreed by the respondents. The limitation and barriers agreed by the respondents are also correlated with the study conducted by Roslan et al. (2019) and Yaakob et al. (2016) which classified the factors which contributes to hindrance of BIM adoption in Malaysian construction industry into four (4) general categories: i) People, ii) Technology, iii) Process, and iv) Policy [28-29]. It can be observed that the challenges faced by industry player in shifting from their common practices into new technological advance constantly revolves around these four main factors including the highway industry.

The results from the interview are consistent with the recent study by Yung et al (2020) which also reveals similar challenges in BIM adoption for road industry in Malaysia [30]. BIM adoption requires upgrading of IT system with BIM enabled software as well as providing relative resources and training. Thus, this would greatly impact the cost and company's return of investment (ROI) will be re-evaluated. BIM incentives such as free training shall be further enhance for highway project which is deemed to increase the level of awareness and intensifies the adoption rate [30].

Respondents has also agreed that limited urges from the relevant authorities are among the BIM adoption challenges for highway project in Malaysia. Public Work Department (PWD) PWD has mandated BIM for any public construction projects worth more than RM 100 million value [31]. However, the requirement for private projects is still lagging as it is commonly depending on client's demand. Thus, this has reasonably contributed to the slow adoption rate as industry players are still reluctant to invest as it is not wholly compulsory yet as per implemented in other BIM leading countries like USA, UK and European countries [32]. Therefore, it is significant for the government and authorities to look into these challenges and further developing a strategic planning in promoting BIM for highway industry.

It is also important for industry players and professionals from all level to understand the concept of BIM and its capabilities in transforming the highway design. In year 2014, Malaysia's authority like Jabatan Kerja Raya (JKR) has established Garis Panduan dan Piawaian BIM JKR and Garis Panduan Pengunaan Template JKR which can suit several of projects. Malaysian Highway Authority (MHA) has also started to look into BIM application for highway project and the midst of data collection but yet to execute any further initiatives. As BIM is relatively new for highway project in Malaysia, there is also the need to introduced more pilot projects to invite more participation from private sector as what has been done in Singapore and Hong Kong [33]. For instance, currently Pan Borneo Highway is one of the projects that has set a benchmark in leveraging the BIM adoption for highway project. Thus, this approach will somewhat gradually increase the numbers of expertise in the field as well as expand the exposure to BIM capabilities among designers. Sinoh et al. (2020) has also recently indicates the roles of government as one of the solutions in undertaking the impediments factors of BIM adoption through expanding BIM competency in the industry, promoting BIM awareness and also BIM deployment in higher learning education [34].

5. Conclusions

This paper reveals that BIM capabilities in changing the direction of highway industry from the highway design consultant's perspective. As the adoption in highway industry is considered quite new, the numbers of consultants who are using BIM tool and software in designing highway projects are very limited. Based on the findings from the interviews conducted, all highway design consultants from various disciplines are aware on the BIM technology and willing to migrate from their current design practices towards BIM design workflow.

The results shows that the top three challenges of the current highway design practices are non-collaborative workflow between multi disciplines, higher likelihood of design changes during construction and risks of design conflicts between disciplines. BIM technology on the other hand has the capabilities in transforming the current highway design process towards providing better visualization through 3D modelling, coordinated and collaborative workflow between multi-disciplines, and ability to perform design optimization and clash analysis prior to construction phase. However, BIM adoption in highway industry in Malaysia from consultant's perspective has been hindered due to the high cost of BIM enabled software and hardware as well as high investment on resources and manpower and limited urges from the government and relevant authorities. These data are deemed to be the significant in understanding the current BIM implementation aspects among highway designers.

In summary, although the utilization is not very common within the highway industry in Malaysia with diverse limitations and barriers, players in the industry are now has starting to recognize its potentials and slowly has steering the industry towards more technological and digitalization advancement. As highway projects nowadays has become more challenging and complex, relevant authorities has also begin to initiate the BIM adoption in for highway projects in Malaysia. Hence, this study is deemed to be as a basis towards further research on BIM strategic implementation planning for highway industry

6. References

[1] Zahrizan Z, Mohamed Ali N, Tarmizi Haron A, Marshall-Ponting A, and Abd. Hamid Z 2014 Exploring the Barriers and Driving Factors in Implementing Building Information Modelling (BIM) in the Malaysian Construction Industry: A Preliminary Study, *The Institution of Engineers, Malaysia* **75** 1

[2] Ahmad Latiffi A, Mohd S, Kasim N, and Fathi M 2013 Building Information Modeling (BIM) Application in Malaysian Construction Industry, *Inter. J. of Const. Eng. and Management 2013*, **2** 1-6

[3] Gerges M, Ahiakwo O, Jaeger M, and Asaad A 2017 An Investigation into The Implementation Of Building Information Modelling In The Middle East, *J. of Info. Tech. in Const. ISSN* 1874-4753

[4] Zakaria S A S and Abdul Rahim N S 2020 Building Information Modelling (BIM) Contribution Towards Sustainability: Awareness Of Malaysian Contractors, *Inter. J. of Property Sci. (E-ISSN: 2229-8568)* **10** 16-25

[5] Costina A, Adibfara A, Hub H and Chenc S S 2018 Building Information Modeling (BIM) for transportation infrastructure –Literature review, applications, challenges, and recommendations, *Automation in Const.* **94** 257–281

[6] Akob Z, Hipni M Z A and Abd Razak A A A 2019 Deployment of GIS + BIM in the construction of Pan Borneo Highway Sarawak, Malaysia *IOP Conf. Series: Mater. Sci. and Eng.* **512** 012037

[7] National Institute of Building Sciences (NIBS) 2007 *United States National Building Information Model Standard* (United States: National BIM Standard)

[8] Azam Haron N, Raja Soh R, and Harun A 2017 Implementation of Building Information Modelling (BIM) in Malaysia: A Review, *Pertanika J. Sci. & Technol.* **25** 661 - 674

[9] Assaf S A and Al-Hejji S 2006 Causes of delay in large construction projects, *Inter. J. of Project Management* **24** 349-357

[10] Thapanont P, Santi C and Pruethipong X 2018 Causes of delay on highway construction projects in Thailand *MATEC Web of Conf.* **192** 02014

[11] Sohu S, Jhatial SS, Ullah K, Lakhiar M T and Shahzaib J 2018 Determining the Critical Success Factors for Highway Construction Projects in Pakistan, *Eng., Tech. & App. Sci. Res.* **8** 2685-2688

[12] Rahman R A, Radzi A R, Saad M S H and Doh S I 2020 Factors affecting the success of highway construction projects: the case of Malaysia *NCWE & ISSCE 2019, IOP Conf. Series:Mater. Sci. and Eng.* **712** 012030

[13] Cooperative Research Centre (CRC) for Construction Innovation 2007 *Adopting BIM for Facilities Management: Solutions for Managing the Sydney Opera House,* Cooperative Research Center for Construction Innovation (Australia)

[14] McGraw-Hill Construction 2012 *The Business Value of BIM for Infrastructure* Smart Market Report (Bedford: McGraw-Hill Construction) pp-14-15

[15] Anderson R 2010 *An Introduction to the IPD Workflow for Vectorworks BIM Users* VP Integrated Practice (Nemetschek Vectorworks)

[16] Shetwi R, Farooq B and Popa C 2015 Optimizing Large-scale Transportation Infrastructure Projects using Building Information Modelling *Canadian Transportation Research Forum 50th 818 Annual Conference-Another 50 Years: Where to From Here* (Montreal)

[17] Strafaci A 2010 *BIM for road and highway design* Visual Technical Position IT pp 55-57 (June)
[18] Marzouk M and Hisham M 2011 Optimizing the layout of bridges construction sites using genetic algorithms *3rd International and the 9th Construction Specialty Conference* (Ottawa) 1-8
[19] Queensland Department of Transport and Main Roads 2017 *Guideline Building Information Modelling (BIM) for Transport and Main Roads A guide to enabling BIM on Road Infrastructure Projects* (Australia)
[20] Findley D J, Schroeder B J, Cunningham C M, and Brown T H 2015 *Highway Engineering: Planning, Design, and Operations* (Butterworth-Heinemann)
[21] Walsh I D, Hunter R N, Darral L, Matthews P, Jameson P and Thorp. J 2011 *ICE Manual of Highway Design and Management* (London: Institute of Civil Engineers)
[22] Anderson S D and Fisher D 1997 *Constructability Review Process for Transportation Facilities* NCHRP Rep. 390, (Washington: National Cooperative Highway Research Program pp 39
[23] Lembaga Lebuhraya Malaysia 2008 *Guidelines for Malaysian Toll Expressway System – Design Standards* (Malaysia)
[24] Nabors D and Soika,J 2013 *Road Safety Audit Case Studies: Using Three-Dimensional Design Visualization in the Road Safety Audit Process* Final Report Doc No: FHWA-SA-14-003 (US: Department of Transportation Federal Highway Administration)
[25] Rajendran P and Gomez C P 2012 Implementing BIM for Waste Minimization in the Construction Industry: A Literature Review *2nd International Conf. on Management* (Kedah)
[26] Wan Mohammad W N S, Abdullah M R, Ismail S and Takim R 2018 Overview of Building Information Modelling (BIM) Adoption Factors for Construction Organizations *IOP Conf. Series: Earth and Envi. Sci.* **140** 012107
[27] Ibrahim F S, Shariff N D, Esa M and A. Rahman R 2019 The Barriers Factors and Driving Forces for BIM Implementation in Malaysian AEC Companies, *J. of Adv. Res. in Dynamical and Control System* **11** 08
[28] Roslan A F, Abd. Hamid Z, Mohd Zain M Z, Mat Kilau N, Dzulkalnine N and Hussain A H 2019 Building Information Modelling (BIM) Stage 2 Implementation Strategy for The Construction Industry in Malaysia, *MCRJ Special Issue* **6** 153-161
[29] Yaakob M, Wan Ali W N A and Radzuan K 2016 Critical Success Factors to Implementing Building Information Modeling in Malaysia Construction Industry, *Inter. Review of Management and Marketing* **6** 252-256
[30] Yung A T B, Aminudin E, Liat C N S, Neardey M, Zakaria R, Hamid A R A, Ahmad F & Yong LY 2020 Adoption of Building Information Modelling in Malaysia Road Construction *IOP Conf. Series: Mater. Sci. and Eng.* **943** 012061
[31] Yusof A and Shah A 2017 *MyBIM promotes modern construction technology* Retrieved on March 20, 2021 from https://www.nst.com.my/business/2017/11/305294/mybimpromotes-modern-construction-technology
[32] Othman I, Al-Ashmori Y Y, Rahmawati Y, Mugahed Amran Y H and Mohammed Al-Bared M A 2021 The level of Building Information Modelling (BIM) Implementation in Malaysia, *Ain Shams Eng. J.* **12** 455–463
[33] Zakaria Z, Mohamed Ali N, Marshall-Ponting A and Haron A T 2012 An Exploratory Study on the Potential of Implementing Building Information Modelling (BIM) in Malaysian Construction Industry: Lesson Learnt from Singapore and Hong Kong Construction, *Industry Inter. J. of Civil Eng. and Geo-Envi.* **3** 21802742
[34] Sinoh S S, Ibrahim Z, Othman F and Muhammad N L N 2020 Review of BIM literature and government initiatives to promote BIM in Malaysia *2nd Inter. Conf. on Mater. Tech. and Energy* **943** 012057

Acknowledgements
Sincere thanks to Universiti Teknologi Malaysia for financial support from High Impact Research Grant (HIR) Q.J130000.2451.04G54 as well as the consultant's participation in this study.

IOP Conf. Series: Earth and Environmental Science **971** (2022) 012004 doi:10.1088/1755-1315/971/1/012004

The Impact of Train Operation Time with Ground Borne Vibration Induced by Railway Traffic

M K A M Lazi[1*], M A Adnan[1] and N Sulaiman[1]

[1] Faculty of Civil Engineering, Universiti Teknologi MARA, Shah Alam, Selangor, Malaysia
*Corresponding author: tmm_afxan@yahoo.com

Abstract. Ground-borne vibration caused by rail transport has detrimental effects on residential areas along the railway tracks. The residents who are subjected to daily ground-borne vibration may also experience more severe health issues over the long term. This involves the mental well-being of the people who live nearby. The efficiency and quality of life of residents may also theoretically be impaired by direct exposure to ground-borne vibration caused by trains. As the government decided to invest further in public infrastructure, especially in the railway transport system, the railway transport system in Malaysia is changing to a higher level. Complaints about human discomfort and irritation caused by the ground-borne vibrations created by rail transport have been reported to increase as a result of the expansion of coverage for rail transport. Many previous studies have investigated the influence of different parameter variables on ground-borne vibrations. The ground-borne vibrations are considered as the dependent variable by some of the inquiries, and the other parameters are classified as independent variables. However, depending on the nature and source of the study performed, the results of the independent variables will differ from the researchers. This analysis aims to examine one of the variables which is the train operating hour in terms of peak and non-peak hours against the ground-borne vibrations caused by the train using local conditions.

1. Introduction

The Malaysian government decided to continue investing in public infrastructure, especially in the rail transport system and make the rail transport system in Malaysia is changing to a higher level. This included the extension of Phase 2 of the Mass Rapid Transit (MRT) with an investment of RM 32 billion after the completion of Phase 1 of the MRT, which opened to the public on 17 July 2017 [1]. In addition, RM 6 billion was invested in the Phase 3 Light Rail Transit (LRT) project, which is now under construction, and a further RM 5.9 billion was invested in upgrading the Keretapi Tanah Melayu Berhad (KTMB) services under the Klang Valley Double Track (KDVT) project. The East Coast Rail Link (ECRL) with an investment of RM 55 billion and the future mega project from Kuala Lumpur to Singapore High Speed Rail is another noteworthy mega rail transport project [2].

Cardona et al. [3] found that concerns about human discomfort and irritation caused by ground-born vibrations generated by rail traffic have been reported to increase due to the expansion of coverage for rail transport. Vibration is produced over a wide range of amplitudes and frequencies by rail traffic. The destabilisation of embankments, settlement of foundations, damage to track components, and damage to surrounding structures may be caused by high vibration amplitudes.

As the wave propagation through soil and the structures, the vibration produced by railway traffic can be elaborated. The induced energy can cause irritability and irritation or annoyance to individuals living in the affected areas and could have a negative effect on sensitive machinery. The vibration generated by the trains' movement will have a noticeable effect on the environment around them. The

IOP Conf. Series: Earth and Environmental Science **971** (2022) 012004 doi:10.1088/1755-1315/971/1/012004

impacts are influenced by the type of trains, type of foundations and the quality of the rail track. These factors determine the level of vibration produces [4].

He et al. [5] highlighted that the environmental effect of railway traffic-induced vibrations has become a concern in the planning and construction of railway networks over the last few years. In addition, with the increase in the standard of living among residents, environmental quality standards and priorities are becoming more important. Vibration pollution refers to ground vibrations, including human movements, from railways and other traffic vehicles, which have an adverse effect on human comfort and the psychological wellbeing of people living near the source of the vibration. Ground-borne vibrations will also impact the protection of structures, the operation of sensitive machinery and equipment and the working conditions in the affected areas.

2. State of Arts

2.1. Ground borne vibration due to rail traffic

It is possible to interpret the phenomenon of vibration caused by train traffic as wave propagation through structures and the ground. The created energy can create irritation for individuals living near the railway tracks and trigger interference with sensitive machinery. Due to the movement of the train at certain locations, a certain degree of response by the residents will be anticipated [4]. Once the railway is in service, the vibration will be produced by trains on the ground. In adjacent structures or buildings, the degree of vibration depends on the forces applied by the train to the rails and the attenuation between the buildings and the rails. This attenuation should be built to achieve the necessary vibration limit for the buildings [6].

The key characteristics of vibration, in general, are frequency and magnitude. The amplitude is the rate of oscillation. There are three distinct ways of describing the magnitude of vibration; velocity, acceleration and displacement. The displacement is measured from the datum to the maximum peak in one direction, known as the peak displacement, or the difference from the maximum peak displacement in the opposite direction. Whereas the magnitude is expressed from one end velocity to another end velocity with the difference between the maximum velocities in one direction and from end to end in the opposite direction. Mm/s is the measure commonly used for speed. On the other hand, acceleration is more related to the vibration energy involved in the movement of oscillatory vibration and is transmitted either through peak acceleration or through peak to peak acceleration. M/s^2 or in the unit of gravity, g, is the unit commonly used to express acceleration. Frequency is the other property of vibration. Frequency is the rate at which the cycles of oscillation are completed over a period of time and measured in inverse second(s). Frequency is expressed in hertz (Hz). Motion usually includes more than one frequency, so determining the frequency of vibration contained is crucial [7].

In general, there are three correlation elements between excessive ground-borne vibration due to rail traffic, i.e. between the source, propagation, and the receiver, as shown in Figure 1. The first component of the relationship is the source where vibrations are produced, followed by the propagation of the vibrations and finally the receiver, which is typically the building. Each variable has its own parameters and properties that influence the levels of vibration. Even though the links or elements function as a sequence, sometimes it can be helpful to study each of it separately when designing the measures of vibration-reduction [8].

IOP Conf. Series: Earth and Environmental Science **971** (2022) 012004 doi:10.1088/1755-1315/971/1/012004

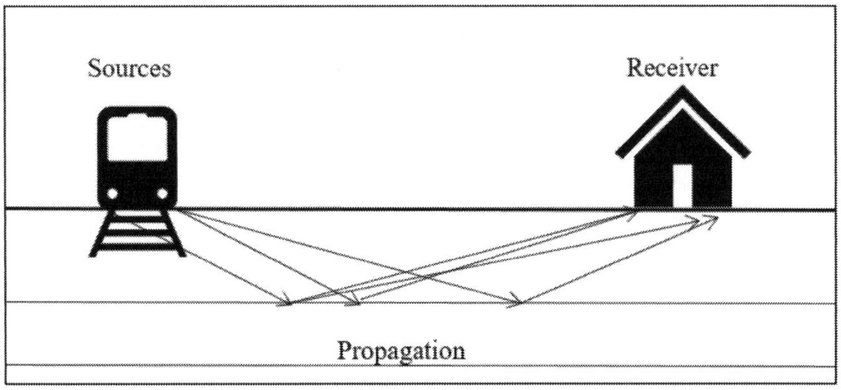

Figure 1. The three correlation elements of the ground-borne vibration.

2.2. Propagation

Typically, after the ground-borne vibrations are produced in the railway, vibration propagates to the surrounding area through the soil media. Propagation characteristics are dependent on the properties of the soil, parameters and distance from the source. The soft earth tends to have a lower frequency relative to stiff or hard ground. The frequency of the ground vibration ranges from 5 to 10Hz in the soft ground profile region. Ground vibration can travel longer distances from its source at these low frequencies. The key factor affecting significant ground vibration is soft soil formation such as silt or soft clay, which may cause discomfort for individuals living 100 to 200 m away from the rail tracks [9]. Persson [8] also reported that the propagation velocity for each wave form is controlled by the conditions of the ground soil as shown in Figure 2 below.

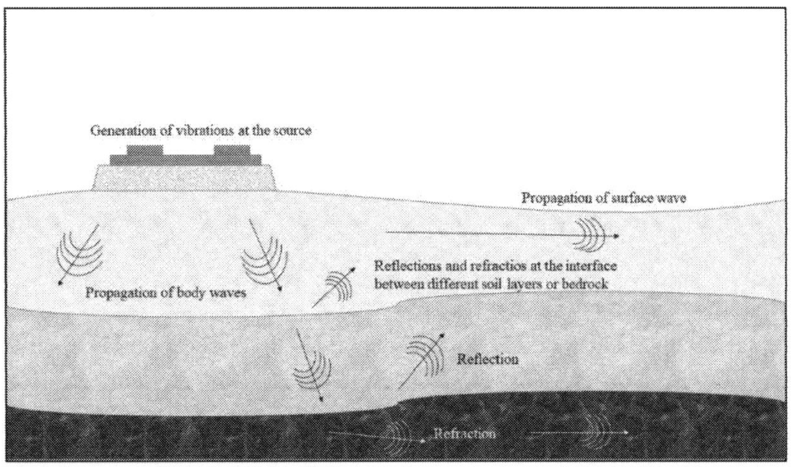

Figure 2. Propagation movement for ground vibration.

The vibration rate is determined by the peak particle velocity (PPV) and it is the most known and used vibration measurement. Vibration limits using PPV are recommended by most guidelines and regulations. The maximum particle velocity over the total recorded time are regarded as the peak particle velocity for each recorded waveform as shown in Figure 3 [10].

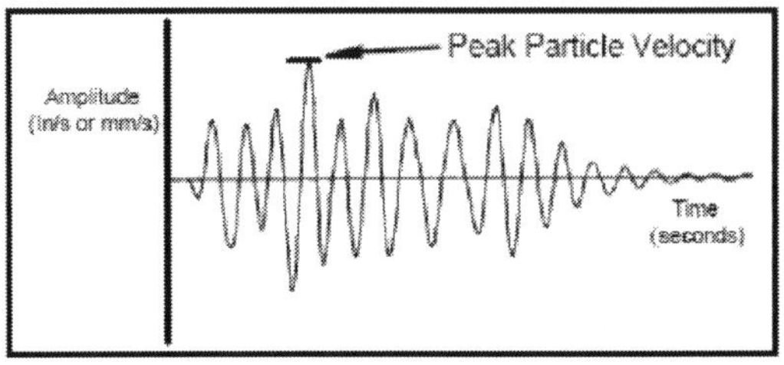

Figure 3. The definition of peak particle velocity [10].

The PPV is the maximum instantaneous velocity of a particle at a point in a particular time interval. Ground particles vibrate with varying particle velocity when the disturbance from the source of vibrations propagates away from the source with a certain amount of wave velocity. The motion at a given position along the propagation path is represented in three mutually perpendicular components (usually transverse, vertical, and radial or longitudinal). All three components must be calibrated at the same time to ensure the PPV is correctly determined [11].

2.3. The impact of train operation time on the ground borne vibrations
To date, research has yet to be performed on the effect of train operating hours on ground-borne vibrations, either at peak or non-peak hours. Nevertheless, research that took into account daytime and night time train activity is already available with regard to the ground vibration. Peris et al. [12] found that, relative to daytime, the concern about irritation and discomfort among the affected population was greater during evening to night time. The survey, however, was focused on the general time frame from 07.00 pm to 07.00 am, whereby night time was taken into account. Furthermore, from this study, no actual vibration data were taken. A research on human discomfort due to noise and ground-borne vibrations has also been conducted by Yokoshima et al. [13]. The processing of data took place from 07.00 am to 10.00 pm. The division of peak and non-peak hours was not considered in the report. In addition, there was no discussion of the relationship between daytime and night time data with respect to the vibration caused by the trains. To explore further on the impact of the train operation time between peak and non-peak hour, this study was conducted by using local conditions. The hypothesis of the train operation time during peak hour would produce higher reading of ground-borne velocity vibrations will be analyse.

3. Methodology

3.1. Case study area
The data collection was carried out along the Keretapi Tanah Melayu (KTM) railway route, from Padang Jawa Station, Shah Alam to Klang Station, Selangor. The railway track under study is a two-way railway with two (2) train routes, one to Kuala Lumpur and the other to Pelabuhan Klang, Selangor. Nine (9) separate sites along Padang Jawa Station, Shah Alam to Klang Station, Klang were involved in this analysis. To discern the spectrum of vibration magnitude caused by the trains running on the railway tracks, these site locations were chosen. Because of the strategic areas, the sites were selected as there were several residential areas along the track that were threatened by the occurrence of ground-borne vibration. This study also concentrated on suburban areas with landed-type houses. Figure 4 shows the map of locations of study in Padang Jawa Station, Shah Alam until Klang Station, Klang.

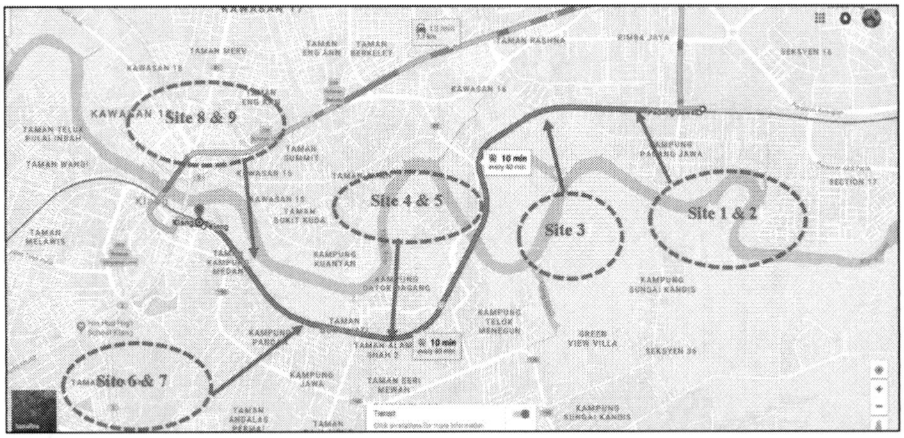

Figure 4. Maps of nine site study located in Padang Jawa until Klang, Selangor.

The route was chosen because to the presence of landed residential buildings in the areas adjacent to the railway tracks. There were no vibration barriers located at the selected locations. The distance range for this analysis from the residential area to the railway tracks was less than 30 metres. This distance was chosen sufficient for this study since there are houses within this range, allowing for the assessment of human annoyance produced by the vibration. Train parameters such as the type of trains and the time conditions that were either during peak hours or non-peak hours were the field data gathered during the site survey. The ground-borne vibration measurements consisting of vertical wave vibration, horizontal and radial wave vibrations were another data collection package. A seismograph mounted at the site under research was used to obtain these measurements. As for the site locations, from Padang Jawa station to Bukit Badak station, three (3) sites were selected. The other six (6) sites chosen were situated between Klang Station and Bukit Badak Station. Hence, nine (9) area were picked altogether. To be as similar as possible to the landed residential areas, the sites were selected. In order to achieve different distances from residential areas to sources, different locations were chosen. Shah Alam and Klang are among the most established in Malaysia and have the largest population. It is possible to overserve the populated area well from aerial views. The railway tracks under study were located near to the residential areas and the tracks run above ground. Figure 5 and Figure 6 demonstrate the aerial view of the positions of sites between Padang Jawa Station, Shah Alam and Klang Station, Klang, from Site 1 to Site 9. The geological conditions of all of the sites chosen were soft soil geological conditions. According to Sulaiman [7], soft soil is more influenced by vibrations, especially those with higher magnitudes, than other types of soil.

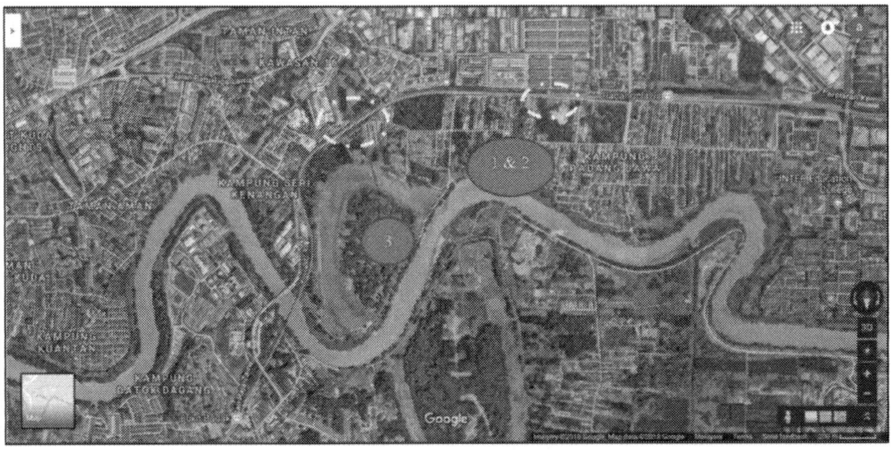

Figure 5. Aerial view of site location for the lane of Padang Jawa Station to Bukit Badak Station consist of sites 1, site 2 and site 3.

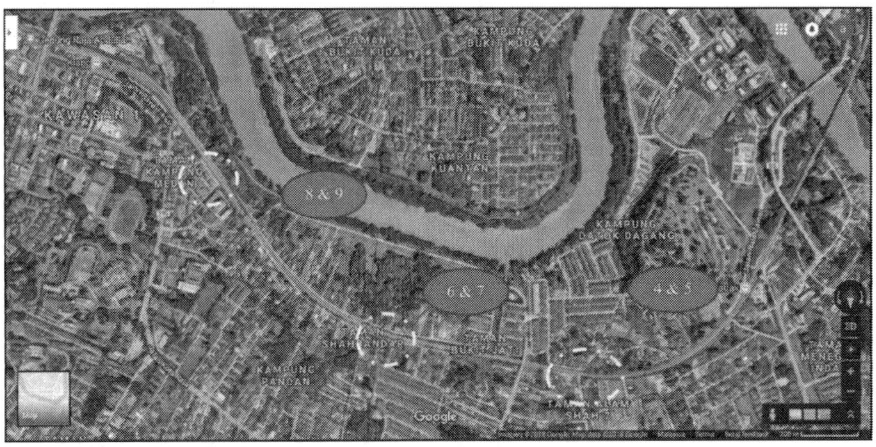

Figure 6. Aerial view for site 4 until site 9 located along the lane of Bukit Badak Station to Klang Station.

3.2. Instrumentation and equipment strategy setting up

For this analysis, using a seismograph metre or also referred to as the Mini-SEIS, five repetitions of data reading and measurements were performed for each site position. Data were registered and grouped into two classifications; at two distinct distances at each site location during peak hours and non-peak hours. This was done because of the fact that the load borne by the trains would vary during peak hours and non-peak hours, since it was anticipated that passenger volumes would be higher during peak hours. Figure 7 and Figure 8 provide a detailed diagram showing the position of the Mini-SEIS placement. Two data collection sessions were carried out with two hours of experimentation for each session for the experiments performed in the morning. This was to gather the data from the peak and non-peak hour sessions. In the evening, another two sessions were performed to collect from the sources of the peak and non-peak hour data from different distance positions of the Mini-SEIS. The peak hour and non-peak hour were classified based on the train interval from the train schedule [14]. Different location of Mini-SEIS was taken to get the different distance within the radius range of 30m.

Table 1. Timeline of data collection based on location of Mini-SEIS.

Train Operation Time		**Location of mini-SEIS and distance taken.**	**Type of Train**
Peak Hour	**Non-peak Hour**		
6.30am – 8.30am	9.00am – 11.00am	1 = d1 & d2	Commuter
5.30pm – 7.30pm	3.00pm – 5.00pm	2 = d3 & d4	Commuter

Note: Location of mini-SEIS 1 & 2 and distance d1,d2,d3 and d4 refer to Figure 7 and Figure 8

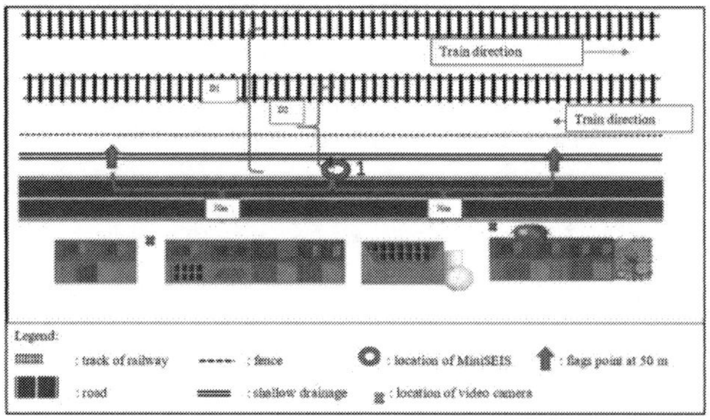

Figure 7. The location of Mini-SEIS at location 1 for morning session of each site locations.

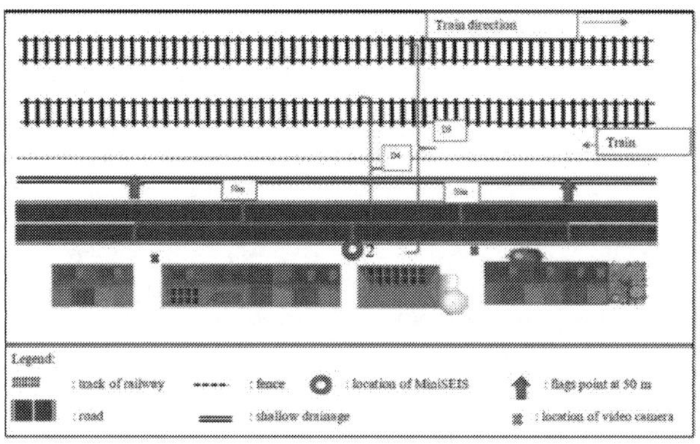

Figure 8. The location of Mini-SEIS at location 2 for evening session of each site locations.

In this data collection, using a Mini-SEIS digital seismograph, ground vibrations were measured. As a data logger, a Mini-SEIS consists of a microphone, a geophone transducer and a display [15]. As shown in Figure 9, the microphone and geophone transducer were mounted at the ground level of the residential areas and were connected to the Mini-SEIS display. Using a GPS metre, the Mini-SEIS coordinates were measured. During data analysis, the times shown by the Mini-SEIS were registered to tally with the actual time of the trains passing through the specified points to denote the data.

IOP Conf. Series: Earth and Environmental Science **971** (2022) 012004 doi:10.1088/1755-1315/971/1/012004

Figure 9. Mini-SEIS installed at the site of study.

4. Result and Analysis

4.1. Descriptive statistical analysis of the empirical peak particle velocity data

An overview of the vibration empirical peak particle velocity (PPV) can be encapsulated in a succinct statistical summary consisting of all variables, including mean, maximum, minimum, skew, kurtosis, and standard deviation information. Table 2 shows the descriptive analysis of the empirical data recorded.

Table 2. Descriptive statistic of empirical data.

Variable	Mean	StDev	Minimum	Median	Maximum	Skewness	Kurtosis
PPV	1.0823	0. 542	0.1588	1.0160	3.4608	0.59	2.27
Train Operation Time	0.59	0.49	0.00	1.00	1.00	-0.37	-1.88

4.2. Correlation analysis of the peak particle velocity and train operation time

The correlation analysis is the method practiced in diagnosing the possible relationship between the dependent and independent variables that are used in the development of regression models. The correlation analysis hypothesis test can be stated as follows:

H_0 = There is no correlation between two variables

H_1 = There is a correlation between two variables

The variable of train operation time and peak particle velocity was evaluated using correlation analysis and the outcome for the variables are displayed as shown in Table 3. Each column stands for r-value and p-value. The p-values are located at the bottom rows while the r-values are located at the top rows.

Table 3. The correlation of train operation time (peak and non-peak hour) with PPV.

Variable	PPV
Train Operation Time	0.038* 0.508#

* r-value
\# p-value

IOP Conf. Series: Earth and Environmental Science **971** (2022) 012004 doi:10.1088/1755-1315/971/1/012004

From the result of correlation above, it shows that the time of the train operation (peak and non-peak hour) did not have significant impact on the PPV as the p-values were more than 0.05. The hypothesis describing that the train operation time during peak hour would produce higher reading of ground-borne velocity vibrations was consequently rejected.

4.3. Scatterplot fundamental of commuter rail traffic vibrations
Figure 10 below shows the scatterplot of data received from the nine sites from this study that being categories between train operation time either peak or non-peak hour.

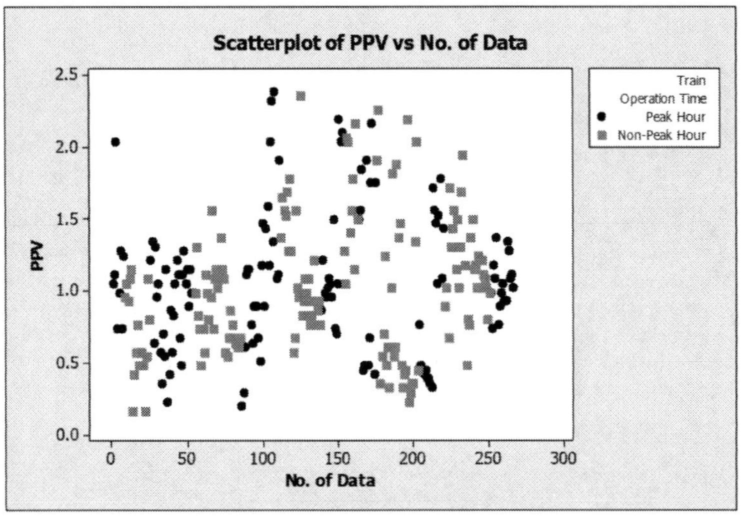

Figure 10. Scatterplot of peak particle velocity.

Figure above shows that the distribution of ground borne vibration induced by train was almost in similar pattern for the two categories which is peak hour and non-peak hour. Figure 11 below shows the scatterplot of data received from the nine sites from this study based on the time of data collection following as in schedule.

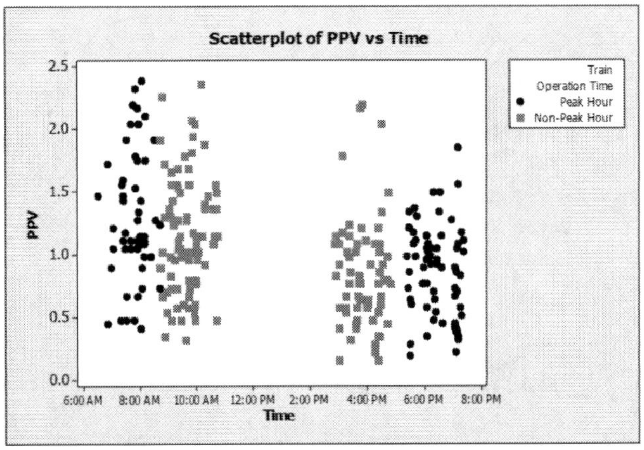

Figure 11. Scatterplot of peak particle velocity vs train operation time.

From the figure above, it shows that the peak particle velocity of ground borne vibrations induced by train was scattered in same pattern between peak hour and non-peak hour for each session. During peak hour of data collection taken from 6.30am until 8.30am, the plot almost similar pattern with non-peak hour data during 9.00am until 11.00am from the same distance of sources to receiver. During evening session with different distance from morning session, data collected from non-peak hour from 3.00pm until 5.00pm also shows the same pattern with data collected for peak hour which is from 5.30pm until 7.30pm. Figure 12 shows the scatterplot of peak particle velocity versus peak and non-peak hour.

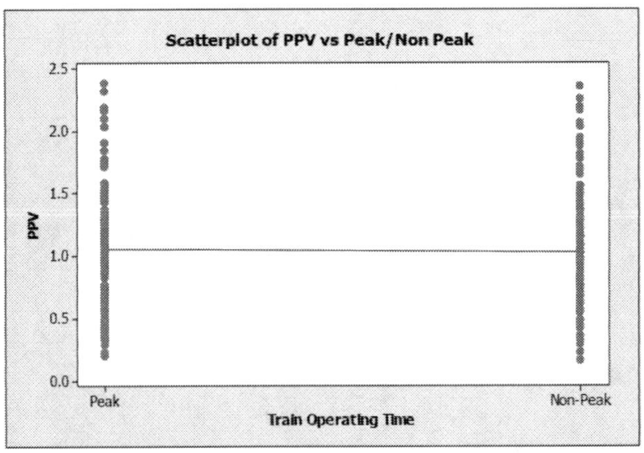

Figure 12. Scatterplot of peak particle velocity versus peak and non-peak hour.

From Figure 12, the trend of the peak particle velocity quantified in mm/s and the train operation time between peak and non-peak shows almost no sloping. Thus, it can be deduced that train operation time did not give an impact towards the ground borne vibrations. The impact of train operation hour (peak and non-peak hour) can be ignored during data collection as being practice by other researchers

[12][13]. From the literature findings, one of the highest factor that allowed the ignorance of train operation time was due to the usage of suspension in the train systems. Therefore, the disparity between the operating hours of the train does not influence the vibration of the ground. A research by Sun, Gong and Zhou [16] concluded that with the suspension systems, the vibrations from the train body could be minimised. The focus is on the resonance frequencies generated by the train or wheel interface. The increase in the usage of slab track in modern railway tracks is due to its lower maintenance cost and higher stability. In reducing ground vibrations, it is also more effective [17]. A study has also shown that a good suspension can minimise ground vibrations by more than 70% [18].

5. Conclusion

This study aims to examine the hypothesis that train running time during peak hours will produce a higher reading of ground-borne velocity vibrations compared to non-peak hours due to higher passenger volumes during peak hours. However, the analysis shows that the time of operation of the train did not influence the peak particle velocity of ground-borne vibrations. To disregard the variables of train operation hours for the respective scope of railway conditions may be a guideline for a future analysis in Malaysia.

6. References

[1] Intan F Z 2018 *LRT3 Cost can be Reduced by RM6bil.* The Star (July 10)
[2] Hemananthani S KTM Double Track Project to be Reviewed, says Loke The Star (June 12)
[3] Cardona J, Romeu J, Arcos R, and Balastegui A 2010 A Ground-Borne Vibration Assessment Model For Rail Systems at Grade *Inter Noise 2010 noise and sustainability* (Portugal)
[4] Monteiro M 2009 Vibration Induced by Trains: Case Study: Extension of Track Platform of Gare do Oriente
[5] He X, Yanmei C, Guido D R, and Geert D 2007 Environmental Problems of Vibrations Induced by Railway Traffic *J. of Railway Sci. and Eng.* **1** 44-51
[6] Arup 2015 *Ground-Borne Noise and Vibration from Rail Tunnels:Impacs of Tunnels in the UK* Report (London: Department for Transport)
[7] Sulaiman N 2017 *Empirical Modelling of Ground-Borne Vibration from Road Traffic* PhD Thesis (Selangor: University Putra Malaysia.)
[8] Persson N 2016 *Predicting Railway-Induced Ground Vibrations* Master Thesis (Sweden: LUND University)
[9] Madshus C, Bessason B, and Harvik L 1996 Prediction Model for Low Frequency Vibration From High Speed Railways on Soft Ground, *J. of Sound and Vibration*
[10] Alcudia A D, Stewart R R, Eliuk N, and Espersen R 2007 *Vibration and Air Pressure Monitoring of Seismic Sources* CREWES research report (Alberta: Compton Petroleum Corporation)
[11] Avellan K, Belopotocanova E, and Puurunen M 2017 Measuring, Monitoring and Prediction of Vibration Effects in Rock Masses in Near-Structure Blasting *Symposium of the International Society for Rock Mechanics* **191** 504-511
[12] Peris E, Woodcock J, Sica G, Moorhouse A T, and Waddington D C 2012 Annoyance due to Railway Vibration at Different Times of The Day *The J. of the Acoustical Society of America* **131**
[13] Yokoshima S, Morihara T, Sato T, and Yano T 2017 Combined Effects of High Speed Railway Noise and Ground Vibrations on Annoyance *Inter. J. of Envi. Res. and Public Health* **14** 845
[14] Lazi M K A F, Mustafa M and Rahman Z A 2016 Assessing Pedestrian Behavior and Walking Speed on Staircase; A Review, *Inter. Civil and Infra. Eng. Conf.* (Singapore: Springer) 1019-1029.
[15] White Industrial Seismology Inc 2009 *Mini-Seis Digital Seismograph: Operating Manual* PbS thin films. Thin Solid Films **519** 4917-4922
[16] Sun Y, Gong D, and Zhou J 2016 Study on Vibration Reduction Design of Suspended Equipment of High Speed Railway Vehicles *J. of Physics: Conf. Series* (IOP SCIENCE) **744** 012212

[17] Kaewunruen, Sakdirat, Remennikov, Alex 2016 Current State of Practice in Railway Track Vibration Isolation: An Australian Overview *Australian J. of Civil Eng.* **14** 63-71
[18] Tuler M V and Kaewunruen S 2017 Life Cycle Analysis of Mitigation Methodologies for Railway Rolling Noise and Groundborne Vibration, *J. of Envi. Management* **191** 75-82

Acknowledgement
The authors would like to thank the authorities from the Faculty of Civil Engineering, Universiti Teknologi MARA (UiTM) for their constant support and encouragement. Appreciation goes to the Research Management Institute (RMI, UiTM) and Ministry of Higher Education Malaysia (MOHE) for the financial supports FRGS Grant scheme: RMI File No: 600-IRMI/FRGS 5/3 (008/2019), Sponsorship File No: FRGS/1/2018/SKK06/UITM/02/5.

IOP Conf. Series: Earth and Environmental Science **971** (2022) 012005 doi:10.1088/1755-1315/971/1/012005

Application of Alternative Filler in Asphalt Mixture: An Overview from Indonesia Perspective

M R A Maha[1], M K Idham[1*], M R Hainin[1,2], S N Naqibah[1] and N A Amirah[1]

[1] School of Civil Engineering, Faculty of Engineering, Universiti Teknologi Malaysia (UTM), 81310 Skudai, Johor, Malaysia
[2] College of Engineering, Universiti Malaysia Pahang (UMP), 26600 Pekan, Pahang, Malaysia
*Corresponding author: khairulidham@utm.my

Abstract. A good quality and proper quantity of filler material is one of the significant factors in producing asphalt mixture. Many studies have been highlighted waste material as an alternative filler to improve asphalt mixture performance. Hence, this paper reviewed the performance of asphalt mixture using rice husk ash (RHA) as filler in terms of the Marshall properties. It shows that 6% RHA by weight of total aggregate was commonly used and depicted better in Marshall properties than the others. This study also found that the RHA can be used to replace up to 100% of the conventional filler. Even though, higher RHA content could significantly reduce the stability yet it still within the acceptable range and could be applied at medium traffic load road. In return, the higher the RHA content the more waste material can be utilized for road construction. Thus, minimized the usage natural source materials.

1. Introduction

Flexible pavement constructed with four layers namely surface course, base course, subbase and subgrade. While, surface course constructed in two layers which are wearing course (WC) and binder course (BC). Proper asphalt mixture design of WC and BC courses is needed to provide a good pavement performance throughout its service life. Improper design and quantity are one of the crucial issues that leads to increasing air void, causing permanent deformation and reducing its performance. In general, asphalt mixture consists of asphalt, coarse aggregate, fine aggregate and filler. Asphalt mixtures is a well graded mixture containing coarse aggregate (50-65%), fine aggregate and filler (35-50%), asphalt (5-8%) of total mass of aggregate [20]. Filler is the component that are dry, free from lumps and passing sieve no 200 (75 Micron). Commonly, materials used as filler are Portland cement, lime ash or fly ash [1]. Among them, filler is one of the most important components of asphalt concrete. It plays a significant role on the properties of asphalt mixtures. Properly designed and constructed surfaces of this general type are capable of carrying almost unlimited volumes of passenger, mixed, or truck traffic, provided only that they are supported by adequate foundation structures. The majority of these surfaces might be expected to have an economic life of 20 years or more[21]. Filler is one of the factors that influenced the asphalt mixture by increasing the density and minimising the mixture's permeability. Filler also affects the elasticity of the mixture and its sensitivity to water [2]. Production of asphalt mixture involves large quantity of natural sources materials and including filler. In order to provide possible alternative for filler, many potential materials have been studied by researchers and one of them is rice husk ash (RHA). It is the result of burning the waste of rice husks.

Since 90% of global production of rice is form Asian countries such as China, India, Indonesia, Bangladesh, Vietnam and Thailand, abundant of rice husk waste can be collected for research purposes.

IOP Conf. Series: Earth and Environmental Science **971** (2022) 012005 doi:10.1088/1755-1315/971/1/012005

As in other countries, Indonesia has also conducted studies on the use of RHA in road construction for the past 20 years. Hence, this paper aimed to review the usage of RHA as a filler in asphalt mixture based on the study conducted in Indonesia.

2. Material and Laboratory Test

2.1. Filler

Filler content affects the process of mixing, spreading and compaction of asphalt mixture. In addition to its size which must be relatively fine, the filler material must have certain properties such as cementitious properties when exposed to water and high adhesion with other aggregates. Among the materials that have cementation properties when exposed to water and are widely used as filler materials are rock ash, fly ash, gypsum, Portland cement (PC) and tiles ash. Previous studies also proved that rice husk ash (RHA) is believed to have good properties as a compactor filler because it has cementation properties, in addition to its relatively small grain size. Besides, the abundance of rice husk has given advantage to be used as alternative filler where the cost is cheaper and easier to obtain.

2.2. Rice Husk Ash (RHA)

Rice husk or grain skin is waste from rice milling process where husk is the second largest part after rice. Milling process produced about 65% rice and 20% rice husk [3]. While rice husk ash (RHA) as shown in Figure 1 is the result of burning the rice husk. It is the 15 % of combustion waste from rice husks, which are usually used as fuel in burning raw bricks.

Figure 1. Rice husk ash.

Rice husk is composed of a network of cellulose fibres that contain high silica content in the form of tough fibres. Based on previous studies, RHA consist of high content of silica (SiO_2) element with a range of 72.29% to 96.7% [1-2,4-6]. High temperature in the combustion process changed the silica structure, affecting the level of pozzolan activity and the smoothness of ash grains. This pozzolan contains cementitious properties when mixed with water [2]. The other chemical composition of rice husk ash was shown in Table 1. While, in term of specific gravity, Rosyidi et al. [7] found that rice hulk ash filler shows lower specific gravity with 0.947 than fly ash filler (as shown in Table 2).

Table 1. Composition of RHA.

Components	Percentage (%)				
	[1]	[2]	[4]	[5]	[6]
SiO_2	86.89	96.7	92.99	96.7	72.28
Al2O_3	1.58	1.01	0.18	1.01	0.37
Fe_2O_3	0.48	0.05	0.43	0.05	0.32
CaO	1.99	0.49	1.03	0.49	0.65
K_2O	5.35	0.91	0.72	0.91	-
Na_2O	0.82	0.26	0.02	0.26	-
MgO	1.29	0.22	0.35	0.22	-
TiO_2	-	0.16	-	0.16	-
P_2O_5	-	0.01	-	0.01	-
MnO	-	0.19	-	0.15	-
SO_3	0.71	-	-	-	-
Water Content	0.52	-	-	-	-

Table 2. Specific Gravity filler [7]

Specific Gravity	Value	Unit
Rice Husk Ash	0.947	g/cm3
Fly Ash	2.688	g/cm3

2.3. Mixture Design & Laboratory Test

In general, there are three layers of asphalt mixture used in flexible pavement in Indonesia. The layers are wearing course (AC-WC), binder course (AC-BC) and base course (AC-Base). The gradation limits used by previous studies complied to Bina Marga published in 2014 [8]. The gradation limits are shown in Table 3. While, amount of filler used is varied based on the respective studies (as shown in Table 4).

Table 3. Gradation limit for asphalt mixture [8].

Sieve size (mm)	Percentage by weight passing	
	AC-WC	AC-BC
37.5	-	-
25	-	100
19	100	90-100
12.5	90-100	75-90
9.5	77-90	66-82
4.75	53-69	46-64
2.36	33-53	30-49
1.18	21-40	18-38
0.6	14-30	12-28
0.3	9-22	7-20
0.15	6-15	5-13
0.075	4-9	4-8

Table 4. Amount of filler for asphalt mixture.

Previous study	Filler Percentage	
	RHA	dust
Syahputra et al. [1]	6	-
Akbar and Wesli [4]	6	-
	-	6
Ismardani and Kasan [5]	1.5	4.5
	6	-
Basuki et al. [9]	6	-
Ferdiana [10]	4	2
Asparini and Sulistiono [11]	6	-
Maulana and Empung [12]	2	4
Ridwan and Nadia [13]	7	-

All mixtures were prepared based on the Marshall mix method and complied the specifications published in Bina Marga 2014 (as shown in Table 5) [8].

Table 5. Marshall properties [7].

Properties	Unit	AC	Modified AC
Density	g/cm^3	2	2
Air voids in mix (VTM)	%	3-5	3-5
Voids in mineral aggregate (VMA)	%	15	15
Voids in aggregate filled with bitumen (VFA)	%	65	65
Stability	N	8000	10000
Flow	mm	2-4	4
Marshall Quotient (Stiffness)	N/mm	2500	3000
Residual Stability	%	90	90

3. Marshall Properties

Recent study conducted by Saepudin [14] in 2021 showed the Marshall parameters using RHA as filler with 6% of optimum binder content (OBC), as presented in Table 6. The data shows that the increase of RHA percentage, increases stability and Marshall quotient but starts to decrease at percentage 7%. The highest stability and Marshall quotient were recorded by 6.5% of rice hulk ash with 9964 N and 2880 N/mm, respectively, followed by 6.0, 5.5, 5.0 and 7.0 %. Only at 6.5% RHA content, all Marshall properties requirement are met. Thus, 6.5% shows the optimum amount of RHA content for the mixture. The reason behind the deviation in optimum binder content (OBC) for RHA be connected to the fact that controls the change in air voids in the mixtures.

Table 6. Marshall Properties using RHA as filler [14].

Mixture requirement	Percentage of rice hulk ash content (%)									
	5.0		5.5		6.0		6.5		7.0	
Stability (N)	8591	✓	9709	✓	9787	✓	9964	✓	9335	✓
Flow (mm)	2.92	✓	3.30	✓	3.00	✓	3.46	✓	3.74	✓
Marshall quotient (N/mm)	2942	✓	2942	✓	3262	✓	2880	✓	2496	✓
VIM (%)	7.56	×	6.34	×	5.06	×	3.65	✓	2.68	×
VFA (%)	56	×	65	✓	70	✓	80	✓	85	×
OBC	6.0%									

IOP Conf. Series: Earth and Environmental Science **971** (2022) 012005 doi:10.1088/1755-1315/971/1/012005

Study by Saepudin [14] was compared with other previous studies to observe any trend in Marshall properties due to varied content of RHA. The properties are Stability, Flow, VTM/VIM, Marshall Quotient and Residual Stability. Results of each property are depicts in Figure 2 to Figure 6. The RHA content used in previous study is from 1.5% to 7% with typical RHA content of 6% was used. In general, no consistent trend can be concluded based on the previous results. From the figure, lower RHA content shows an increasing on the stability (MS) value compared to higher RHA content. For instance, the MS of the mixture with RHA containing 1.5% and 2% are increased to 18kN and 16kN respectively compared to the mixture with RHA containing 6%, 6.5% and 7% are decreased to 8kN, 9kN, 9.5kN and 8.5kN. However, an increase in RHA content will reduce the stability value but still within the acceptable range outline by the specification. This trend is consistent with the finding by Rosyidi et. al, [7] where low specific gravity of RHA could be one of the factor contribute to lower stability value. Furthermore, low specific gravity mixture still applicable at the medium traffic load area. This improvement in Marshall stability for RHA mixtures can be attributed to the irregular particles of RHA [15] that may enhance the shear strength and stiffness of the modified binder.

In the case of flow values, most of the prepared mix lies between the specified limit of 2-4mm [16]. Also, the voids in mineral aggregate (VMA) and voids filled with bitumen (VFB) values did not cross the specified limits. From Figure 3, most of the percentage of RHA content used as filler in asphalt mixtures lies between the specified limit which are 3.8mm, 3.3mm, 2.4mm, 4mm, 2mm and 3.3mm except for some cases from the previous study that shows the flow values exceed 4mm. The flow value reflects the plasticity and flexibility properties of asphalt mixtures. Marshall Samples corresponding to the deformation of the load are broken, which represents a measure of the flow and also the value of mixing and flow with the value of internal friction. Flow has a linear inverse relationship with internal friction [17-18].

As it is known, increment in bitumen content slowly decreases the air void and all the studied fillers follow the said pattern individually. However, with increment in filler percentage, the air void in the mixture having RHA continuously is increasing respective to same bitumen content for all the filler percentages as shown in Figure 4. This phenomenon may be occurred due to the presence of more fine particles of RHA in the mixture pushes the coarse aggregates. As a result, the coarse aggregates suspend in the fine particles. All the alternative fillers like RHA nearly follow the same path as HL except 7% RHA content, where air void values are more than that of 5% maximum specified value (Figure 4). This trend may be credited to the fact that RHA, due to its high SSA and porous structure, has a propensity to absorb bitumen binder, making the bitumen film on aggregate thicker, which efficiently reduced the air voids of the mixture [19].

Furthermore, the change in Marshall Quotient (MQ) values may explain the nature of the obtained Marshall test results effortlessly. Marshall quotient (MQ) is defined as the proportion of stability (kN) to flow (mm), and as a hint of the stiffness of mixtures. It is well accepted that the MQ is a quantifier of the materials' resistance to shear stresses, permanent deformation, and hence rutting. The outcome of all the different RHA content on MQ values is shown in Figure 5. As a whole, the MQ values for 6% of RHA shows an increasing values which are 4.5 kN/mm and 3.6kN/mm, respectively. Nevertheless, with more added in filler quantity for RHA mixes, the MQ values gradually escalate and surpass the maximum permissible specified limit of MQ of 5 kN/mm [16]. For this reason, despite having the lowest OBC among all the mixtures, more than 6% RHA should not be used for field application. The increasing tendency in MQ values may be accredited to the fact that the higher concentration of studied waste fillers absorbs the light constituent of bitumen which upsurge the viscosity and stiffness of bituminous binder. This improves the adhesion between mastic and aggregates and boosts the overall ability of the HMA to bear loads.

Last but not least, Residual Marshall stability is another index used for evaluating the water stability of an asphalt mixture (Figure 6). Large residual Marshall stability value indicates good water stability [22]. With the addition of RHA, the residual stability of the asphalt mixture was improved. It can be seen as in Figure 6 that the statistical results show that the residual Marshall stability of the asphalt mixtures was significantly showing high values for high RHA content in asphalt mixtures.

IOP Conf. Series: Earth and Environmental Science **971** (2022) 012005 doi:10.1088/1755-1315/971/1/012005

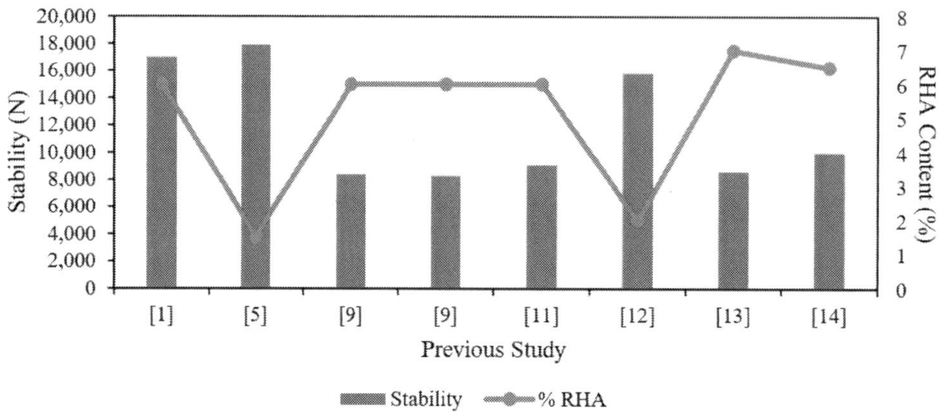

Figure 2. Marshall Stability with different RHA content from previous studies.

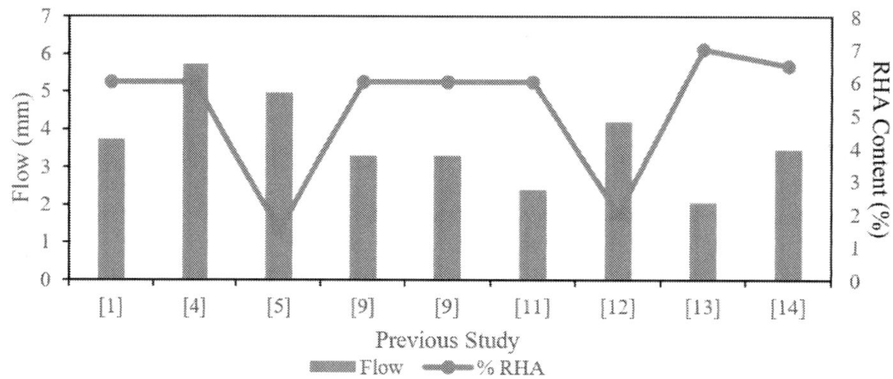

Figure 3. Flow with different RHA content from previous studies.

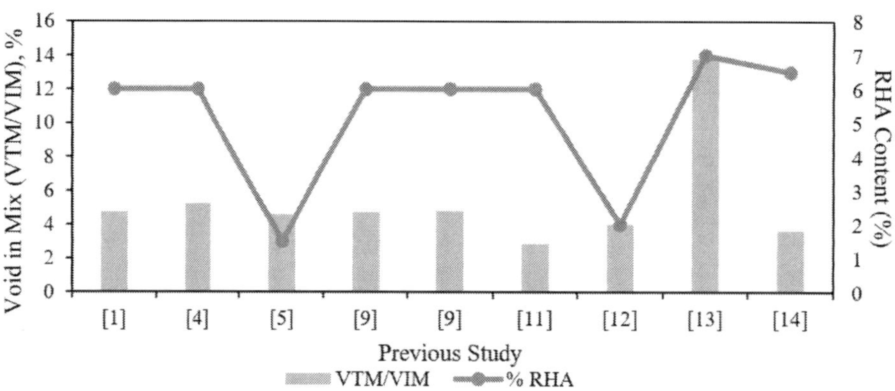

Figure 4. Void in mix (VIM / VTM) with different RHA content from previous studies.

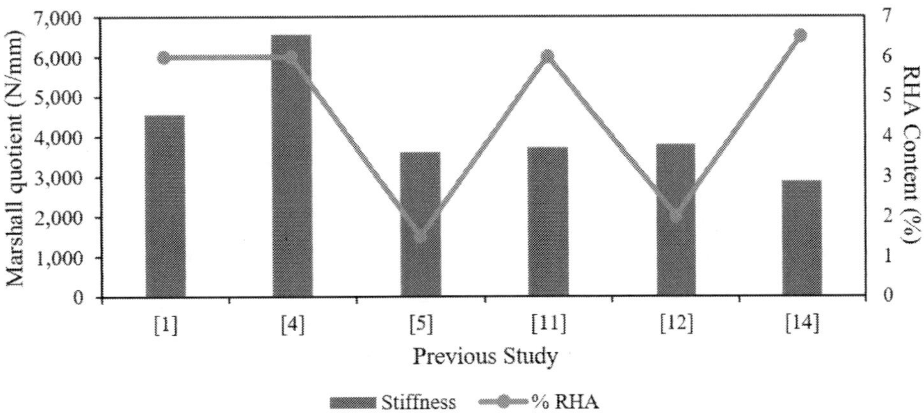

Figure 5. Marshall quotient with different RHA content from previous studies.

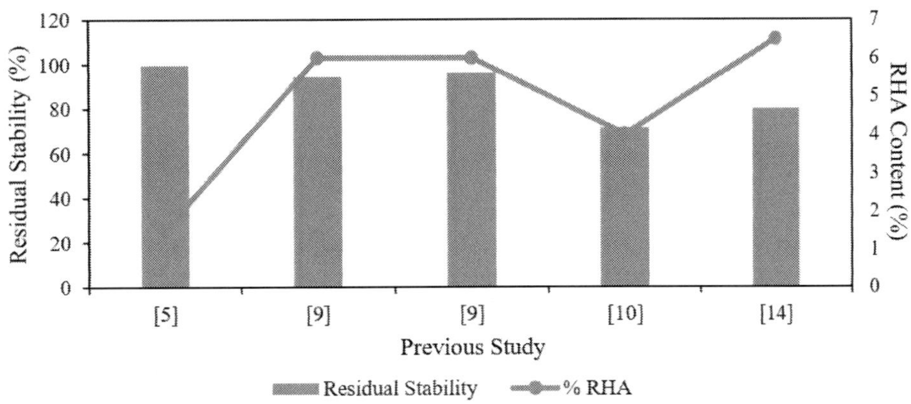

Figure 6. Residual stability with different RHA content from previous studies.

4. Conclusion
Application of rice husk ash (RHA) as a filler in asphalt mixture showed comparable performance than conventional filler. In contrast, higher RHA content could significantly reduce the stability yet still within the acceptable range. The higher the RHA content the more waste material could be utilized for road construction. Thus, minimized the usage natural sources materials.

5. References

[1] Syahputra N, Malik A and Sebayang M 2019 Penggunaan Abu Sekam Padi Sebagai Bahan Pengisi (Filler) Dalam Campuran Aspal Jenis AC-WC Dengan Pengujian Marshall *Jom FTEKNIK* **6**

[2] Rianto R H 2007 *Pengaruh Abu Sekam Sebagai Bahan Filler Terhadap Karakteristik Campuran Aspal Emulsi Bergradasi Rapat (Cebr) Terhadap Karakteristik Campuran Aspal Emulsi Bergradasi Rapat (Cebr)* Thesis (Semarang: Universitas Diponegoro)

[3] Ginanjar R R, Ma'ruf A and Mulyadi A H 2014 Ekstraksi Silika Dari Abu Sekam Padi/Rice Hulk Ash Menggunakan Pelarut/Solvent NaOH *Prosiding Seminar Nasional Hasil - Hasil Penelitian dan Pengabdian LPPM* (Universiti Malaysia Pahang) 306-312

[4] Akbar, Said J and Wesli 2012 Stability of AC-WC Asphalt Concrete Layer Using Rice Husk Ash, *Teras J.* **2**(4) 310-320
[5] Ismardani, Risman and Kasan M 2013 Karakteritik Beton Aspal Lapis Pengikat (AC-BC) Yang Menggunakan Bahan Pengisi Pengisi (Filler) Abu Sekam Padi, *Majalah Ilmiah Teknik* **15**(2) 94-102
[6] Bakri 2009 Komponen Kimia Dan Fisik Abu Sekam Padi Sebagai SCM (Supplement Cement Material) Untuk Pembuatan Komposit Semen, *J. Perennial* **5**(1) 9-14
[7] Rosyidi S A P, Fachriani E and Purwanto A 2012 Pengaruh Penggunaan Abu Sekam Padi sebagai Bahan Pengisi pada Campuran Hot Rolled Asphalt terhadap Sifat Uji Marshall, *J. Ilmiah Semesta Teknika* **15**(2) 98-107
[8] Sugiyartanto 2018 *Spesifikasi Umum 2018 untuk Pekerjaan Konstruksi Jalan dan Jembatan* Surat Edaran Dirjen Bina Marga Nomor 02/SE/Db/2018 (Jakarta: Kementerian Pekerjaan Umum dan Perumahan Rakyat)
[9] Basuki A, Syahrul and Wuaten H M 2016 Perencanaan Campuran Aspal Beton Ac-Bc Dengan Filler Abu Sekam Padi, Pasir Anggana dan Split Palu *e-journal untag-smd* (Samarinda: Universitas)
[10] Ferdiana A 2013 *Penggunaan Filler Abu Sekam Padi Pada AC-WC Halus Spesifikasi Jalan Bina Marga 2010* Thesis (Bandung: Universitas Pendidikan Indonesia)
[11] Asparini A and Sulistiono D 2008 Pemanfaatan Bahan Limbah Untuk Perkerasan Lentur Jalan Raya, *J. Purifikasi* **9**(1) 9-18
[12] Maulana R, Kurnia A and Empung 2019 Bandingan Abu Sekam Padi Dan Abu Kapur Pada Campuran Lataston Perkersan Jalan, *J. Akselerasi Teknik Sipil* **1**(1) 9-15
[13] Ridwan F S and Nadia 2017 Analysis Of The Effect Of The Utilization Of Husk Ash As A Filler On Asphalt Concrete Mixture, *J. Konstruksia* **8**(2) 1-8
[14] Saepudin U 2021 Analisis Perbandingan Kualitas Lapis Aspal Beton (Laston) Dengan Menggunakan Filler Abu Batu Dan Abu Sekam Padi, *J. Ilmu Sipil Program Studi Teknik Sipil Fakultas Teknik Universitas Galuh* 5(1) 12-18
[15] Mistry R, Karmakar S and Roy T 2018 Experimental evaluation of rice husk ash and fly ash as alternative fillers in hot-mix asphalt, *Road Mater. and Pave. Design* 20(4) 979-990
[16] Ministry of Road Transport and Highways (MORT&H) 2013 Specifications for road and bridge works *Indian Roads Congress* (New Delhi)
[17] Serin S, Morova N, Saltan M and Terzi S 2012 Investigation of usability of steel fibres in asphalt concrete mixture *Constr. Build. Mater.* **36** 238–44
[18] Umar F and Ag˘ar E 1991 Pavement structure (Instabul:Technical University)
[19] Arabani M and Tahami S A 2017 Assessment of mechanical properties of rice husk ash modified asphalt mixture *Constr. Build. Mater.* **149** 350-358
[20] Asohakt U 1993 *Mix Design Methods for Asphalt Concrete and other Hot Mix Types* ed A H Abedali (Baghdad: Mustansiriyh University)
[21] Rocksana A and Md K H 2017 Influence of Rice Husk Ash and Slag as Fillers in Asphalt Concrete Mixes, *American J. of Eng. Res. (AJER)* **6** 303-311
[22] Bo L, Jinyu Y, Xiaohui L, Feng H and Liangying L 2015 Effect of Short-Term Aging Process on the Moisture Susceptibility of Asphalt Mixtures and Binders Containing Sasobit Warm Mix Additive, *Advances in Mater. Sci. and Eng.* **5** 1-8

Acknowledgments

The authors would like to thank Universiti Teknologi Malaysia (UTM) for providing financial support under the Research Grant No. Q.J130000.2651.16J15. The financial assistance is highly appreciated.

Durability of POFA-modified dense-graded cold mix asphalt

K R Usman[1,3], M R Hainin[1,2*], M K I M Satar[1], M N M Warid[1], M A Bilema[4], A Usman[5] and A J Zulfiqar[1,6]

[1]School of Civil Engineering, Universiti Teknologi Malaysia, 81310 Johor Bahru, Johor, Malaysia.
[2]Department of Civil Engineering, College of Engineering, University Malaysia Pahang, 26600 Pekan, Pahang, Malaysia
[3] Department of Civil Engineering Technology, School of Engineering Technology, Nuhu Bamalli Polytechnic Zaria, P.M.B 1016, Zaria, Kaduna State, Nigeria.
[4]Department of Civil Engineering, Faculty of Engineering, Benghazi University, Benghazi, Libya.
[5]Civil and Environmental Engineering Department, Universiti Technologi PETRONAS, Bandar Seri Iskandar, Perak, 32610, Malaysia.
[6]Department of Civil Engineering, Balochistan University of Engineering and Technology, Khuzdar, Pakistan.
*Corresponding author: mrosli@utm.my

Abstract. Cold mix asphalt (CMA) is a versatile and energy efficient mixture often prepared at ambient temperature. Efforts to improve the performance of CMA mixtures recorded appreciable success. Nonetheless, the critical performance measures of moisture and abrasive resistance still baffles researchers. Additionally, CMA's design, production, testing, and site application is bedevilled by slow rate of strength gain, high void content, and absence of a globally acceptable mix design. This study aimed at assessing CMA's durability measures using the modified Lottman (moisture damage) and Cantabro tests by adding a 3% palm oil fuel ash (POFA) as filler replacement and 3% cement. Fast-curing mixtures fabricated with a cationic quick-set (CQS-1h) and a rapid-set (RS-1K) emulsion both of medium viscosities with a nominal maximum aggregate size (NMAS) of 4.75 mm were evaluated. In addition to improving the Marshall volumetric properties, POFA's inclusion is aimed at improving adhesion thereby enhancing the durability of the fine dense-graded cold mix asphalt (FGCMA-4.75 mm). The modified Marshall as stated in the asphalt institute (AI) manual series (MS-19) for dense-graded CMA was used for design. Results indicated that POFA reduces the void, improves stability, and enhance durability. Moreover, the CQS-1h has better durability for both control (FGCMA-C) and POFA-modified mixtures (FGCMA-P). While RS-1K has higher tensile strength in both dry and moisture-conditioned states with a higher tensile strength ratio (TSR) for both FGCMA-C and FGCMA-P mixtures. POFA's pozzolanic capability proves effective in improving the durability and moisture damage resistance of FGCMA-P.

1. Introduction

Asphalt mixtures for pavement construction, rehabilitation and maintenance are mainly produced as Hot Mix Asphalt (HMA) in most part of the globe for its durability, and safe riding quality. However, release of greenhouse gases, other harmful fumes, and influence of weather on mixing and laying encourages exploring of viable alternatives to HMA by stakeholders, especially researchers [1,2]. The recent global sustainability campaign aimed at curbing the negative effect of global warming, eschew unhealthy environmental practices, and minimise the use of non-renewable resources led to the introduction of low-temperature paving technologies in form of warm mix asphalt (WMA) [3]. WMA can save up to 70% energy when compared to a conventional HMA, and its performance is often reported to be superior than that of HMA, but WMA additives are costly while its production needs skill and heating too [4]. Other low-temperature paving mixtures include half-warm mix, and semi-warm mix asphalt. The mixing temperature requirement for WMA between 100 – 140 °C still emits harmful fumes affecting both the atmosphere and the workers alike with attendant handling difficulty [2]. Consequently, cold mix asphalt (CMA) promises to be a better alternative mixture to both HMA and WMA since it is produced without heating, thus releases zero fumes.

CMA has gained recognition in recent years for its superior energy saving, zero fumes emission, ease of production, minimal equipment and skill requirement. CMA is produced from a mixture of well-graded aggregates, water, emulsified asphalt, and additives mixed and placed at a temperature in the range of 0 – 60 °C, but mostly at ambient temperature [5]. Its versatility made it applicable in subbases, base courses and wearing courses of pavements as new construction and as a maintenance mixture for medium to heavy-trafficked roads [6,7]. However, challenges associated with CMA of slow strength development rate, and weak early strength has relegated the use of CMA to maintenance works. Furthermore, other most critical challenges inhibiting CMA's use include high susceptibility to moisture damage due to its inherent high voids and low abrasion resistance due to its peculiar lower base binder content.

Several efforts by researchers were made to solve the aforementioned challenges especially, the moisture resistance of CMA. The typical method, which was developed by Hughes and adapted with minor modifications by the Asphalt Institute (AI) requires measuring the tensile strength of wet CMA specimens subjected to three (3) hours vacuum saturation at room temperature in comparison to the tensile strength of the dry samples [8]. Boiling test and Coating ratio are the two related common tests utilised for testing CMA's and warm mix asphalt's moisture damage resistance [9,10]. However, coating test does not truly mimic site in-service conditions and because these tests are conducted on the component materials, some aggregates may perform worse in those tests individually, but do great in a mixture. This could be explained by the interaction of the aggregate's chemical component with other materials in a mixture [11].

Therefore, this study set to test moisture damage in CMA by subjecting samples to the same 60 °C total water immersion for 24 hours similar to that for HMA to simulate worst case scenario. On the other hand, the solutions proffered towards improving CMA abrasion resistance include the inclusion of pozzolanic fillers to which POFA belongs. These pozzolanic fillers of either biomass or industrial origins and notably rich in calcareous and siliceous bases improves CMA performance [12]. The use of pozzolanic fillers serves a dual purpose of reducing void and improving mixture's adhesion. POFA is a dark-grey ash by-product obtained by burning parts of a palm tree. A recent study by the author revealed that the addition of 3% POFA (0.075mm size) by aggregate weight is effective in improving the mechanical performance of dense-graded CMA [7]. Moreover, POFA's inclusion in CMA will ameliorate environmental hazards posed by its landfill disposal whilst upholding sustainable engineering practice.

2. Materials and test methods

The detailed properties of the materials used in this study and the methodology employed are explained in section 2.1 to 2.5.2.

2.1. Binders

Two categories of binders are used for this study namely; primary and secondary binders. The primary binders are the emulsified asphalts while the secondary binder is the cement.

2.1.1. Emulsified Asphalt. Emulsified asphalt is a homogenous mixture of an oil-based and water-based liquids. The two liquids are dispersed into tiny globules held in position by differing electrical charges with the help of a 'SURFace ACTion AgeNT' (Surfactant). Depending on the category of the emulsion, the surfactant also imparts the setting characteristics of the emulsion, either rapid, slow, medium, or quick setting. The higher the concentration of the charges around the bitumen droplets, the slower will be the setting rate. Emulsion viscosities are designed to enable low temperatures application as cold mixtures are mostly applied in slurry form [13]. Traditionally, medium to slow setting emulsions is often used for CMA production to allow for ample time for mixing and placing on site. However, to solve the lingering challenge of slow strength development a quick and rapid set cationic emulsions are introduced in this study.

Cationic emulsions are compatible with both alkaline and acid-nature aggregates which ensures greater adhesion and aggregate stone-on-stone lock [14]. The durability of CMA is dependent on the emulsion performance which is translated by the film-thickness on the aggregate. But the asphalt film thickness on the aggregate relies on the amount of water and breaking speed of the emulsion. Thus, the addition of prewetting water during CMA mixing hampers adequate adhesion and proper coating, hence exclusion of prewet water is considered in this study. Enhanced adhesion reduces the high void content peculiar with CMA, and a reduced void improves the moisture damage resistance component of CMA's durability [15]. Moreover, the durability performance of dense-graded CMA is more compared to other gradations, nonetheless, the samples preparation and testing for moisture damage test differs [16]. The emulsion types used are a quick and a rapid set emulsion both of medium viscosities.

The quick set emulsion is polymer-modified cationic (CQS-1h) emulsion specifically meant for emulsified asphalt slurry seals, allowing faster curing of the mixture and early opening to traffic. Both emulsions were tested of mechanical performance whilst their residues further evaluated for rheological behaviour. Conditioning of emulsions before each test was done at 25 °C according to ASTM D244 while adjustments were made where necessary to replicate on-site service conditions as reported by Usman [18]. In a nut shell, quick and rapid set emulsions is pertinent towards early opening of the road to traffic, especially in the absence of solar radiation, and low humidity [19].

Table 1 present the result of the emulsions and their residues' tests in accordance to the American Society for Testing and Materials (ASTM) and American Association of State Highway and Transport Officials (AASHTO). The result revealed that the requirements set by Jabatan Kerja Raya (JKR) – Malaysian Public Works Department for cold mixes were satisfied [20].

Table 1. Physical and mechanical properties of RS-1K and CQS-1h.

Test Description	Test Method		Result		Specification (CQS/RS)	Requirement
	AASHTO	ASTM	RS-1K	CQS-1h		
TEST ON EMULSIFIED ASPHALT						
Physical appearance	-	-	brownish	brownish		
Storage stability (%)	T 59	D6930	1.0	0.9	1% max.	
Settlement test (%)	T 59	D6930	1.0	1.0	1% max.	JKR, 2008, MS 161, & ASTM D2397
Residue by evaporation (%)	T 59	D 6934	63	65	57/50 min.	
Particle charge	-	D 7402	Positive	positive		
Saybolt Furol viscosity (Sec)	-	D 7496	22	20	15 – 50	
TEST ON EMULSIFIED ASPHALT RESIDUE						
Penetration at 25 °C (0.1 mm)	D5	T 49	40	68	40/60 – 90/200	
Softening point (°C)	D36	T 53	69	-	54 – 57°C min	JKR
Solubility in Trichloroethylene (%)	D2042	-	95	100	97.5min	

2.1.2. Cement. An ordinary Portland cement conforming to ASTM C150/C 150M - 11 was used as part of the aggregate gradation. It is added as a supplementary binding agent in CMA mixtures [21]. Usually, 1-2% by aggregate weight is added to a CMA mixtures [22,23]. A crucial role of cement in this study was to trigger the cementitious properties of the POFA, which is a pozzolana, and pozzolanas only contribute their cementing or binding properties in the presence of SiO_2 (lime) which provided by cement [24]. In addition to the dual role of binding aggregate and supplementing pozzolanic reactivity, cement do assist in the early strength development and curing (water expulsion) of CMA by heat of hydration. For this study, a fixed 3% cement by total weight of aggregate was used for the control and POFA-modified mixtures alike. The cement was mixed with the aggregate dry and subsequently added to the emulsified asphalt.

2.2. Granite Aggregate

The aggregate for this study was dried at a temperature of 110 °C in the oven and sieved according to the various sizes in the gradation. The choice for granite aggregate was in consonance to the fact that it is an acidic nature aggregate. Acidic aggregates were reported to be compatible with cationic emulsions, and most preferred than anionic emulsions in terms of enhanced adhesion [14].

The aggregate's mechanical performance was tested according to specified standards and result presented in Table 2.

The result in Table 2 indicated that the aggregate has a negligible water absorption, hence, the decision to eliminate addition of premix water did not affect the mix's performance. The aggregate also has a high angularity well above the minimum limit of 45% in addition to an excellent abrasion resistance. The absence of deleterious substances is key to proper aggregate-binder bonding in CMA, the sand equivalent result revealed that the aggregate of being free of clay and other organic particles capable of reducing bonding.

Table 2. Mechanical Properties of fine and Coarse aggregate.

Aggregate	Properties		Granite	Limit	Relevant Specification
Fine aggregate (< 4.75mm)	Specific Gravity	-	2.89	-	ASTM C 128
	Sand Equivalent	(%)	96.0	> 45	ASTM D 2419
	Water Absorption	(%)	0.48	< 2	ASTM C 128/MS 30
	Bulk Density	(Kg/m^3)	1815	-	ASTM C 128
	Fine aggregate angularity (FAA)	(%)	54.13	> 45 %	ASTM C 1252
Coarse aggregate (> 4.75mm)	Los Angeles abrasion value (LAAV)	(%)	23.00	< 45 %	ASTM C131/AASHTO T96
	Flakiness Index	(%)	17.40	< 20	BS 812-21:1989

The plot of the aggregate gradation used in this study with a nominal maximum aggregate size of 4.75 mm is presented in Figure 1.

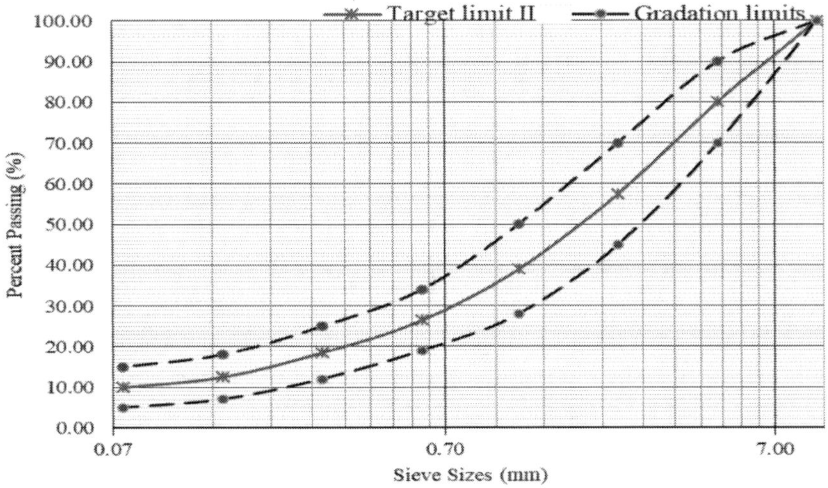

Figure 1. Aggregate Gradation for CMA.

2.3. Palm oil fuel ash

Palm oil fuel ash (POFA) is used in this study as a cementitious pozzolanic material that aid CMA adhesion. It is a by-product obtained by burning palm tree biomass, including empty palm fruit bunches, kernels/shells, fibres, fronds, and palm leaves in palm oil mills, and power generating plants. POFA is typically grey but turns black when the proportion of unburnt carbon becomes high. The black POFA reverts grey by further burning the black POFA at 1000°C. Aside of fly ash, POFA was opined as the second most attractive filler used for construction [25]. POFA is inert however, in the presence of water act as a pozzolana in construction, especially, concrete in which it acts as a supplementary cementing material with excellent flowability and anti-segregation ability [26].

Many studies had proven POFA as a pozzolanic material [27]. In asphalt concrete and bitumen modification, POFA has gained increasing attention with attendant good performance [10]. However, not much was achieved in the utilisation of POFA in asphalt concrete mixtures, especially, cold mixtures. Recently, the inclusion of POFA in CMA was studied by the author and 3% by aggregate weight in a fine dense-graded aggregate matrix was discovered to yield optimal performance [7]. The chemical composition of POFA as revealed by a number of researches signifies that it contains cementitious chemical compounds like silica, alumina, and oxides of magnesium, iron and calcium [27]. The chemical properties of POFA as reported by various researchers is presented in Table 3.

Table 3. Chemical Properties of POFA reported in the literature.

Percentage Chemical Constituent	Researchers											
(%)	[28]	[29]	[30]	[31]	[7]	[32]	[33]	[34]	[35]	[36]	[37]	[38]
SiO_2	65.01	43.60	65.30	64.17	62.60	55.20	71.67	59.62	79.31	63.41	55.50	64.20
Al_2O_3	5.72	11.40	2.50	3.73	4.65	4.48	0.94	2.54	3.15	5.55	9.20	4.25
Fe_2O_3	4.41	4.70	1.90	6.33	8.12	5.44	2.77	5.02	7.12	4.19	5.60	3.13
CaO	8.19	8.40	6.40	5.80	5.70	4.12	5.61	4.92	2.79	4.34	12.40	10.20
MgO	4.58	4.80	3.00	4.87	3.52	2.25	4.91	4.52	1.21	3.74	4.60	5.90
K_2O	6.48	3.50	5.70	8.25	9.02	2.28	7.89	7.52	3.23	6.33	-	8.64
SO_3	0.33	2.80	0.40	0.72	1.16	2.25	1.03	1.28	0.45	0.91	2.30	-
MnO	0.11	-	-	-	-	-	0.12	-	-	0.17	-	-

P_2O_5	4.69	-	-	5.18	-	-	-	3.58	2.32	3.78	-	-
Na_2O	0.07	0.39	0.30	0.18	-	0.10	0.12	0.76	-	0.16	-	0.10
TiO_2	0.25	-	-	-	-	-	0.09	-	-	0.33	-	0.09
C	0.09	-	-	-	-	-	-	-	-	-	-	-
SiO_2 $+Al_2O_3+$ Fe_2O_3	75.14	59.70	69.70	74.24	75.37	76.12	95.15	67.18	89.58	92.91	70.30	96.51

From Table 3, it is glaring that all POFA samples irrespective of their origin or production parameters, has high SiO_2, Al_2O_3, and Fe_2O_3 with silicone dioxide having the highest percentage. Generally, most POFA samples satisfied to be classified as 'class F' pozzolana according to ASTM C-618 [39]. The high silicone dioxide content common with most POFA samples makes it idle pozzolanic candidate which enhances aggregate-binder bonding aided by cement hydration [38]. The incorporation of POFA in CMA will improve its early strength via increased adhesion from cement-pozzolan-water hydration which will in turn improve durability. Furthermore, the durability of CMA based on ITSR result was reported to improve significantly compared to control mixtures with the inclusion of pozzolanic ash (Fly ash and Ground Granulated Blast Furnace Slag) [40]. Furthermore, the geometric increase in crude palm production in Malaysia suggested a matching quantity of waste generation as over 1000 tons of POFA are dumped in landfills [29,41]. Incorporating POFA waste in construction will ameliorate the detrimental consequences of its landfill disposal as the waste has reached worrying levels in Malaysia [38].

2.4. Test Methods
The constituent materials were tested for compliance to relevant specifications and compatibility with each other. The mixtures are designed employing the standard Marshall design method using 75 blows and sample conditioning before testing in a water bath at 60 °C. It is worth mentioning that the usual CMA Marshall design employs 50 blows/face and a sample conditioning at ambient temperature which ranges from 10 °C to 40 °C [42,43]. Absence of a universally accepted mix design led to the customisation of CMA mix design, in USA superpave approach is used while in Africa and Asian countries a modified Marshall is used [4]. The modified Marshall method based on Illinois study was used for this study and samples were further tested for moisture damage and Cantabro performance measures. Subsequent section detailed the material preparation, mixture design, and performance testing.

2.4.1. Mix Proportion and Sample Preparation. All the mixtures were prepared in accordance to the revised practice for the preparation of Marshall samples outlined in ASTM D6926-10 [44]. Regarding the gradation, CMA could be prepared with a wide-ranging aggregate gradation including dense-graded, gap-graded, open-graded, and uniformly-graded mixes. A total combined aggregate weight including filler, cement and POFA of 1000g was batched as against the usual 1200g for a standard Marshall sample. The reason was to curb the swelling effect of the mix experienced especially with the addition of premix water, and to allow for adequate compaction. Thus, premix water was eliminated because samples swell above the mould and needed to be trimmed to size. This was to ensure proper adhesion thereby increasing moisture damage resistance and improve durability, as moisture present in CMA hampers good performance [4].

The initial emulsion content (IEC) was established by employing the AI empirical formulae which depends on the gradation type base on percentages passing the 4.75 mm, 2.36 mm and 0.075mm test sieves. An IEC of 8% was obtained and varied by adding and subtracting 0.5% in two steps above and below the 8% to form an emulsion range of 7 – 9%.

The aggregate was manually mixed dry in a bowl after which the required emulsion percentage was added and further mixed to obtain a minimum of 50% coating. Emulsions were conditioned in a water bath at 25 °C, mixed thoroughly for homogeneity before being added to the aggregate and further mixed at ambient temperature. The mixture was then transferred into pre-oiled moulds in three (3) layers. Each layer was given fifteen (15) light strokes by the mould's periphery and ten (10) blows at the centre with a 10 cm long and 5 mm wide rod, as spelt out in ASTM D6926 –10 [44].

Considering the higher void levels obtained with the standard CMA compaction level of 50 blows as specified by the Malaysia's public works department – Jabatan Kerja Raya (JKR) and India's ministry of road, transport and highways (MoRTH), adjustment were made [20,43]. This study adopted 75 blows per face for all samples in order to keep void levels to a practical minimum and to achieve maximum density. However, some researches highlighted that 3 - 5% air void is hard to achieve even at compaction levels above 75 blows [45]. Conversely, a 3 - 5% void in the total mix (VTM) was attained in a related study with the addition of 3% cement and POFA in a fine dense-graded mix. Nevertheless, concerns of aggregate crushing for coarser mixtures at higher compaction levels is imminent, yet some studies employ it but with similar or even higher void levels [42]. Unlike with other CMA mixes which collapses at early curing ages or immediately after compaction due to weak early strength, the FGCMA mix was stiff enough to withstand shocks immediately after being casted [46].

Samples were then cured in their moulds at 40 °C in an oven for 24 hours before being demoulded and subjected to a further curing at 40 °C for 72 more hours [47]. Some studies have employed similar curing procedure though at a higher temperature of 60 °C [48].

2.4.2. Mix Design. A modified Marshall design method adopted by Illinois design method as outlined in asphalt institute (AI) manual series MS-19 as elucidated in MoRTH specification was employed for designing the FGCMA-C and FGCMA-P [43]. Sample testing was done according to the revised Marshall flow and stability test contained in ASTM D6927-15 [49]. The revised test specifically suits dense-graded mixtures made from bitumen or emulsified asphalt. Additionally, theoretical maximum specific gravity of all mixtures was evaluated in accordance to ASTM D2041/D2041M – 11 [50]. Fully cured Marshall samples were evaluated for density and specific gravity measurements and subsequently transferred into a thermostatically controlled water bath set at 60 °C for forty (40) minutes prior to testing. Each of the three (3) samples for each binder content were tested one at a time by installing onto the Marshall machine mould and subjected to a gradual continuous uniform loading at a deformation rate of 50 mm/min. The Marshall stability in kilo Newton (kN) is recorded as the maximum load at failure. While the deformation in millimetre (mm) recorded by a dial gauge represents the flow of the sample.

Result of the Marshall test was plotted to obtain the optimum binder content. The determined OEC was re-confirmed by fabrication and testing samples at the established OEC. Performance testing was conducted on samples fabricated at the determined OEC as well.

2.5. Performance Testing

Durability performance tests were conducted in addition to modified Marshall test. The durability of asphalt mixtures especially, CMA is influenced by environmental conditions during their service life, whilst moisture and frost damage were identified as critical amongst other factors [39]. A similar research evaluated the durability performance of cold mix patching mixtures by employing Marshall stability, Cantabro, and wheel tracking tests [16]. Moreover, other studies employed Cantabro only for evaluating CMA durability [51]. Thus, considering the fact that moisture damage is considered a critical durability measure for CMA, utilising the modified Lottman test (moisture damage) and Cantabro test should provide adequate indication of their performance.

2.5.1 Cantabro Test. Cantabro test is considered as a cheap, fast, simple, and efficient alternative to the complex cracking durability testings for asphalt mixtures [52]. For this study, the Cantabro durability test was conducted in accordance to the Texas Department of Transportation's specification (Tx-DOT-Tex-245) or ASTM D7064 [16,53]. Notwithstanding the fact that Cantabro test was originally meant for open-graded friction courses especially at the design stage, while it is used here for production and performance assessment [53]. Cantabro test measures the resistance of the combined effect of abrasion and impact of a compacted bituminous specimen, which in turn translates to the asphalt mixture durability performance. It serves as an indication for binder-aggregate bond or adhesion properties of an asphalt mix [51].

The test was conducted by subjecting Marshall samples casted at OEC to an impact and abrasive force in a Los Angeles abrasion machine. Samples for this study were subjected to 300 revolutions in the LAAV machine at 30 rpm (revolutions per minute) without steel charges. Triplicate samples were prepared and their weights in gram (g) was measured before and after the complete revolution to the nearest whole number. The mass loss experienced by each sample is expressed as the weight difference between the final and initial weights of the sample under testing. The value expressed in percentage gives the Cantabro loss of the Marshall sample.

The percentage loss experienced by a sample represents the sample's durability; thus, the smaller the Cantabro loss, the more durable the sample is expected to be.

2.5.2. Moisture Damage Test (Modified Lottman test). This study adopted the standard HMA moisture damage protocol which involves conditioning the dry and moisture-conditioned samples in a water bath at 25 °C and 60 °C respectively with a subsequent vacuum saturation for the former. This is drastically different with the usual ambient temperature (10 – 40 °C) conditioning of CMA samples for longer duration [39]. As such the samples for dry indirect tensile strength (ITS) were conditioned at 25 °C while the moisture-conditioned samples were conditioned at 60 °C for 24 hours as in Figure 2. The air void level for all samples was kept to 7% + 1% and the durability assessment was achieved by evaluating the tensile strength ratio (TSR). The TSR is a value in percentage (%) determined by taking the ratio of the conditioned sample's ITS to those of the unconditioned samples.

The ITS test procedure for both conditioned and unconditioned samples was the same. It entails applying a constant diametral compressive force at a uniform deformation rate of 50 - 53 mm/min. The standard 101.6 mm and 63.8 mm thick samples with a nominal maximum aggregate size of 4.75 mm were loaded between the two-halves mould of a standard compression testing machine. Failure of the sample is caused by the splitting tensile stresses developed within the vertical diametric axis of each sample.

Figure 2. Moisture-conditioned samples in water bath at 60 °C.

3. Results and Discussion

3.1. Results of optimum emulsion content for the two mixtures

The optimum emulsion contents (OEC) for CQS-1h and RS-1K of the designed and control mixtures determined from Marshall testing is presented in Table 4. It is evident from Table 4 that CQS-1h has lower emulsion demand than RS-1K for both control and modified mix. This could be linked to the result of residue by evaporation in Table 1, which revealed that CQS-1h has higher base binder than do RS-1K. Therefore, as higher binder is required to coat aggregate, hence the need for higher emulsion content. The addition of POFA to the mixture has increased the emulsion demand for control and modified mixtures of both emulsions.

Table 4. Optimum emulsion contents for modified Marshall test on samples.

Percentage POFA Replacement	Mixture Designation	Average Marshall OEC (%)	
		CQS-1h	RS-1K
Control	FGCMA-C	7.7	8.1
3% POFA	FGCMA-P	8.1	8.4

The increased surface area of the POFA in the mix leads to higher emulsion absorption which in turn increases the OEC. All samples were verified for Marshall volumetric parameters at the designed OECs. Triplicate samples each for the two emulsions were casted using the established OEC and the result is presented in Table 5.

The four notable Marshall volumetric parameters of stability, flow, VTM, and voids filled with binder (VFB) were selected for all mixture types as presented in Table 5. The obtained results were compared with the MoRTH standard specifications for CMA mixtures [43]. From the values in Table 5, it is glaring that the POFA modified mixes of both emulsions have higher stability, lower voids, and higher voids filled with asphalt. Moreover, there is a significant void reduction with the addition of 3% POFA in the mixture. Equally, CQS-1h POFA-modified mixes have the highest stability of 15.35 kN as against the 13.28 kN for RS-1K. However, the lowest flow was obtained with the RS-1K emulsion for both FGCMA-C and FGCMA-P. From Table 5, tick marks were used to denote samples that passed the MoRTH requirement, while the cross marks indicate the opposite.

With the exception of the control mixtures for both emulsions, all other mixtures have satisfied the minimum criteria set by India Ministry of Road Transport and highways' (MoRTH) specifications for emulsified asphalt cold mixtures [43].

The significant void reduction achieved by POFA addition is due to enhanced adhesion achieved by the increase in hydration products from POFA's pozzolanic reaction with water and cement altogether. Similar void level was achieved with standard cold mixtures of asphalt concrete (AC) gradations and stone mastic concrete (SMA) using gyratory compactor at 40 gyrations [45].

Table 5. Summary of Marshall Verification of Result for CQS-1h and RS-1K Mixtures.

MIX TYPES	CMA MIX REQUIREMENTS & MARSHALL RESULT VERIFICATION TO MoRTH [43]							
	STABILITY (Minimum) 2.2 (kN)		FLOW (Minimum) 2.0 (mm)		VTM (Range) 3.0 - 5.0 (%)		VFB (Minimum) 15.0 (%)	
			CQS-1h					
FGCMA-C	9.70	✓	2.40	✓	10.50	✗	60.20	✓
FGCMA-P	15.35	✓	2.40	✓	4.00	✓	82.40	✓
			RS-1K					
FGCMA-C	10.60	✓	2.10	✓	10.20	✗	62.10	✓
FGCMA-P	13.28	✓	2.16	✓	3.80	✓	83.90	✓

3.2. Durability test results

3.2.1. Results of the Cantabro test. Plot of the Cantabro loss result is presented in Figure 3. The plot visually signified that POFA addition in CMA improved the durability of the mixtures by 50% for both CQS-1h and RS-1K. This is deduced from the Cantabro loss percentage, which means that lower Cantabro losses suggest better durability. Generally, the result indicates that CQS-1h exhibits lesser losses than RS-1K mixtures; thus, CQS-1h has higher adhesion properties and durability.

The result proves the efficacy of POFA's pozzolanic attribute in enhancing aggregate-binder bond thereby improving cracking/abrasion resistance, and durability of asphalt mixes. Even in concrete mixes, despite the extended duration of pozzolanic reaction and cement hydration of up to a year or more, yet, the significant strength attainment occurs within the first 28 days. Similarly for CMA, full strength is attained within a week or less of casting. The durability of FGCMA-P mixtures was improved for CQS-1h and RS-1K by 55.2% and 47.4%, respectively, compared to FGCMA-C. This result is significantly appreciable considering the fact that the test was specifically designed for HMA mixtures. The result in Figure 3 further explained the desirability of the polymer-modified CQS-1h emulsion over the RS-1K. Furthermore, the durability of dense-graded CMA mixtures in terms of Cantabro loss was reported to be influenced by void content, presence of polymer, and aggregate gradation and type [43]. Thus, the polymer matrix in CQS-1h improved the binder-aggregate adhesion, enhancing the resistance of CMA mixture to impact and abrasive forces better than RS-1K.

Therefore, the addition of POFA in CMA improves abrasive and impact damage of the mixture by roughly 50% based on the result of Cantabro test presented in Figure 3.

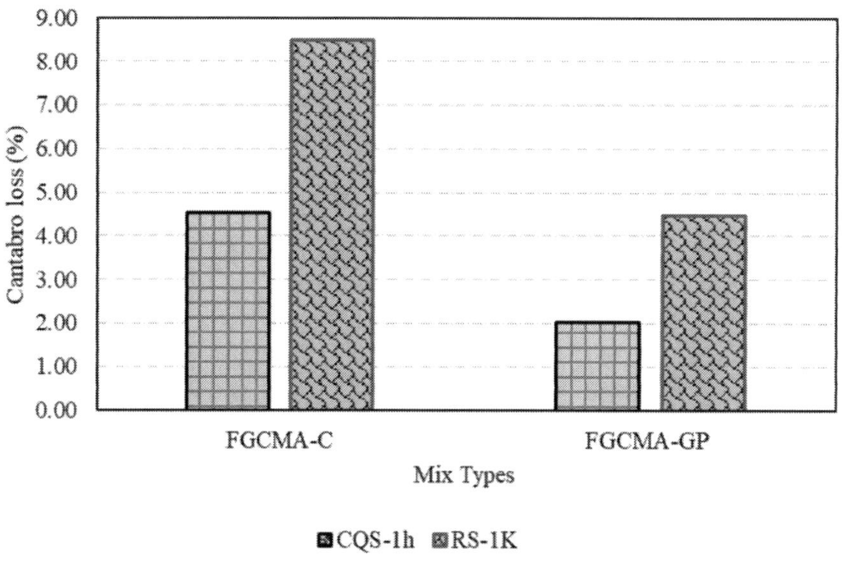

Figure 3. Cantabro loss plot for CQS-1h and RS-1K.

3.2.2. Results of the Modified Lottman test. The result of the modified Lottman's test is presented in Figure 4. From Figure 4 it can be deduced that dry ITS is higher compared to wet ITS for all mix types, reason being that moisture reduces bond strength in asphalt mixtures and in turn impairs mixture durability [54]. ITS values could be used to define an asphalt mixture's cracking resistance – higher ITS values denote enhanced resistance to fatigue cracking. However, this fact is to be treated with caution as specific tests are available for evaluating the cracking resistance of asphalt mixtures.

The addition of POFA in the mix improves both the ITS and TSR of both CQS-1h and RS-1K emulsions. The TSR limit for HMA was plotted as a basis for comparison to that of the FGCMA-4.75 mm mixes. The plot is represented with a straight line at the 80% mark of the TSR axis as in Figure 4. In spite the fact for their lower TSR compared to the standard HMA mix, yet, the FGCMA-P has higher TSR than FGCMA-C for CQS-1h and RS-1K by 7.09% and 11.09% respectively suggesting improved resistance to moisture damage. Hence, the POFA modified mix (FGCMA-P) has improved durability and moisture damage resistance compared to the control mixture (FGCMA-C).

Unlike CQS-1h optimised mix, the margin of improvement in TSR between FGCMA-P and FGCMA-C for RS-1K is relatively small. Furthermore, contrary to expectations that the polymer-modified emulsion mixtures will yield better moisture damage resistance, the FGCMA-GP of RS-1K has higher TSR than that of CQS-1h emulsion.

It is worth mentioning that the aggregate type and gradation affects the moisture damage resistance of CMA more than does emulsion type. It was reported that limestone aggregate in CMA has superior moisture damage resistance than granite aggregate for the same emulsion type [9]. To this end, there is a need to explore the efficacy of POFA in improving CMA durability of mixtures made with varying types of aggregate to arrive at a robust conclusion.

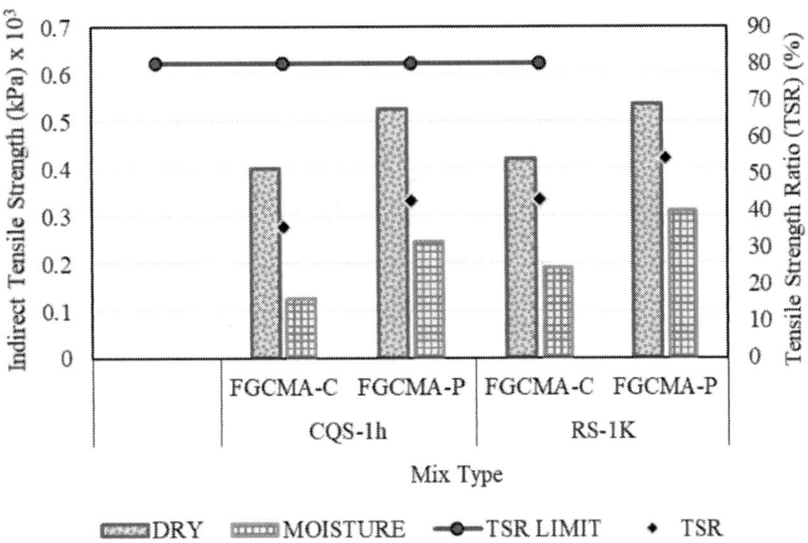

Figure 4. result of the Indirect Tensile Strength and Tensile Strength Ratio.

4. Conclusion

Based on the data analysed and result evaluated, the following conclusions were drawn:

- Despite of CMA's attractive features of being cost-effective and energy efficient mixture, yet, it has a number of setbacks, prominent of which are weak early strength and durability.
- The modified Marshall design for CMA is adequate in designing dense-graded mixes, however, the Marshall verification of result revealed the FGCMA-C to fail in terms of void content as it has higher VTM above the 3-5% stipulated by MoRTH. Thus, the need to devise an encompassing universal design approach for CMA to cater for this concern.
- Fine-graded cold mix asphalts (FGCMA) have better durability compared to other gradation types, and the influence of POFA filler on the durability of FGCMA-4.75 mm was evaluated in terms of Cantabro and moisture damage tests.
- POFA addition has improved the abrasion resistance of FGCMA-4.75 mm by 50% compared to unmodified mixes based on the Cantabro result.
- The modified Lottman test result in addition to revealing the efficacy of POFA in improving the cracking resistance of CMA, also suggest its ability in enhancing CMA's resistance to moisture damage.

5. References

[1] Nur Naqibah Kamarudin S, Rosli Hainin M, Khairul Idham Mohd Satar M and Naqiuddin Bin Mohd Warid M 2018 Comparison of Performance between Hot and Warm Mix Asphalt as Related to Compaction Design *J. Phys. Conf. Ser.* **1049**

[2] Thives L P and Ghisi E 2017 Asphalt mixtures emission and energy consumption: A review *Renew. Sustain. Energy Rev.* **72** 473–84
[3] Rubio M C, Martínez G, Baena L and Moreno F 2012 Warm Mix Asphalt: An overview *J. Clean. Prod.* **24** 76–84
[4] Jain S and Singh B 2021 Cold mix asphalt: An overview *J. Clean. Prod.* **280** 124378
[5] Usman K R, Hainin M R, Satar M K I, Warid M N M and Giwangkara G G 2020 Rheological criteria assessment of a rapid setting emulsion as compared to quick set for emulsified asphalt cold mixes *IOP Conference Series: Earth and Environmental Science* vol 476 p 2
[6] Dulaimi A, Al Nageim H, Ruddock F and Seton L 2017 High performance cold asphalt concrete mixture for binder course using alkali-activated binary blended cementitious filler *Constr. Build. Mater.* **141** 160–70
[7] Usman K R, Hainin M R, Mohd Satar M K I, Mohd Warid M N, Kamarudin S N N and Abdulrahman S 2021 Palm oil fuel ash application in cold mix dense-graded bituminous mixture *Constr. Build. Mater.* **287** 123033
[8] Khalid H A and Monney O K 2009 Moisture damage potential of cold asphalt *Int. J. Pavement Eng.* **10** 311–8
[9] Ling C, Hanz A and Bahia H 2014 Evaluating Moisture Susceptibility of Cold-Mix Asphalt *Transp. Res. Rec. J. Transp. Res. Board* **2446** 60–9
[10] Abdulrahman S, Hainin M R, Satar M K I M, Hassan N A and Usman A 2019 Rutting and Moisture Damage Evaluation of Warm Mix Asphalt Incorporating POFA Modified Bitumen *Int. J. Eng. Adv. Technol.* **9** 90–8
[11] Kumar P and Anand P 2012 Laboratory Study on Moisture Susceptibility of Dense Graded Mixes *J. Transp. Eng.* **138** 105–13
[12] Al-Merzah S, Al-Busaltan S and Nageim H Al 2019 Characterizing Cold Bituminous Emulsion Mixtures Comprised of Palm Leaf Ash *J. Mater. Civ. Eng.* **31** 04019069
[13] Usman K R, Hainin M R, Satar M K I, M Warid M N and Abdulrahman S 2020 Modified Marshall Test assessment for emulsified asphalt cold mixes *IOP Conf. Ser. Earth Environ. Sci.* **498** 012023
[14] Ronald M and Luis F P 2016 Asphalt emulsions formulation: State-of-the-art and dependency of formulation on emulsions properties *Constr. Build. Mater.* **123** 162–73
[15] Thanaya I N A 2003 *Improving the performance of cold bituminous emulsion mixtures (CBEMs) : incorporating waste*
[16] Huang C W, Yang T H and Lin G B 2020 The Evaluation of Short- And Long-Term Performance of Cold-Mix Asphalt Patching Materials *Adv. Mater. Sci. Eng.*
[17] Oluwasola E A, Hainin M R and Aziz M M A 2016 Comparative evaluation of dense-graded and gap-graded asphalt mix incorporating electric arc furnace steel slag and copper mine tailings *J. Clean. Prod.* **122** 315–25
[18] Usman K R, Hainin M R, Idham M K, Warid M N M, Yaacob H, Hassan N A, Azman M and Puan O C 2019 Performance evaluation of asphalt micro surfacing – a review *IOP Conf. Ser. Mater. Sci. Eng.* **527** 012052
[19] Yaacob H, Chang F L, Rosli Hainin M and Jaya R P 2015 Curing of asphalt emulsified tack coat subjected to malaysian weather conditions *J. Mater. Civ. Eng.* **27** 1–10
[20] JKR 2008 JKR/SPJ/2008-S4 Standard Specification for Road Works Part4 Flexible Pavement *JKR Specification for Road Works Part4 Flexible Pavement* pp 1–187
[21] American Society for Testing and Materials 2016 ASTM C150/C151 - 11 Standard Specification for Portland Cement (ASTM International, West Conshohocken, PA.) pp 1–14
[22] Du Y, Kong L and Wei T 2019 Laboratory investigation into early-age strength improvement of cold recycled asphalt mixture containing asphalt emulsion and cement *Adv. Civ. Eng.* 1–10
[23] Jiang J, Ni F, Zheng J, Han Y and Zhao X 2020 Improving the high-temperature performance of cold recycled mixtures by polymer-modified asphalt emulsion *Int. J. Pavement Eng.* **21** 41–8
[24] Bueno E T, Paris J M, Clavier K A, Spreadbury C, Ferraro C C and Townsend T G 2020 A review of ground waste glass as a supplementary cementitious material: A focus on alkali-silica reaction *J. Clean. Prod.* **257** 120180

[25] Hafizah N, Khalid A, Hussin M W, Mirza J, Farhayu N, Ismail M A, Lee H, Mohamed A and Putra R 2016 Palm oil fuel ash as potential green micro-filler in polymer concrete *Constr. Build. Mater.* **102** 950–60
[26] Salam M, Safiuddin M and Jumaat M 2018 Durability Indicators for Sustainable Self-Consolidating High-Strength Concrete Incorporating Palm Oil Fuel Ash *Sustainability* **10** 2345
[27] Aldahdooh M A A, Bunnori N M and Johari M A M 2013 Development of green ultra-high performance fiber reinforced concrete containing ultrafine palm oil fuel ash *Constr. Build. Mater.* **48** 379–89
[28] Megat Johari M A, Zeyad A M, Muhamad Bunnori N and Ariffin K S 2012 Engineering and transport properties of high-strength green concrete containing high volume of ultrafine palm oil fuel ash *Constr. Build. Mater.* **30** 281–8
[29] Awal A S M A and Hussin M W 1997 The effectiveness of palm oil fuel ash in preventing expansion due to alkali-silica reaction *Cem. Concr. Compos.* **19** 367–72
[30] Tangchirapat W, Jaturapitakkul C and Chindaprasirt P 2009 Use of palm oil fuel ash as a supplementary cementitious material for producing high-strength concrete *Constr. Build. Mater.* **23** 2641–6
[31] Ranjbar N, Mehrali M, Alengaram U J, Simon H and Metselaar C 2014 Compressive strength and microstructural analysis of fly ash / palm oil fuel ash based geopolymer mortar under elevated temperatures *Constr. Build. Mater.* **65** 114–21
[32] Abdullah K and Hussin M 2010 Fire Resistance Properties of Palm Oil Fuel Ash Cement Based Aerated Concrete *Concr. Res. Lett.* **1** 1–5
[33] Momeen M, Islam U, Mo K H and Alengaram U J 2016 Durability properties of sustainable concrete containing high volume palm oil waste materials *J. Clean. Prod.* **137** 167–77
[34] Awal A S M A and Shehu I A 2013 Evaluation of heat of hydration of concrete containing high volume palm oil fuel ash *Fuel*
[35] Adebayo Mujedu K, Ab-Kadir M A and Ismail M 2020 A review on self-compacting concrete incorporating palm oil fuel ash as a cement replacement *Constr. Build. Mater.* **258** 119541
[36] Islam A, Alengaram U J, Jumaat M Z, Bashar I I and Kabir S M A 2015 Engineering properties and carbon footprint of ground granulated blast- furnace slag-palm oil fuel ash-based structural geopolymer concrete *Constr. Build. Mater.* **101** 503–21
[37] Tangchirapat W, Khamklai S and Jaturapitakkul C 2012 Use of ground palm oil fuel ash to improve strength , sulfate resistance , and water permeability of concrete containing high amount of recycled concrete aggregates *Mater. Des.* **41** 150–7
[38] Hamada H M, Jokhio G A, Yahaya F M, Humada A M and Gul Y 2018 The present state of the use of palm oil fuel ash (POFA) in concrete *Constr. Build. Mater.* **175** 26–40
[39] Nassar A I, Mohammed M K, Thom N and Parry T 2016 Mechanical, durability and microstructure properties of Cold Asphalt Emulsion Mixtures with different types of filler *Constr. Build. Mater.* **114** 352–63
[40] Nassar A I, Mohammed M K, Thom N and Parry T 2018 Characterisation of high-performance cold bitumen emulsion mixtures for surface courses *Int. J. Pavement Eng.* **19** 509–18
[41] Abdullah N and Sulaim F 2013 The Oil Palm Wastes in Malaysia *Biomass Now - Sustainable Growth and Use* (InTech)
[42] Joni H H 2018 Studying the effect of emulsified asphalt type on cold emulsified asphalt mixtures properties *IOP Conf. Ser. Mater. Sci. Eng.* **433**
[43] MoRTH 2013 Specifications for Road and Bridges Works *Indian Roads Congress on behalf of the Govet. of India, Ministry of Road Transport & Highway* vol 1 pp 234–683
[44] American Society for Testing Materials 2015 ASTM D6926. Standard Practice for Preparation of Bituminous Specimens Using Marshall (ASTM International, West Conshohocken, PA.)
[45] Dash S S and Panda M 2018 Influence of mix parameters on design of cold bituminous mix *Constr. Build. Mater.* **191** 376–85
[46] Ling C and Bahia H U 2018 Development of a volumetric Mix Design Protocol for Dense-Graded Cold Mix Asphalt *J. Transp. Eng. Part B Pavements* **144** 04018039

[47] Thanaya I N A, Zoorob S E and Forth J P 2009 A laboratory study on cold-mix, cold-lay emulsion mixtures *Proc. Inst. Civ. Eng. - Transp.* **162** 47–55
[48] Du S 2015 Performance Characteristic of Cold Recycled Mixture with Asphalt Emulsion and Chemical Additives *Adv. Mater. Sci. Eng.* **2015** 2–8
[49] American Society for Testing Materials 2015 ASTM D6927. Standard Test Method for Marshall Stability and Flow of Asphalt Mixtures (ASTM International, West Conshohocken, PA.)
[50] American Society for Testing and Materials 2015 StandardTheoretical Maximum Specific Gravity and Density of Bituminous Paving Mixtures *D2041/D2041M – 11.* (ASTM International, West Conshohocken, PA.)
[51] Maleka A M, Hamad A W and Jaya R P 2015 Effect of Palm Oil Fuel Ash (POFA) on the Durability of Asphaltic Concrete *Appl. Mech. Mater.* **744–746** 1560–5
[52] Cox B C, Smith B T, Howard I L and James R S 2017 State of Knowledge for Cantabro Testing of Dense Graded Asphalt *J. Mater. Civ. Eng.* **29** 04017174
[53] Texas DoT M and T D 2014 Cantabro loss *TxDoT* pp 3–5

Acknowledgments
The authors would like to thank Universiti Teknologi Malaysia (UTM) and Ministry of Higher Education (MoHE) for providing financial support under the Research grant No. FRGS/1/2019/TK06/UTM/02/1. In addition, the main Author wish to appreciate the support from the Federal Government of Nigeria's Tertiary Education Trust (TETFund) intervention/sponsorship through Nuhu Bamalli Polytechnic.

Evaluating the Road Construction Feasibility Performance of the Rigid Pavement in the Flood-Prone Area: A Case Study of Indonesia

R Mudiyono[1*]

[1] Civil Engineering Department, Universitas Islam Sultan Agung Semarang, Indonesia
*Corresponding author: rachmat@unissula.ac.id

Abstract. Rigid pavement is one of several pavement types often used in road construction. This pavement has sufficient flexural strength to transmit the stresses of the wheels over a wider area underneath. During construction, this road uses concrete slabs with or without steel bars. This slab can be installed either on the sub grade of foundation, the layers of foundation or on the sub grade. Planning must examine the thickness level of concrete pavement required to handle the load with the concrete strength, taking into account the characteristics of the pavement that are capable of dispersing the load widely to the sub grade. This study aimed at evaluating the road construction performance of the rigid pavement in the flood-prone area of Kaligawe Road Semarang Indonesia. Data were collected by testing the pavement under certain specifications during the rainy season, particularly in the flooding time, and analyzed using the software of plaxis 3D foundation. Result of the study showed the pavement on Jl. Kaligawe Road Semarang Indonesia was in the "safe" category, as a result, the design of which could serve as a model for constructing roads with similar qualities elsewhere.

Keywords: *rigid pavement, design, plaxis 3d foundation*

1. Introduction

Kaligawe Road Semarang Indonesia is a land transportation infrastructure made of a rigid pavement. This road connects the city of Semarang with the cities of Demak and Kudus. Located on the coastal area, the road is often flooded with sea water as the result in high tide. Inundation which continuously occurs has caused a decrease in the quality of the pavement. The decrease happens as water from the inundation gets into the pavement layers and makes the pavement easier to crack, thus less durable. To cope with the recent issue, it is necessary to have a more effective and efficient design of road structure capable of handling not only the issue of damages due to the entry of water into the pavement layers, but also the issues of damages due to the traffic loads.

This paper aimed at evaluating the road construction feasibility performance of rigid pavement in the flood - prone area of Kaligawe Road Semarang and finding out possibility whether the construction design can be reference to construct road with similar characteristics elsewhere.

Flood is one of the most common and most destructive natural disasters worldwide [1]. It is a condition where the dry land is inundated by water overflowing from water bodies such as a river, lake, or ocean. In Indonesia, flood often occurs in the rainy season around October to January. While the intensity of rain at the time is considerably high, around 2000 to 3000 mm/year, condition and management of the existing 600 big rivers across Indonesia are less optimal, thus, flooding is inevitable [2].

Rigid pavement is one among the existing pavements. This pavement is made of cement functioning as a concrete binder or often referred to as cement concrete pavement. Seeing its characteristics, this pavement is deemed suitable for road conditions with high traffic volumes dominated by heavy vehicles [3]. Although it has several advantages compared to the other pavements, this pavement is also decreased in term of quality due to various factors.

The decrease in quality in general can be indicated by deformation such as collapse, fracture, pumping and rocking; and by cracking such as block cracks, corner cracks, diagonal cracks, longitudinal cracks, irregular cracks and transverse cracks.

2. Methodology

The Methodology used in this study is related to software relevant Plaxis road pavement modeling.

2.1. Research Stage

A quantitative approach to data collection was employed to the study, consistent to the purpose of the research which was to evaluate the road construction performance of a rigid pavement. To obtain data, series of evaluation were conducted to the pavement on Kaligawe Road Semarang Indonesia during the rainy season, particularly in the flooding time to discover the dimensions, the quality of concrete, the quality of concrete pavement layers, the quality of thin concrete layers, and the quality of foundation layers, as well as the quality of sub grade and CBR.

Completed with data collection, data were analyzed using Plaxis 3D Foundation Software to generate the magnitude of deflection/displacement, stress, and forces in the pavement structure. Data analyses in the study took place in several stages; (1) Geometry, it was a process of inserting dimension (size) of the rigid pavement, (2) Creating an Entry, it was a process where data were inserted and parameters were managed as required based on the analysis program, (3) Calculating, once data entry was completed, the program would automatically count the inserted data to generate an element network and a complete element model, (4) Output, after doing all processes from geometry to calculation, the results would appear, thus output of the program could be evaluated. In the output, the displacements, the stresses in either all parts or in certain pieces of geometry, and the internal forces such as axial, shear and moment were observable.

2.2. Plaxis Analysis Software

Plaxis is an engineering simulation software developed in particular for deformation and stability analysis of soil and rock [4]. Recognized for ease of use and accuracy, this software helps optimize design more effectively than the conventional calculation methods. Bokko et al. [5] explains different functions of the main parts of the entry program in the Plaxis analysis software; Main menu, it is where users can find the all entry and operational options; General toolbar, it is where users can find specific buttons related to filing, printing, or selecting objects in running programs; Geometric toolbar, it is a part in the software where users can find and use special buttons to create geometric model; Ruler, this part shows dimension of geometry; Image area, it is a worksheet where the drawing model can be created and modified; Axis, it is a part to show coordinate of x and y axes; manual entry, it is an alternative if the use of mouse cannot provide accuracy level as required; and indicator of cursor position, it is where mouse can show either in physical unit (x and y coordinates) or in pixel unit of display.

2.2.1. Pavement Structure Analysis Stage. At this stage, an analysis of the rigid pavement structure design will be carried out to determine the amount of deflection/displacement, stress, internal forces, and pavement structure stability based on the level of safety against deformation and stress that occurs. This data analysis includes:

- Pavement Structure Analysis with Plaxis 3D Foundation
  The stages of analysis with Plaxis 3D Foundation are carried out in the following order:

IOP Conf. Series: Earth and Environmental Science **971** (2022) 012007 doi:10.1088/1755-1315/971/1/012007

- Geometry

  The geometric model is a rigid pavement structure with dimensions of 6 m x 3 m, as shown in Figure 1.

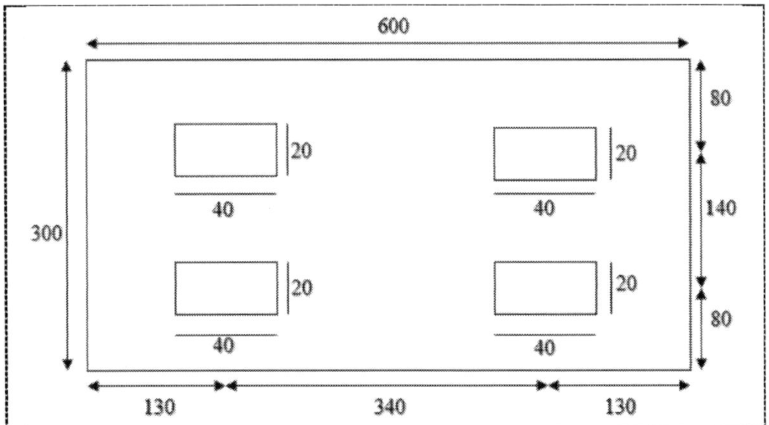

Figure 1. Top of Simple Geometry of Pavement Structure.

- Making Feedback

  To run Plaxis it is necessary to input data and set some required parameters. The input data are in the form of work planes, geometric contours, loading, drill holes, and input of material data sets. After inputting the data and setting the parameters, the next step is to arrange the 2D and 3D element networks.

- Doing Calculations

  After the element net is compiled, the finite element model is complete, then all elements and data that have been input into PLAXIS are calculated.

- Showing Output

  After the calculation is complete, the result can be evaluated in the Output program. In the Output window you can see displacements and stresses that occur throughout the geometry or in certain sections as well as internal forces on structural elements, if necessary. The computational results can also be seen in the form of curves and tables.

- Stage of Comparison of Analysis Results

At this stage, a comparison of the results of the analysis of the output magnitude of displacement/deflection, stress, and internal forces on the rigid pavement structure will be carried out to determine the difference in output of these quantities and the stability of the pavement structure against deformation and stress that occurs. The results of this comparison will be used as a basis for evaluating the output results of pavement structures that have been analyzed with Plaxis 3D Foundation.

2.2.2. Analysis Parameter of Road Sub grade Structure. In an analysis, parameter is a valuable component. Analysis parameter is useful component in an analysis. It refers to characteristics that are used to define a given object of a research. It is used to describe a specific characteristic of the whole researched objects. In order to conduct the present study, the followings parameters of analysis are used; (1) Modulus of sub grade reaction (Ks), it is defined as a value of Ks which can be determined based on soil CBR data, considering the presence of a non-linear correlation between ks and CBR [6]; (2) Elastic modulus of soil (Es), it is a value of the modulus which can be measured from its correlation between sub grade resilient modulus and CBR; (3) Poisson's Ratio (v), it is a value magnitude of Poisson's Ratio

(v) based on type of soil [7]; (4) Ultimate soil bearing capacity (qu), this capacity can be calculated based on the formula of J.E. Bowles (qu = Ks / 40); (5) Deflection limit (δ), it refers to allowable maximum deflection of the pavement structure on the sub grade.

The axle load used for road pavement design refers to the Bina Marga regulation [8] regarding the Standard Single Axle Load, which is 8.16 tons. Based on Government Regulation no. 43 of 1993 concerning Road Infrastructure and Traffic states that the Heaviest Axis Load (MST) permitted for class IIIA arterial roads is 8 tons. In the analysis of the pavement structure, the MST is determined to be 8 tons as a static load. Based on Equation 1, the simplified tire contact area in the form of a circle with radius r is:

$$L = \sqrt{\frac{Pd}{0.5227q}} = \sqrt{\frac{2000}{0.5227x\ 5.95}} = 25.36 \qquad (1)$$

$$\pi.r^2 = (2 \times 0.5227 \times L^2) + ((Sd - 0.6) \times L)$$

$$\pi.r^2 = (20.4454 \times 25.36^2) + (34.4 \times 25.36)$$

$$\pi.r^2 = 1158.83$$

$$r = \sqrt{\frac{1158.83}{\pi}} = 19.21 \text{ cm} \approx 20 \text{ cm}$$

Thus, the contact area of the wheel load can be seen in Figure 2, as shown below:

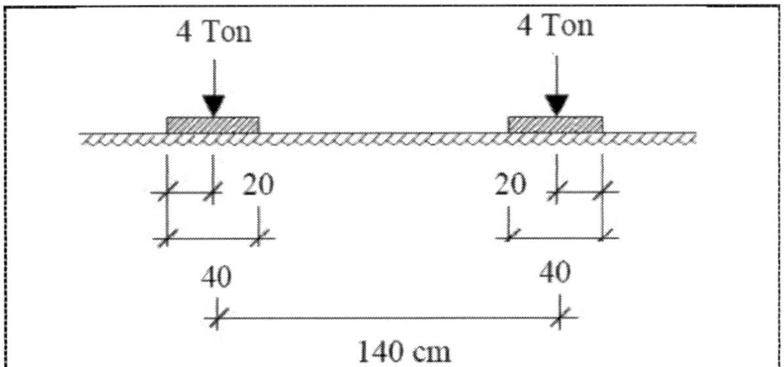

Figure 2. Wheel Load Contact Area.

2.2.3. Parameters of Road Sub grade Structure Analysis

The following will describe the parameters that affect the structure of the subsoil of the road which is the basis for calculations using the Plaxis software.

- Sub grade Reaction Modulus (Ks)
  The coefficient of Modulus of Sub grade Reaction (Ks) used for the analysis of the pavement structure can be calculated based on the CBR value of the sub grade. According to Oglesby [6], the value of ks can be determined based on soil CBR data because there is a nonlinear correlation between Ks and CBR.
  The subgrade in the analysis of this road pavement structure is clay with a CBR value of 6 %, so the Ks value is 160 psi/in = 160 x 6895/0.0254 (N/m^2)/m = 43,433 .071 N/m3 = 43,433 kN/m^3.
  From the data above, the value of the Subgrade Reaction Modulus (Ks) to be used in the analysis of the pavement structure is 43,433 kN/m^3.
- Soil Elasticity Modulus (Ice)
  The value of the soil elasticity modulus can be measured from the correlation between the sub grade resilience modulus and the CBR as follows:

MR of sub grade (MPa) = 10 x CBR (%) = 10 x 6 = 60 MPa.
So, the value of the modulus of elasticity of the soil is 60 MPa or equivalent to 60 MPa = 60 N/mm^2.

$$\frac{10^{-3}}{10^{-6}} = 60000 \text{ kN/m}^2$$

- Poisson's Ratio (v)
According to Bowles [7], The value of Poisson's Ratio (v) based on the type of soil is presented in Table 1.

Table 1. Range of Poisson's Ratio Values [7].

Soil Type	N value
Saturated Clay	0.40 - 0.50
Unsaturated Clay	0.10 - 0.30
Sandy Clay	0.20 - 0.30
Silt	0.30 - 0.35
Sand (solid) Pebbled sand	0.10 - 1.00
Commonly used	0.30 - 0.40
Rock	0.10 - 0.40
Land of Lus	0.10 - 0.30
Ice	0.36
Concrete	0.15

The sub grade in the analysis of the pavement structure is clay, so the value of based on the table above lies in the range of values 0.10 - 0.50. For the analysis of the pavement structure, it is determined that the value of is taken on an average of 0.30.

- Ultimate Soil Bearing Capacity (qu)
The ultimate soil bearing capacity can be calculated based on the approximate formula [7]. From the Ks data, it is known that the ks value is 43,433 kN/m^3, so the value of the soil bearing capacity can be calculated using Equations 2 and 3 as follows:

$$Ks = 40 \times qu \tag{2}$$
$$qu = Ks/40 \tag{3}$$

$$qu = 43433/40$$
$$qu = 1085.825 \text{ kN/m}^2$$

From the above calculation, it is known that the carrying capacity of the soil is 1,085.825 kN/m^2.

- Permit Deflection (δ)
The maximum allowable deflection of the pavement structure above the subgrade can be calculated by the formula. From the Ks data, it is known that the value of qu is 1,085.825 kN/m^2 and ks is 43,433 kN/m^3, so the allowable deflection value (Equation 4) is:

$$\delta = qu/Ks = 1085.825/43433 = 0.025 \text{ m} = 2.5 \text{ cm} \tag{4}$$

So, the allowable deflection of the pavement structure located above the sub grade is a maximum of 2.5 cm.

3. Analysis

The followings are the analysis results of the deflection/displacement, stresses and forces in the pavement structure.

3.1. Plaxis Analysis

At this stage, the data from the field are analyzed using Plaxsis software.

3.1.1. Displacement. The magnitude of the displacement values, both horizontal displacement (x), horizontal displacement (z) and vertical displacement (y) that occur in rigid pavement layers can be seen in Table 2. The table shows that the extreme value of horizontal displacement (Ux) is 239.72 x 10^{-9} m, the extreme value of horizontal displacement (Uz) is 34.16 x 10^{-9} m and the extreme value of deflection or vertical displacement (Uy) is - 12.04 x 10^{-6} m. The maximum vertical displacement value of bending pavements was taken at the nodal point having the maximum vertical displacement value. The maximum vertical was located in the short-span direction of the plate of the bending pavement which lied under the rear axle

Table 2. The values of displacements on the rigid pavement layers obtained from Plaxis 3D Foundation Software

Horizontal Displacements [Ux] (m)	Horizontal Displacements [Uz] (m)	Vertical Displacements (deflections) [Uy] (m)
239.72 x 10^{-9}	34.16 x 10^{-9}	- 12.04 x 10^{-6}

3.1.2. Stress. The magnitude of the stress values, both effective stress and total stress that occurs in each layer of rigid pavement can be seen in Tables 3 and 4. Table 3 shows that the extreme value of horizontal effective stress (Sig'-xx) is - 19.55 kN/m^2, the value of vertical effective stress (Sig'-yy) is - 45.63 kN/m^2, the value of shear stress (Sig'-xy) is - 2.87 kN/m^2, the value of shear stress (Sig'-yz) is - 2.37 kN/m^2, and the value of shear (Sig'-xz) is - 818.31 x 10^{-3} kN/m^2.

Table 3. The values of effective stress and shear stress on the rigid pavement layers obtained from Plaxis 3D Foundation Software.

Effective Stresses					
Horizontal Effective Stresses (Sig'-xx) kN/m^2	Vertical Effective Stresses (Sig'-yy) kN/m^2	Horizontal Effective Stresses (Sig'-zz) kN/m^2	Shear Stress (Sig'-xy) kN/m^2	Shear Stress (Sig'-yz) kN/m^2	Shear Stress (Sig'-xz) kN/m^2
- 19.55	- 45.63	- 19.55	- 2.87	- 2.37	- 818.31 x 10^{-3}

Based on Tabel 4 the value of horizontal total stresses (Sig'-xx) is – 36.28, the value of vertical total stresses (Sig'-yy) is – 62.35 dan the value of horizontal total stresses (Sig'-zz) is - 36.28.

IOP Conf. Series: Earth and Environmental Science **971** (2022) 012007 doi:10.1088/1755-1315/971/1/012007

Table 4. The values of total stresses on the rigid pavement layers obtained from Plaxis 3D Foundation Software.

Total Stresses		
Horizontal Total Stresses (Sig'-xx) kN/m^2	Vertical Total Stresses (Sig'-yy) kN/m^2	Horizontal Total Stresses (Sig'-zz) kN/m^2
- 36.28	- 62.35	- 36.28

3.1.3. Forces. Based on the values of internal forces, both the axial force N2, the shear force Q23 and the bending moment M22 that occur in the rigid pavement layer are shown in Figures 3, 4 and 5 below. Based on all figures below, the results of the evaluation of the rigid pavement structure can be concluded that the axial force, shear force, and maximum moment occur at the top layer of the pavement structure, namely cement concrete pavement.

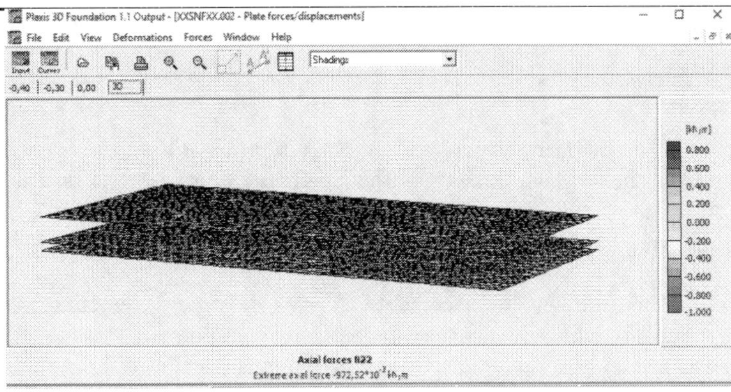

Figure 3. The Axial Force of 972.52 x 10^{-3} kN/m occurred in the pavement layer.

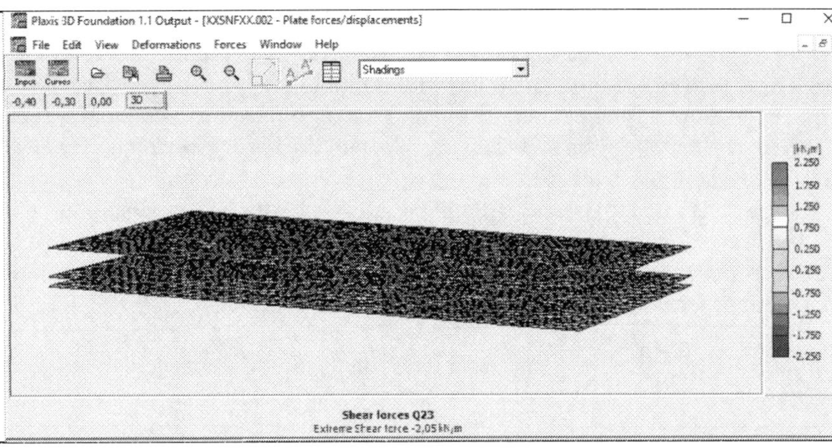

Figure 4. Shear force of - 2.05 kN/m occurred on the pavement layers

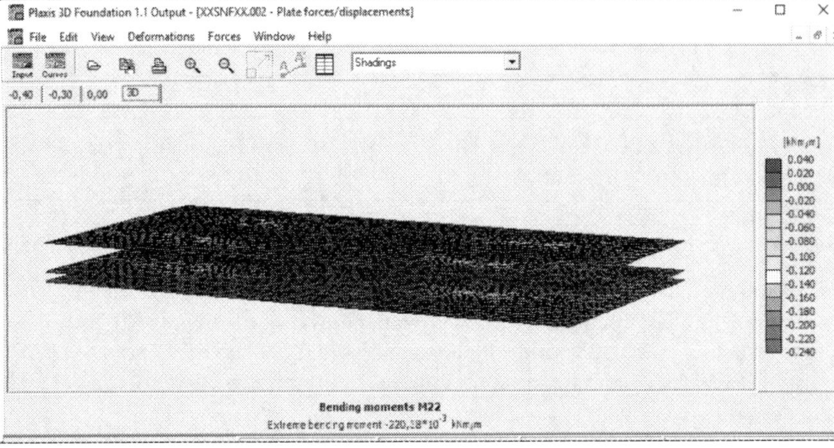

Figure 5. Bending moment of -220,38 x 10^{-3} kN/m occurred on the pavement layers

3.2. Discussion

The internal forces on the pavement structure are the magnitudes of the axial forces, shear forces, and moments resulting from the workload acting on the pavement structure. The internal forces that occur in the pavement structure provide an overview of the values of the internal forces that occur in each layer of the pavement which can affect the strength of the pavement structure itself.

3.2.1. Displacements. The following displacement values on the rigid pavement explains that deflection limit was a critical boundary which occurred on the pavement structure and was regarded as a parameter if the displacement values exceed the values of displacement limit, thus the structure was considered a failure and not feasible of design.

Source of the deflection limit was based on J.E. Bowles Formula (Equation 5) below:

$$\frac{qu}{Ks} = \frac{1085.825}{43433} = 0.025 \text{ m} = 2.5 \text{ cm} \tag{5}$$

3.2.2. Stresses. The ultimate soil bearing capacity in the pavement structure describes the critical stress limit supported by the pavement structure, so that if this ultimate stress has been exceeded, the pavement structure is considered not strong enough to support the rigid pavement structure. From the previous calculation, it is known that the ultimate bearing capacity of the pavement structure is 1,085.825 kN/m^2.

The evaluation of the stresses that occur in the rigid pavement structure is presented in Table 5. The data shows that the values of the stresses belong to "safe" category, which means "feasible of design". Ultimate bearing capacity of pavement structure in this case defines the critical limit of stresses. If the ultimate stress does not exceed the critical limit, the pavement structure is considered strong in term of supporting the rigid pavement structure, and vice versa.

Table 5. The values of effective stress and shear stress on the rigid pavement layer obtained from Plaxis 3D Foundation Software.

Rigid Pavement Stresses		Ultimate bearing capacity	Analysis result
Horizontal Effective Stresses (Sig'-xx) kN/m^2	- 19.55	1,085.825 kN/m^2	Safe
Vertical Effective Stresses (Sig'-yy) kN/m^2	- 45.63		Safe

Rigid Pavement Stresses		Ultimate bearing capacity	Analysis result
Horizontal Effective Stresses (Sig'-zz) kN/m^2	- 19.55		Safe
Shear Stress (Sig'-xy) kN/m^2	- 2.87		Safe
Shear Stress (Sig'-yz) kN/m^2	- 2.37		Safe
Shear Stress (Sig'-xz) kN/m^2	- 818.31 x 10^{-3}		Safe
Horizontal Total Stresses (Sig'-xx) kN/m^2	- 36.28		Safe
Vertical Total Stresses (Sig'-yy) kN/m^2	- 62.35		Safe
Horizontal Total Stresses (Sig'-zz) kN/m^2	- 36.28		Safe

Source of the ultimate soil bearing capacity was based on J.E. Bowles formula (Equation 5) as follow:

$$qu = 43433/40$$
$$qu = 1085{,}825 \text{ kN/m}^2$$

3.2.3. Forces. There are several types of forces occurred on the pavement structure in the study; axial, shear and moment. These forces happened due to the loads working on the pavement structure. From table 7, it can be inferred, the forces took place on the top layer of the pavement structure (cement concrete pavement).

4. Conclusion

Based on the findings and results of discussion, rigid pavement on Jl. Kaligawe Raya Km. 4 belongs to "safe" category. As design with the existing data obtained from the field is considered optimal, it can be as a reference for other road constructions under similar characteristics elsewhere.

The findings and outcomes of the discussion indicate that the rigid pavement on Jl. Kaligawe Raya Km. 4 is in the "safe" category. Because the design based on existing field data is deemed best, it can be used as a model for subsequent road constructions with similar features elsewhere.

5. References

[1] Nordbeck R, Steurer R and Löschner L 2019 The future orientation of Austria's flood policies: from flood control to anticipatory flood risk management, *J. of Envi. Planning and Management* **62(11)** 1864 - 1885
[2] Findayani A 2018 Kesiap siagaan masyarakat dalam penanggulangan banjir di Kota Semarang, *J. Geo.: Media Informasi Pengembangan dan Profesi Kegeografian* **12(1)** 102-114
[3] Rachardi R and Kurniawan R 2019 Analisis Tebal Perkerasan Jalan Rigid Di Kecamatan Sinar Peninjauan, *J. Deformasi*, **3(2)**, 74-83.
[4] Kajian J, Sipil T, Jl D I, Pelabuhan R, Priok T, Sipil T, Agustus U, Permai J S, Utara J and Jakarta D K I 2019. 04(2), 118 - 127.
[5] Bokko J 2019 Analisis Kelongsoran Jalan Poros Sangalla-Batualu Dengan Program Plaxis, *J. Dynamic Saint*, **4(1)**, 764-772
[6] Oglesby C, Parkir H, and Hole G 1989 Productivity Improvement In Construction (Mcgraw-High: New York)
[7] Bowles J E 1989 Sifat-Sifat Fisis Dan Geoteknis Tanah, Erlangga, Jakarta

IOP Conf. Series: Earth and Environmental Science **971** (2022) 012007 doi:10.1088/1755-1315/971/1/012007

[8] Direktorat Jenderal Bina Marga 2003 Perencanaan Perkerasan Jalan Beton Semen. Jakarta: Departemen Permukiman dan Prasarana Wilayah Ferad Puturuhu. (2015). Mitigasi Bencana Dan Penginderaan Jauh. Cetakan I. (Yogyakarta: Graha Ilmu).

Acknowledgments

The author would like to thank the UNISSULA Institute for Research and Community Service (LPPM) which has provided support through funding for this research. The author would also like to thank the Transportation Laboratory, Faculty of Engineering, UNISSULA, which contributed a lot of data during the research.

IOP Conf. Series: Earth and Environmental Science **971** (2022) 012008 doi:10.1088/1755-1315/971/1/012008

Effectiveness of and Public Perception about the Installation of Transverse Rumble Strips with Road Studs

M F Hazim[1], N Mashros[1*], S A Hassan[1], N A Hassan[1] and A Y Kurnia[2]

[1] Department of Geotechnics and Transportation, School of Civil Engineering, Faculty of Engineering, Universiti Teknologi Malaysia, 81310 Skudai, Malaysia
[2] Department of Civil Engineering, University of Sriwijaya, 30662 Inderalaya Sumatera Selatan, Indonesia
*Corresponding author: mnordiana@utm.my

Abstract. The installation of transverse rumble strips (TRS) is a common practice to reduce vehicle speed and alert drivers on the roadway. However, the combination of TRS with road studs to enhance traffic safety is still not widely used by local authorities. This paper aims to evaluate the effectiveness of and the public perception about the installation of TRS with road studs. A field observation was carried out on the TRS road section with and without road studs for speed evaluation. A questionnaire survey was also conducted to assess the public perception towards the installation of this measure. The results indicate that the mean speed at the TRS road section with road studs was lower than the speed at the TRS road section without a road stud and its effectiveness was evaluated using a statistical test. Responses from the questionnaire survey showed that respondents agree that both TRS with and without road studs can reduce speed effectively. Respondents also believed that the installation of road studs might cause damage to the vehicle. This paper concludes that TRS installed with road studs effectively reduces speed which is consistent with the results of the public perception survey.

1. Introduction

Road traffic injuries are the leading cause of death globally with approximately 1.3 million fatalities each year. In 2018, the World Health Organization (WHO) [1] estimated the global rate of road traffic deaths to be 18.2 per 100,000 people. In South-East Asia alone, the regional rates are the second highest with road traffic deaths at 20.7 deaths per 100,000 people. The huge number of yearly deaths from road traffic accidents has been recognized as one of the main challenges caused by unsustainable transport, hence, making road safety relevant and a key priority [2].

Motorization in Malaysia is growing rapidly due to which, the country has had the highest fatality risk in the world since 1996 [3]. Speeding, which is defined as driving too fast for road conditions or exceeding the posted speed limit, has been a significant contributing factor in traffic fatalities nationwide. Many scientists pointed out that the impact of speed on the number of accidents and injury severity is greater than almost any other known risk factors [4–6]. Thus, controlling vehicle speed can prevent accidents, reduce the impact when they occur, and lessen the severity of injuries sustained by the victims.

Managing speed has been a central issue in the last decade. One of the engineering practices to combat speeding and other unsafe behaviour is the use of traffic calming devices. Traffic calming uses physical design and other measures to improve safety for motorists, pedestrians and cyclists, especially in urban and residential areas. Speed humps, raised pedestrian crossings, raised junctions (plateau junctions), rumble strips, narrowing the road width, and speed zones [7] are among the traffic calming devices that force road users to go at the permissible speed on roads and streets. Nevertheless, the effectiveness of road calming devices depends on the geometric design [8].

Rumble strips are an effective countermeasure for reducing roadway departure crashes [9]. Among the types of rumble strips, transverse rumble strips (TRS) are a common choice to warn drivers of a need to slow down or stop or of other upcoming adverse situations that may not be anticipated by an inattentive driver. This is done so engaging their auditory and tactile senses in addition to their visual senses through traffic signals. However, continuous exposure to noise and vibration by transverse rumble strips often results in diminished effectiveness and erratic behaviours leading to additional safety challenges [10]. TRS were also installed as a countermeasure along roadways to alert drivers to lane departures or changes in roadway environment, such as the nearing of stop-controlled intersections [11].

TRS is commonly used in Malaysia [12] at interchange ramps, climbing and downhill road sections, and approaching intersections. However, in practice, TRS does not always ensure compliance with the speed limit. Thus, it encouraged some local authorities to combine TRS with road studs on the travel lane. The effectiveness of TRS in improving road safety can be measured using two distinct methods; direct and indirect. Direct method analyzes by taking the accident data into account while the indirect method describes potential treats [13]. To provide some basic insight, the present study evaluated the effectiveness of TRS when applied with road studs. The effectiveness of this practice is assessed through speed reductions since the impact of speed management measures on road safety can be made using intermediate criteria such as changes in vehicular running speed caused by a specific measure [14]. The study also considered public perception through a questionnaire survey. The findings of this study are expected to assist road planners and engineers in providing better solutions in reducing speed in urban and residential areas.

2. Methodology

In the following sub-sections, the methodology for experimental research and questionnaire survey are presented.

2.1. Experimental Research on the Effect of TRS with Road Studs

Experimental research was aimed at determining the extent of vehicle speed reduction resulting from TRS with road studs. The research was conducted at three sites; one site has a TRS section installed with road studs and the other two sites have a road section with TRS only as control sites. The sites' photos are shown in figure 1. All sites are two-lane dual carriageway roads and located in residential areas where the speed limit is 60 km/h. The effect of TRS with and without road studs was determined based on instantaneous speed and the vehicle type. Study sites with a 210 m long section were divided into five points as shown in figure 2; Point A as the upstream end and Point E as the downstream end. Points D and E have TRS, Points B and C are the transition while Point A is free from the influence of TRS. The placement of Point A requires the consideration of stopping sight distance. Research on instantaneous speed was carried out at all points using radar gun and the speed data were recorded together with the type of vehicles. The measurements were taken on a workday during the off-peak daylight period.

(a) TRS with road studs (b) TRS without road studs

Figure 1. Photos of the study site.

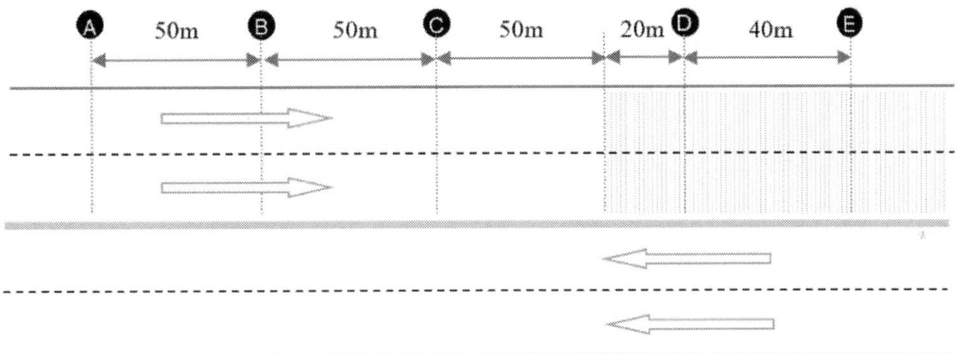

Figure 2. Layout of typical setup at the study site.

To produce reliable experimental results, the quantity of data collected must be assured. The minimum sample size of the test data that satisfies the research requirements can be determined by the following equation [15]:

$$N = \frac{Z^2 p(1-p)}{E^2} \quad (1)$$

where, N is the minimum number of samples required; Z is the confidence coefficient, E is the maximum allowable error in data measurement and p is the ratio of the number of measured samples to the total traffic flow during the measurement period. Note that Z is used to characterize the reliability. Thus, $Z = 1.65$ is utilized for 90% confidence level and $Z = 1.96$ is for 95% confidence level. Using the maximum allowable error of 5%, and the confidence level of 95%, the minimum sample size is 384. The observed data in this research at each site is much larger than the minimum sample size, so the sample size of this research satisfies the requirements.

2.2. *Questionnaire Survey of Public Perception*

A survey of public perception was based on a simple questionnaire developed for this study. The questionnaire was divided into three sections. Section A covered demographics, Section B was related to travel information and Section C was about the effectiveness and the impact on the vehicle. The items in Sections A and B were multiple-choice questions while items in Section C were scored based on a five-point Likert-type response scale (1 = strongly disagree, 2 = disagree, 3 = neutral, 4 = agree, and 5 = strongly agree). The questionnaire was distributed randomly to respondents through an online survey. The link to the questionnaire was shared through a WhatsApp. 397 responses which met the minimum requirement of the study were obtained over three months with a majority completing the survey within the first month. Respondents were also informed that the submission of completed questionnaires implied their consent to participate in the study. The responses on the questionnaire were kept anonymous for confidentiality and privacy of information. Data preparation and analysis were carried out using IBM SPSS (Statistical Package for the Social Sciences) Statistics version 22.

3. Results and Discussion

The following sub-sections present the results of the experimental study and the survey.

3.1. *Effect on Installation of TRS with Road Studs*

3.1.1 Speed analysis. Table 1 shows some statistics describing vehicle speed. The mean speed, maximum speed, and 85th percentile speed at Point E were slower than at Points A to D. Thus, the usage of TRS either with or without road studs could capture the driver's attention and alert them of upcoming roadway changes. Consequently, it could reduce the occurrence of traffic accidents. When the vehicle passes through Point A to Point E, the general trend of all speed types, including standard deviation, decreases when approaching the TRS except the maximum speed at Point B. The maximum speed at Point B is however not much different from Point A.

Table 1. Speed statistics.

Site	Parameter	Speed (km/h)				
		Point A	Point B	Point C	Point D	Point E
		TRS Without Road Studs				
	Mean	71.30	60.79	51.65	47.16	43.19
	85th Percentile	86.25	69.00	58.00	53.00	49.00
1	Min	38.00	34.00	29.00	24.00	21.00
	Max	97.00	100.00	74.00	67.00	63.00
	Std. Dev	12.59	10.10	7.68	7.24	7.24
	Mean	69.74	60.68	51.35	46.70	44.08
	85th Percentile	84.55	68.00	59.00	53.00	50.00
2	Min	40.00	34.00	26.00	24.00	23.00
	Max	96.00	100.00	74.00	73.00	67.00
	Std. Dev	12.52	9.75	8.38	7.40	7.09
		TRS With Road Studs				
3	Mean	71.35	60.01	50.13	43.66	40.02
	85th Percentile	85.00	68.00	58.25	49.00	47.40
	Min	35.00	34.00	27.00	22.00	20.00
	Max	98.00	88.00	79.00	73.00	60.00
	Std. Dev	12.42	9.64	9.24	7.46	7.10

Figure 3 shows a scatterplot of vehicle speeds at five points. Point A is located at 0 m while Point E is located 210 m from Point A. In between these two points are Points B, C and D. The scattered points at Point A were generally higher in which the maximum speed was around 100 km/h. The speed of vehicles at Point B seemed similar at sites 1 and 2 with the range of speed between 34 and 100 km/h, but slightly lower at Site 3 with the speed spread from 34 to 88 km/h. The presence of road studs together

IOP Conf. Series: Earth and Environmental Science **971** (2022) 012008 doi:10.1088/1755-1315/971/1/012008

with TRS 100 m ahead of the drivers really motivated those going at high speed to slow down. At Point C, the distribution of speed was lower than Points A and B. However, due to the speed of one vehicle, the maximum speed at Site 3 was found to be higher than other sites. The range of speed at Point D at Site 1 was lower than Sites 2 and 3, but at the last point, it was evident that the range of speed at Point E was lower than Sites 1 and 2. It can be understood that when a vehicle is passing through a road section with TRS and road studs, the strips and road studs will produce extra noise and vibration to encourage drivers to reduce speed and increase their alertness.

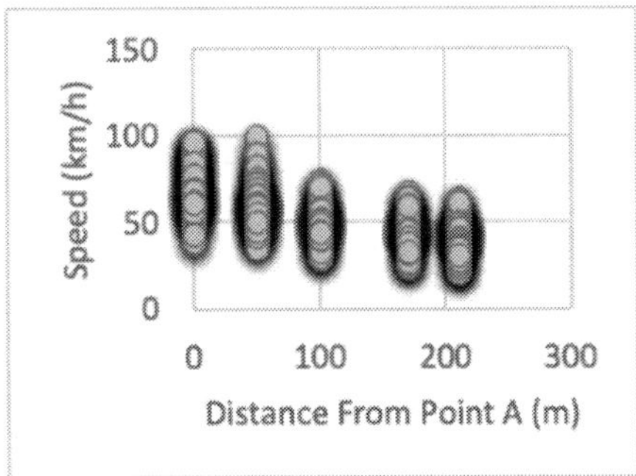

(a)Site 1

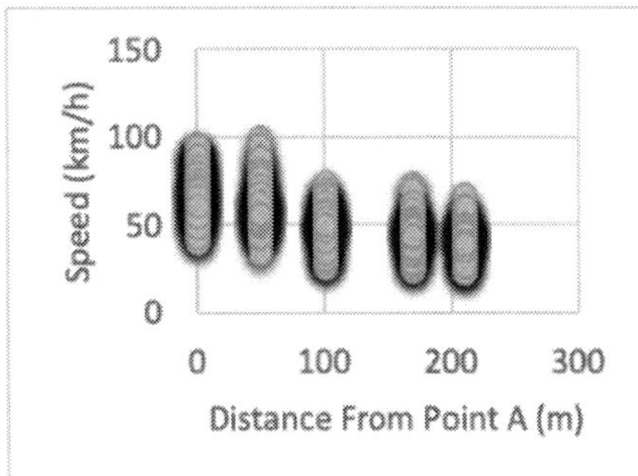

(b) Site 2

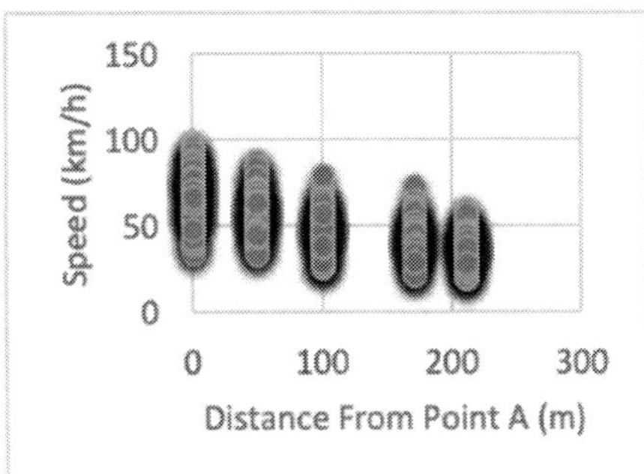

(c) Site 3

Figure 3. Scatterplot of speed (black shadow shows the repetition of data).

3.1.2 Statistical Analysis. As vehicle speed decreased from Point A to Point E, the speed changes were evaluated using a statistical test to determine whether TRS with road stud significantly affected the mean speed. T-test was applied in this study for comparing the mean speed of two sample populations. The hypothesis for checking two different speeds is as follows:

H_0: The mean speed for one set of data is the same as the second set of data
H_a: The mean speed for one set of data is different from the second set of data

The p-value for the t-test is the probability that the null hypothesis (H_0) is true. It is interpreted as when the p-value is less than 5%, the H_0 is rejected and the alternative hypothesis (H_a) would be accepted at 95% confidence level. On the contrary, when the p-value is greater than 5%, the H_a is rejected and the H_0 would be accepted at a 95% confidence level [16].

Table 2 presents the mean speed and t-test results for road sections with only TRS as well as TRS with road studs. The results in table 2(a) suggested that the difference in mean speeds between Site 1 and Site 3 were not statistically significant at Point A and Point B, respectively. It might be that on approaching the TRS section, the vehicles were still not affected by TRS either with or without road studs. However, the mean speed comparison results at Point C, Point D and Point E were statistically significant with a reduction of 1.52 km/h, 3.50 km/h, 3.17 km/h, respectively. T-test results for mean speed between Site 2 and Site 3 in table 2(b) also showed the same results as in table 2(a) except at Point C where the difference in mean speed was not significant. This finding suggests that speed decreases with the existence of road studs on a TRS section.

Table 2. t-test for mean speed.

(a) Comparison between Site 1 and Site 3.

Point	A	B	C	D	E
Hypothesis	H0: $\mu 1 = \mu 3$; Ha: $\mu 1 \neq \mu 3$				
Mean speed at Site 1 (km/h)	71.30	60.79	51.65	47.16	43.19
Mean speed at Site 3 (km/h)	71.35	60.01	50.13	43.66	40.02
Change (km/h)	+0.05	-0.78	-1.52	-3.50	-3.17
p-value	0.959	0.326	0.028 [a]	0.000 [a]	0.000 [a]

[a] significant different, $p < .05$ (Reject H0)

(b) Comparison between Site 2 and Site 3.

Point	A	B	C	D	E
Hypothesis	H_0: $\mu_2 = \mu_3$; H_a: $\mu_2 \neq \mu_3$				
Mean speed at Site 2 (km/h)	69.74	60.68	51.35	46.70	44.08
Mean speed at Site 3 (km/h)	71.35	60.01	50.13	43.66	40.02
Change (km/h)	+1.61	-0.67	-1.22	-3.04	-4.06
p-value	0.111	0.391	0.089	0.000 [a]	0.000 [a]

[a] significant different, $p < .05$ (Reject H_0)

The analysis of speed difference was also conducted at each point based on the vehicle category. In this study, vehicles were classified into six categories as presented in figure 4. According to the figure, it can be found that a larger proportion of vehicles consists of cars with around 55%, followed by motorcycles with the proportion near to 20%. Van is the third higher proportion, and the rest of the vehicle categories share almost similar proportion. All sites present the same trends for six vehicle categories. Table 3 and table 4 show the summary of the results of the t-test on mean speeds based on vehicle categories between Site 1 and Site 3 and mean speed between Site 2 and Site 3, respectively. Result obtained from the analyzes indicate that the changes in mean speed for all categories of vehicles at Point A and Point B was found not statistically significant. Point A and Point B, which were located more than 100m from TRS, indicated that the mean speed of vehicles was not influenced by the existence of TRS with or without road stud.

The results in both tables also revealed that only cars at Point C were affected by the application of TRS with road stud. On the other hand, other categories of vehicles at the same point have a p-value more than 0.05 which indicated that there was no significant difference in the mean speed. Point D and Point E in both table 3 and table 4 proved that the difference in mean speeds for car, van, motorcycle, and small lorry were statistically significant except at Point D for the small lorry in table 3.

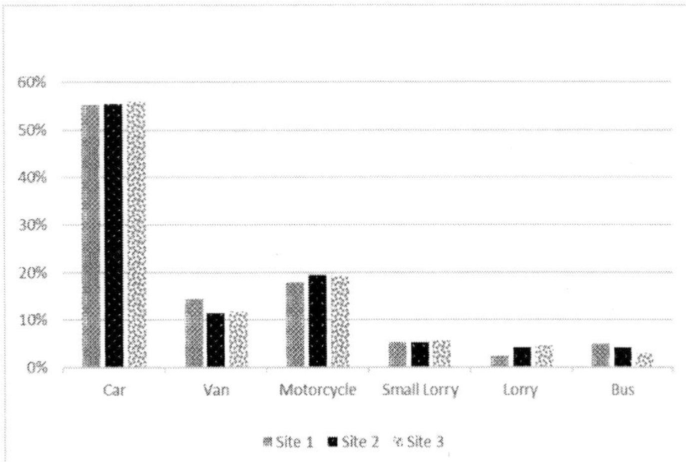

Figure 4. Vehicle composition.

Table 3. t-test on mean speed based on vehicles categories between Site 1 and Site 3.

Point	Vehicles Class	Car	Van	Motorcycle	Small Lorry	Lorry	Bus
	Hypothesis	H_0: $\mu_1 = \mu_3$; H_a: $\mu_1 \neq \mu_3$					
A	Change (km/h)	-1.50	5.25	-3.11	6.61	-0.87	4.24
	p-value	0.205	0.055	0.161	0.426	0.739	0.07
	Reject H_0?	No	No	No	No	No	No
B	Change (km/h)	0.56	-1.60	-4.99	0.84	-1.57	2.22
	p-value	0.599	0.21	0.168	0.599	0.581	0.465
	Reject H_0?	No	No	No	No	No	No
C	Change (km/h)	-2.70	0.63	-2.15	4.57	-0.40	0.45
	p-value	0.001	0.529	0.237	0.472	0.848	0.855
	Reject H_0?	Yes	No	No	No	No	No
D	Change (km/h)	-3.98	-3.00	-4.80	-2.08	-0.20	-1.58
	p-value	0.000	0.000	0.001	0.594	0.916	0.302
	Reject H_0?	Yes	Yes	Yes	No	No	No
E	Change (km/h)	-3.82	-2.09	-3.61	-3.44	2.57	0.13
	p-value	0.000	0.002	0.002	0.040	0.244	0.917
	Reject H_0?	Yes	Yes	Yes	Yes	No	No

Table 4. t-test on mean speed based on vehicles categories between Site 2 and Site 3.

Point	Vehicles Class	Car	Van	Motorcycle	Small Lorry	Lorry	Bus
	Hypothesis	H_0: $\mu_2 = \mu_3$; H_a: $\mu_2 \neq \mu_3$					
A	Change (km/h)	2.10	5.36	-2.62	-0.38	-4.24	3.28
	p-value	0.094	0.064	0.308	0.882	0.119	0.237
	Reject H_0?	No	No	No	No	No	No
B	Change (km/h)	0.08	2.00	5.90	-0.13	-5.99	-0.06
	p-value	0.930	0.152	0.058	0.955	0.055	0.986
	Reject H_0?	No	No	No	No	No	No
C	Change (km/h)	-2.94	0.38	0.99	1.47	-0.83	1.67
	p-value	0.001	0.729	0.602	0.376	0.720	0.514
	Reject H_0?	Yes	No	No	No	No	No
D	Change (km/h)	-5.09	-3.88	-2.71	-2.76	3.02	-2.12
	p-value	0.000	0.000	0.015	0.023	0.150	0.130

	Reject H_0?	Yes	Yes	Yes	Yes	No	No
	Change (km/h)	-5.09	-3.88	-2.71	-2.76	3.02	-2.12
E	p-value	0.000	0.000	0.015	0.023	0.150	0.130
	Reject H_0?	Yes	Yes	Yes	Yes	No	No

3.2. Public Perception

Figure 5 shows the attributes of the respondents. Of the 397 respondents, 206 (52.14%) were male and 191 (47.86 %) were female. The respondents aged 19–25 comprised 17% of the total proportion while those between 26–30 years comprised 28%, those between 31–35 years comprised 29%, and those between 36–40 years and more than 40 years comprised 15% and 11%, respectively. The percentage of respondents aged 26 to 35 was found to be larger than the other age groups. In this survey, the car was the most frequent vehicle used and it made up 69% of the total proportion, followed by motorcycle at 23%. The percentage of van and lorry were evenly distributed at 4% each. In terms of driving experience, 156 (39%) respondents out of 397 respondents had a driving experience between 6–10 years. 100 (25%) respondents out of 397 respondents have a driving experience between 11–15 years. Four percent of the respondents have a driving experience of more than 20 years, thus showing that this survey also considered drivers with a wide experience on the road.

The respondents were asked on how frequently they were traveling on road sections of TRS with and without road studs per day. Results shown in figure 6 indicated that only 17 (4%) of the respondents traveled on the road without TRS while the rest have experience traveling on TRS every day. Most of the respondents (38%) met four to six TRS per day. 72 (18%) of respondents met TRS more than nine times per day. For TRS with road studs, it was found that 200 (51%) of the respondents met TRS between one to three times per day. This result revealed that TRS with road studs had been used in rural areas to alert drivers that they are approaching a lower speed zone or a stop-controlled intersection. On the other hand, only 96 (24%) of the respondents encountered TRS with road studs for about four to six times daily. The number of respondents who do not meet TRS with road stud per day was higher than the number of respondents who met TRS without road studs.

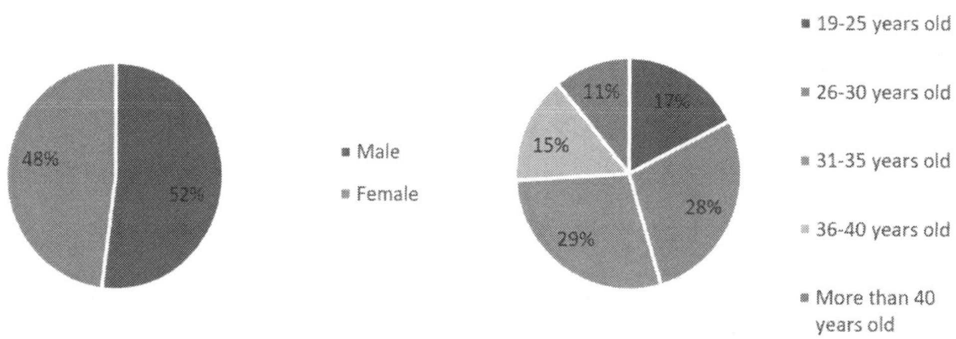

(a) Gender (b) Age

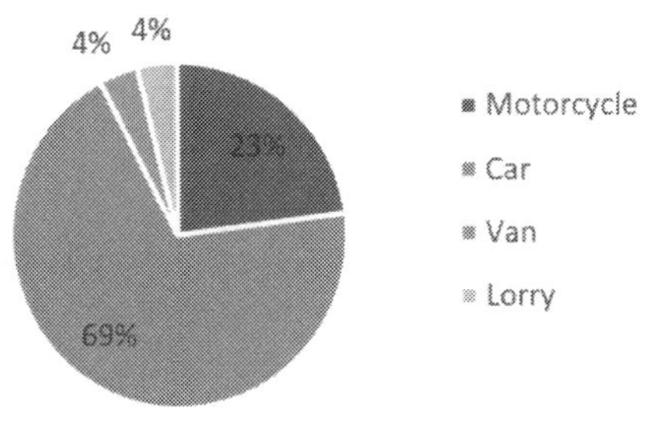

(c) Frequent vehicle used

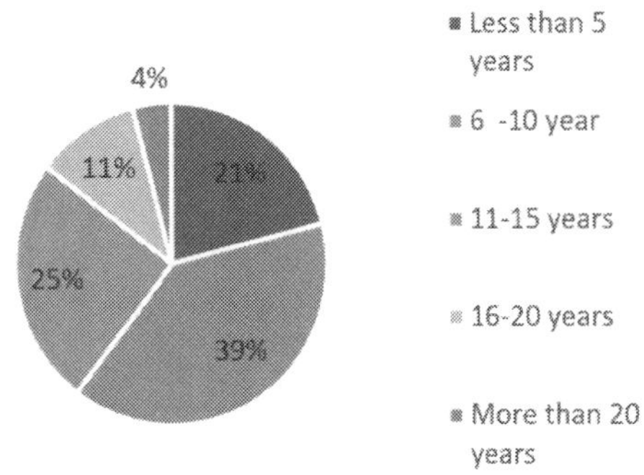

(d) Driving experience

Figure 5. Respondents' attributes.

The respondents were also asked about their average vehicle speed on the main residential road. As presented in figure 7, most respondents (35 %) had an average speed between 46 and 60 km/h. These speeds were almost similar to the spot speed observed at the site when the drivers were approaching TRS with or without road studs. About 131 (33%) drove with an average speed between 30 and 45 km/h. Surprisingly, there were a few respondents who drove more than 90 km/h on the main residential road. This speeding behaviour might lead to road accidents in the residential area.

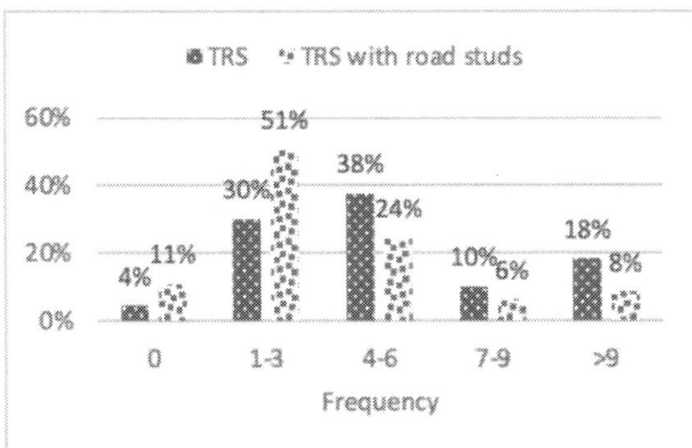

Figure 6. Respondents' travelling frequency.

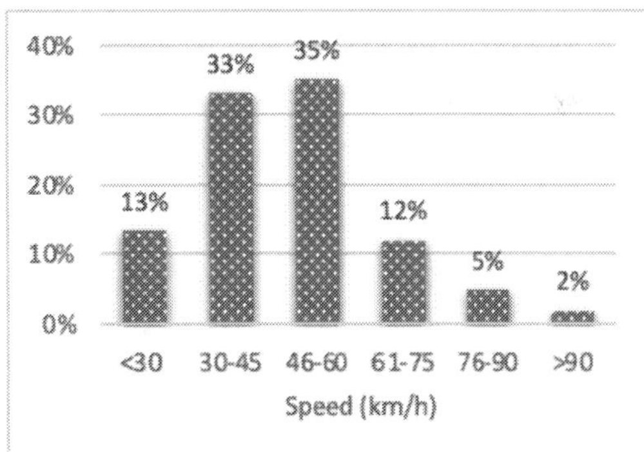

Figure 7. Respondents' operating speed.

Public perception towards TRS with and without road studs were assessed using a Likert scale and the results are as shown in table 5. As can be seen from the table, most of the respondents agreed that TRS with and without road studs could effectively reduce vehicle speed. It shows that these devices which were placed across the travel lanes of the roadway and were perpendicular to the flow of vehicles were able to alert respondents about the upcoming hazard. Respondents had neutral responses when asked about vehicle damage due to TRS. However, the results showed that a majority of the respondents believe that TRS with road studs can cause damage to the vehicle. The application of road studs might cause monetary losses in the form of bent wheels, damaged tires and suspension componentry, broken wheel joints and even more so, the loss of lives that may result from the aforementioned.

Table 5. Public perception towards TRS with and without road stud.

No.	Questions	Mean	Level of agreement
1.	Transverse rumble strip reduce speed effectively.	3.616	Agree
2.	Transverse rumble strip installed together with road stud reduce speed effectively	3.929	Agree
3.	Transverse rumble strip cause damage to vehicles	2.671	Neutral
4.	Transverse rumble strip installed together with road stud cause damage to vehicles	3.621	Agree

4. Conclusion

This study was conducted by focusing on the effectiveness of and public perception about the installation of TRS with road studs. Three locations were selected as study sites to determine the effectiveness of TRS with road studs on speed reduction. A simple questionnaire was also developed to know public perception towards TRS with and without road studs. The key findings obtained in this study are:

- The speed analysis conducted indicates that the operating speed was reduced at all sites when vehicles traveled from Point A to Point E. However, the speed reduction was lower with the presence of road studs. In addition, the speed reduction was more obvious on TRS with road studs.
- In terms of statistical analysis, the difference in mean speed was statistically significant at Point D and Point E. Cars, vans, motorcycles, and small lorries were significantly different in speed at these two points.
- Assessment of public perspective indicates that respondents agreed that TRS with and without road studs could reduce speed effectively. Respondents have neutral responses when asked about vehicle damage due to TRS. However, a majority of the respondents believe TRS with road studs can cause damage to vehicle.

5. References

[1] Paprika 2018 *Global status report on road safety 2018* Licence: CC BY- NC-SA 3.0 IGO. (Geneva: World Health Organization)

[2] Anila P 2017 *Road Safety - Considerations in Support of the 2030 Agenda for Sustainable Development* UNCTAD/DTL/TLB/2017/4 (Geneva: United Nations)

[3] Jacobs G, Thomas A A, Astrop A 2000 *Estimating global road fatalities* TRL Report 445 (London: Transport Research Laboratory)

[4] Nilsson G 1981 The effects of speed limits on traffic accidents in Sweden *Proc. of the Inter. Symp. on the Effects of Speed Limits on Traffic Accidents and Transport Energy Use* Dublin, Ireland (Paris: OECD) 1-8

[5] Letty A, Charlotte B, Ragnhild D, Charles G, Theo J, Boudewijn v K, René M, Peter M, Ingrid v S, Chris S, Divera T, Willem V, Fred W and Paul W 2006 Advancing Sustainable Safety *National Road Safety Outlook for 2005−2020* ed F Wegman, L Aarts (Leidschendam: SWOV) pp 215

[6] Elvik R 2009 *The Power Model of the Relationship between Speed and Road Safety* Update and New Analyses Report 1034 (Oslo: Institute of Transport Economics)

[7] Abdul M MM, Alvin P WH, Ahmad F MS 2010 Evaluation of Traffic Calming Scheme Around the Schools in Malaysia *8th Malaysian Road Conf. (MRC8)* (Kuala Lumpur)

[8] Brindle R E 1999 Towards guidelines for LATM in different speed environments, *Road and Trans. Res.* **8(1)** 92-95

[9] Dadashova B, Wu L, and Dixon K. 2018 Evaluating the impact of rumble strips on fatal and injury freeway crashes *Trans. Res. Record* **2672(30)** 131-141

[10] Hossen M S, Kappes C, Trabia M, Morris B, Park J and Paz A 2019 Design and preliminary testing of demand-responsive transverse rumble strips *Advances in Mechanical Eng.* **11(9)**
[11] Miles J D, and Finley M D 2007 Factors that influence the effectiveness of rumble strip design *Trans. Res. Record* **2030(1)** 1-9
[12] Othman M H, Haron Z, Hainin M R, Yahya K, Yaacob H and Sanik M E 2015 Malaysian transverse rumble strips: A review and recommendations for practice, *J. Tech. Sci. & Eng.* **73(4)** 117–123
[13] Gaca S, and Pogodzińska S 2017 Speed management as a measure to improve road safety on Polish regional roads *Archives of transport* **43**
[14] Gargoum S A, and El-Basyouny K 2016 Exploring the association between speed and safety: A path analysis approach *Accident Analysis & Prevention* 93 32-40
[15] Pan F, Yang Y, Zhang L, Ma C, Yang J and Zhang X 2020 Analysis of the impact of traffic violation monitoring on the vehicle speeds of urban main road: Taking China as an example *J. of Advanced Transportation* **2020** 11
[16] Montgomery D C, Runger G C and Humble N F 2009 *Engineering Statistics* (New York: John Wiley and Sons) pp 544

Acknowledgment
This research was supported by Ministry of Higher Education (MOHE) through Fundamental Research Grant Scheme (FRGS/1/2020/TK02/UTM/02/5) and Universiti Teknologi Malaysia through Research University Grant Scheme (R.J130000.2651.17J34).

Impact of Ageing and Moisture Damage on the Fracture Properties of Plastic Waste Modified Asphalt

H R Radeef [1,2], N A Hassan[1*], A R Z Abidin[1], M Z H Mahmud[1], C R Ismail[1], H F Abbas[1], Z H Al-Saffar[3] and S Redha[1],

[1] School of Civil Engineering, Universiti Teknologi Malaysia, Skudai, 81310, Johor, Malaysia
[2] Department of Civil Engineering, College of Engineering, University of Kufa, Najaf Governorate, Iraq
[3] Building and Construction Eng. Technical College of Mosul, Northern Technical College of Mosul, Northern Technical University, 41002, Iraq
*Corresponding author: hnorhidayah@utm.my

Abstract. The utilisation of plastic waste as a modifier in asphalt mixtures has inflicted significant impacts on the cracking resistance of the produced mixtures. Whilst many studies have evaluated the cracking resistance of asphalt mixtures incorporating plastic waste using wet method, limited studies have used the dry method are available. The current study aims to evaluate the effect of plastic waste incorporation on the fracture properties of conventional asphalt mixture. In addition, the impact of ageing and moisture damage on the fracture properties of modified asphalt was also investigated. Indirect tensile strength test was carried out to assess the CT-index, tensile strength, and fracture energy of the asphalt mixtures before and after exposure to ageing and moisture conditioning. The finding revealed that the asphalt mixture incorporating plastic waste demonstrated superior resistance to thermal and fatigue cracking compared to the control mixture, thus proving the capability of plastic waste in increasing the resistance of asphalt against ageing and moisture damage.

1. Introduction

The incorporation of plastic waste into asphalt mixtures has demonstrated outstanding economic and environmental benefits. Indeed, the increasing amount of plastic waste such as plastic bags and plastic bottles is considered as the most critical environmental challenge in Malaysia and the world, likewise. It was found that thirteen percent of the solid waste generated in Malaysia is of plastic-based, of which 55% is being mismanaged [1]. The Malaysian Plastics Manufacturers Association has disclosed that every Malaysian litters 300 plastic bags a year on average, where 30% of the plastic waste might have been thrown into the ocean [2]. This calls the attention to utilise plastic waste as a sustainable material for construction. Accordingly, an increase in traffic loading reduces the asphalt pavement service life, and as a result, this increases the pavement maintenance work. A high asphalt pavement maintenance cost implies the need for relevant materials utilisation to improve the performance of asphalt mixtures [3]. Thus, there is an increasing interest in plastic waste incorporation into industrial asphalt pavement for construction [4–10]. Numerous studies have focused on plastic waste integration into asphalt mixtures using wet method [5,11,12]. Polymer modification is a widely utilised wet process for modifying asphalt mixtures through rheological properties improvement in asphalt binder prior to mixing with aggregate [9,13–15]. However, the wet method has some vital limitations regarding the mixing process and storage. This method requires the mixing of asphalt binder with plastic waste at high temperatures for full disintegration of the plastic waste into the binder. Moreover, modified asphalt binder requires a special storage tank to ensure its homogeneity. On the other hand, a very limited number of studies is available on the effect of plastic waste incorporation into asphalt binder using dry method on the cracking performance of asphalt mixtures. This method includes a pre-mixing process between plastic waste and hot aggregate before a further mixing with asphalt binder [4,16]. In this study, 1 wt.% of shredded plastic bags by total weight of mixture was used for the modification of conventional asphalt mixture. The plastic content was selected by recommendation of a previous study conducted by Radeef et al. [17]. The modified and control mixtures were evaluated using tests of CT-index, tensile strength, total fracture energy, and fracture energy at peak load. Moreover, these test methods included the evaluation of the impact of two ageing levels and moisture conditioning on the performance of the produced mixtures. According to several prior researches, the IDEAL-CT test has demonstrated encouraging results, as it closely simulates the mixture resistance to crack propagation [18–20].

2. Materials

Asphalt binder of 60/70 penetration grade with an average softening point of 48°C was utilised in this study. Table 1 illustrates the physical properties of the asphalt binder. Asphalt mixtures were prepared using crushed granite of nominal maximum aggregate size of 14 mm in accordance to the Malaysian Public Works Department specification [21]. 2% hydrated lime was added as an anti-stripping agent to reduce the moisture susceptibility of the produced asphalt mixtures [22]. Figure 1 shows the aggregate gradation of AC 14 mixture.

Table 1. Physical properties of the asphalt binder

Parameter	Unit	Recommended Value	Experimental Value	Reference Method
Penetration at 25°C	mm	60–70	66.7	ASTM D5
Softening Point	°C	47 Min	48	ASTM D36
Ductility	cm	> 100	137	ASTM D113
Specific Gravity at 25°C	kg/m^3	—	1.033	ASTM D70
Viscosity at 135°C	Pa·s	—	0.4	ASTM D4402/D4402M
Viscosity at 165°C	Pa·s	—	0.2	ASTM D4402/D4402M

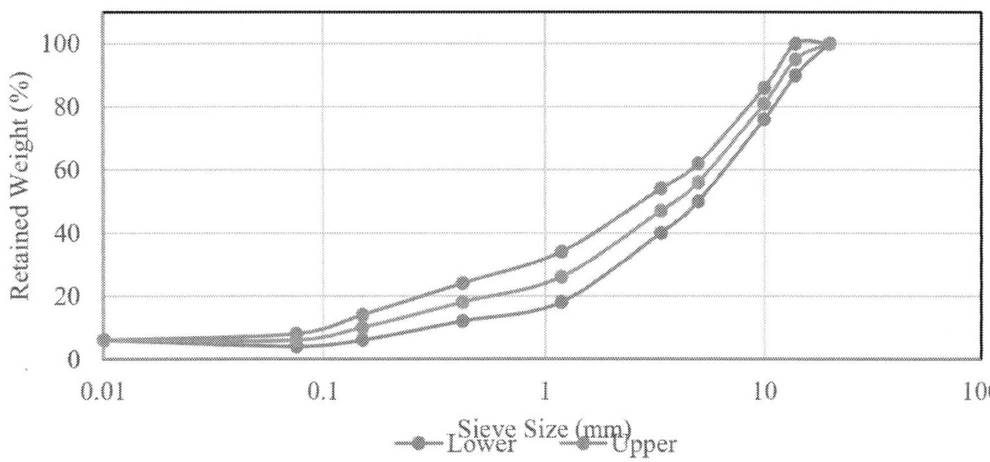

Figure 1. Aggregate gradation of AC 14 mixture.

Low-density polyethylene of 5-10 mm particle size acquired from shredded plastic bags was utilised as the plastic waste. In this material preparation, cleaned plastic bags were mechanically shredded at ambient temperature. This range of small particle size was chosen to increase the homogeneity of the mixture during the blending process [17]. 1 wt.% of the plastic waste by weight of the total mixture was incorporated into asphalt mixture using dry process method, as recommended in a previous study conducted by Radeef et al. [23].

2.1. Mixture preparation

To examine the impact of plastic waste incorporation on the performance of asphalt mixture, two mixture types were considered in this study; conventional and plastic waste modified asphalt mixtures. The Marshall mix design method was employed for the asphalt mixtures preparation. The plastic waste modified asphalt mixture was produced by adding 1% plastic waste by weight to aggregate retained on 3.35 mm mesh sieve. Mixing process was carried out at 160°C for the control mixture and 180°C for the modified mixture based on the melting point of the shredded plastic utilised in this study. The mixing process for the modified mixture involved the blending of the shredded plastic with hot aggregate to create melted plastic coat on the aggregate particles, as shown in Figure 2. Subsequently, the modified aggregate was mixed with hot binder prior to the addition of fine aggregate and compaction using a Marshall hammer to produce the modified mixture.

Figure 2. Plastic waste coated aggregate particles

3. Testing method

To fulfil the objectives of this study, a two-phase test method was implemented. The first phase included mixtures preparation and volumetric properties evaluation. Two types of mixture were prepared, namely conditioned and unconditioned samples. For sample conditioning, samples were exposed to long-term ageing and moisture conditionings prior to testing. Meanwhile, the second phase involved the examination of the effect of moisture conditioning and long-term ageing on the cracking properties of the control and plastic waste modified mixtures via fracture or indirect tensile stress test (IDEAL-CT). In addition, the fracture behaviour of the plastic waste modified mixture was assessed using three different fracture parameters; fracture energy, CT-index, and tensile strength. The asphalt mixtures were coded into eight different blends as shown in Table 2.

Table 2. Mixture type coding.

Mixture Description	Mixture Coding
Control conditioned	C
Control Moisture conditioned	C-MD
Control Long-Term Aged	C-LTA
Control Short-Term Aged	C-STA
Plastic Waste modified	PLW
Plastic Waste Moisture conditioned	PLW-MD
Plastic Waste Long-Term Aged	PLW-LTA
Plastic Waste Short-Term Aged	PLW-STA

3.1. Marshall properties
The Marshall design method as specified by the Malaysian Public Works Department [21] was implemented to determine the optimum binder content for the control and modified asphalt mixtures. The dry, wet, and saturation weights of the compacted samples were then calculated in accordance to the ASTM D 2726 [24]. For loose mixtures, the maximum theoretical density was determined based on ASTM D2041/D2041M-11 [25]. Subsequently, the samples were conditioned in 60°C hot water for 45 minutes to evaluate the flow and stability of the mixtures via Marshall stability test [26]. Table 3 displays the results on the stability and volumetric properties of the asphalt mixtures.

Table 3. Stability and volumetric properties of asphalt mixtures.

Properties	Control Asphalt Mixture	Plastic Waste Modified Asphalt Mixture (1% PLW)
Stability (N)	16 410	19 319
Flow (mm)	3.30	3.63
Stiffness (N/mm)	4972	5322
Void filled with binder (%)	79.9	71.9
Air void (%)	3.10	4.30
Asphalt binder content (%)	5.00	4.93
Bulk density (kg/m^3)	2351	2330

3.2. Ageing and moisture conditionings
The ageing conditions adopted in this study included the long-term ageing (LTA) and short-term ageing (STA) as presented in the AASHTO R30 [27]. The LTA procedure involved the exposure of samples to heat at 85°C in oven for 120 hours. While the STA required the exposure of loose asphalt mixture to heat at 145°C in oven for 2 hours including mixture stirring at 1-hour interval to ensure uniform heat distribution within the loose mixtures. Moisture damage conditioning was carried out by immersing partially saturated samples in water of 60°C temperature for 24 hours, as recommended in D4867M 2014 [28]. Successively, the samples were transferred from the hot water into 25°C water bath and left for 2 hours prior to testing.

3.3. Fracture energy
Several critical parameters were investigated using a load-displacement curve in assessing the fracture properties of the asphalt mixtures, including fracture energy, fracture energy at peak load, and displacement at peak load. These parameters provide valuable information on the behaviour of asphalt mixtures at fracture. As shown in Figure 3, the fracture energy at maximum denotes the cumulative energy consumed up to the peak load as represented by the shaded area, whilst the total area below the load-displacement curve represents the total fracture energy.

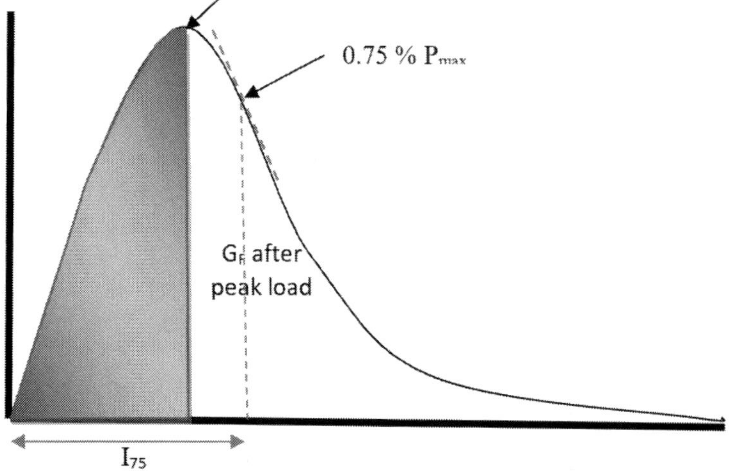

Figure 3. Typical load-displacement curve [18].

Zhou [18] utilised Equation (1) for determining fracture energy, G_f.

$$G_f = \frac{W_f}{A_f} \tag{1}$$

where;
G_f, A_f, and W_f respectively represent fracture energy (J/m^2), fracture area (m^2), and the work of fracture (J).

3.4. IDEAL-CT

The CT–index parameters developed by Zhou [18] were utilised to investigate the fracture properties of the control and modified mixtures. IDEAL-CT was conducted using the recommended procedure followed by conventional indirect tensile strength test. In IDEAL-CT, continuous measurement of load and displacement data throughout a fracture test is required to construct a load-displacement curve. The CT-index parameters obtained from this plot are proven appropriate and valuable for evaluating the crack propagation rate. The CT-index combines the effect of three essential parameters (load, displacement, and fracture energy) extracted from the load-displacement curve as shown in Figure 3. A good correlation between CT-index and the performance of field cracking has also been proven and confirmed by the National Centre for Asphalt Technology [29]. In the CT-index determination, control and modified mixtures of air void content of 7±0.5% were prepared. The test was conducted by applying a vertical loading along the diametrical direction of the samples. The established CT-index parameter was determined using Equation (2).

$$\text{CT indix} = \frac{G_f}{|m_{75}|} \times \frac{l_{75}}{D} \tag{2}$$

where;
I_{75} is the displacement at the 75% post point of peak load (mm); $|m_{75}|$ is the absolute value of the slope at the 75% inflexion point of the peak load (kN/mm); G_f is the fracture energy (J/m^2); and D is the diameter of the sample (mm).

3.5. Indirect tensile strength test

The indirect tensile strength test (ITS) was carried out in accordance with ASTM D6931-12 [30] as represented by Equation (3). The test aims to evaluate the tensile strength of unconditioned and conditioned samples. In this test, a vertical loading was applied along the diametrical axis of the Marshal compacted samples at a constant rate of 50mm/min to failure.

$$St = 2000P/\pi Dt \quad (3)$$

where;
St is the indirect tensile strength (kPa); D and t are the diameter and height of the sample (mm), respectively; and P is the peak load (N).

4. Results and Discussion

4.1. Fracture energy

The load-displacement curves for the conditioned and unconditioned control and modified mixtures are displayed in Figure 4. As shown in the figure, the control mixture exhibited a more flattened curve than that of the plastic waste modified mixture. In contrast, the modified mixture demonstrated a greater maximum load compared to the control asphalt mixture, for both conditioned and unconditioned samples. The results indicate that the plastic waste modified mixture behaves in a linear elastic way due to the increase in the stiffness value. In addition, a similar effect was observed for the aged control and modified mixtures [6].

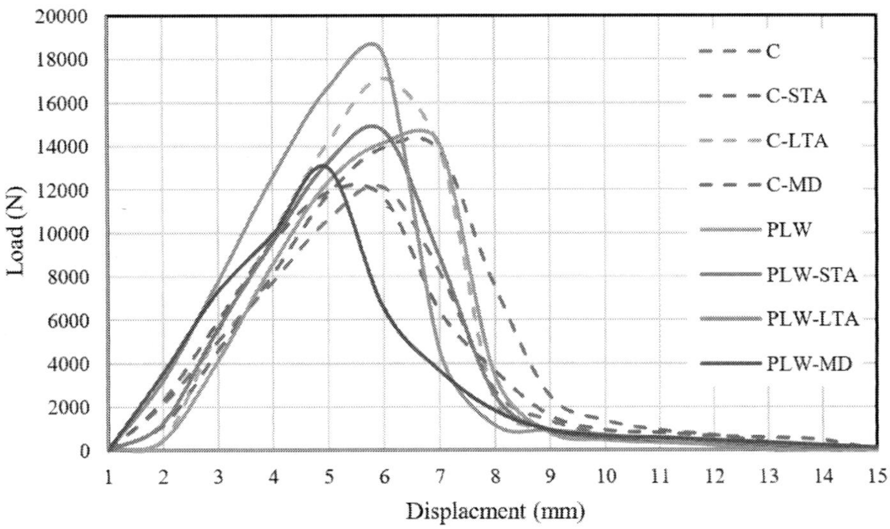

Figure 4. Load–displacement curves for conditioned and unconditioned mixtures.

The results of displacement at maximum load, total fracture energy, and fracture energy at maximum load of the control and plastic waste modified mixtures are presented in Figure 5. Both conditioned and unconditioned plastic modified mixtures demonstrated lower displacement at peak load than that of the control mixture. The results of displacement at maximum load as shown in the load-displacement curve have effectively justified the impact of plastic waste incorporation in improving the stiffness of the mixtures. This finding compliments a previous study which also revealed that the incorporation of plastic waste into asphalt mixture could increase the stiffness of the mixture [4]. In terms of fracture energy, all the pre- and post-conditioned PLW mixtures displayed superior performance to the control mixture, excluding the short-term aged sample. Short-term ageing has contributed in the total fracture energy enhancement in the sample by 33%, while the PLA-STA mixture exhibited an increase by only 5%. This is possibly due to the effect of the short-term ageing in decreasing the workability of the plastic waste modified mixtures [12]. Low workability may increase the air void content in a mixture, which consequently decreases its fracture energy [31]. The fracture energy at maximum load of all plastic waste modified asphalt mixtures was higher than that of the control mixture. It is worth mentioning that the effect of plastic waste incorporation in increasing the fracture energy at maximum load was higher compared to the total fracture energy for the conditioned samples. Nonetheless, moisture conditioning has decreased the total fracture and fracture energy at maximum load of the mixtures by 7.5% and 1%, accordingly. The displacement at maximum load for the plastic modified mixture also exhibited a decrement by 22%. On the other hand, the control mixture demonstrated a higher susceptibility to moisture damage, where lower reduction in total fracture energy, fracture energy, and displacement at maximum load by 20%, 7%, and 12% respectively was observed. Such finding indicates the capability of plastic waste in increasing the resistance of asphalt to cracking and environmental effect.

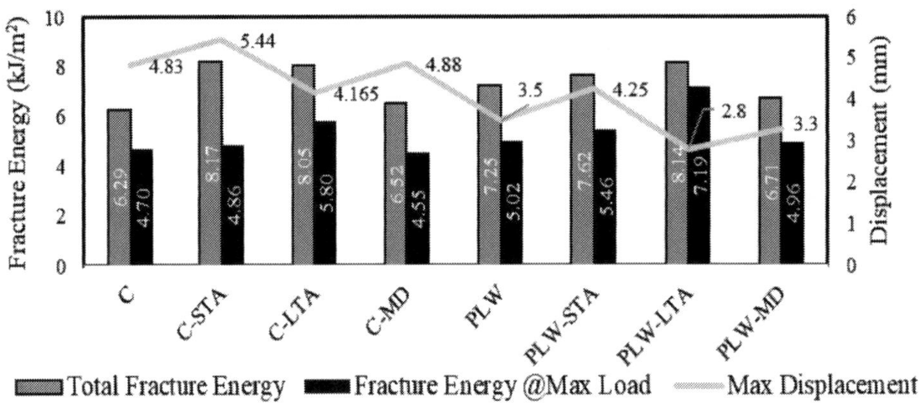

Figure 5. Fracture energy and maximum displacement results of various asphalt mixtures.

4.2. IDEAL CT

Figure 6 presents the results of CT-index for the various asphalt mixtures under study. The IDEAL-CT revealed comparable results between both the control and PLW mixtures, which is significant, as the CT-index parameters combine the effect of post-peak load and fracture energy for determining the value of CT-index. The PLA mixture demonstrated superior performance compared to the control mixture based on its higher CT-index, where a higher CT-index indicates higher mixture resistance against fatigue crack. A similar finding was manifested in a previous study conducted by Modarre and Hamedi [6] on the impact of plastic incorporation (2-10 % by weight of asphalt binder) into asphalt mixture via dry process. In the study, fatigue test was conducted on plastic modified mixtures to reveal the effect of plastic waste incorporation in increasing the fatigue resistance of asphalt mixture. Moreover, in the current study, the CT-indexes of the short-term aged PLA-STA and C-STA mixtures increased by 23% and 17%, respectively, and decreased by about 40% after long-term ageing. These results were expected as observed in the load-displacement curve in Figure 5. Long-term ageing has increased the stiffness of the mixture and resulted in significantly steeper post-peak load slope and lower CT-index. Whilst, moisture damage has slightly decreased the CT-indexes of the control and PLA mixtures. On the other hand, the PLW-MD mixture exhibited a higher CT-index than that of the C-MD mixture, differed by about 12%. This indicates the capability of plastic waste in increasing the resistance of asphalt to moisture damage.

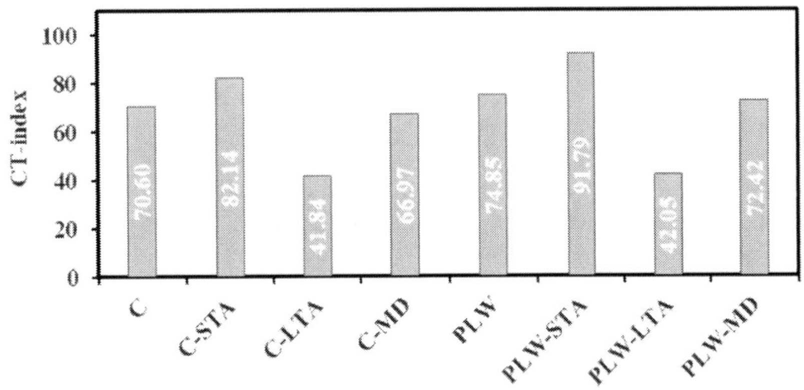

Figure 6. CT-index results of various asphalt mixtures.

4.3. Indirect tensile strength

The indirect tensile strength test results for the control and plastic waste modified asphalt mixtures are presented in Figure 7. It was found that the addition of plastic waste has improved the tensile strength of the conditioned and unconditioned asphalt mixtures. The increase in tensile strength is attributed to the contribution of plastic waste in increasing the bonding between aggregate and mastic (asphalt-aggregate paste). The finding is in agreement with several previous studies [9,31]. For instance, Abdo [32] whom investigated the utilisation of plastic waste in the production of asphalt mixtures disclosed that plastic waste modified mixtures yielded higher indirect tensile strength by 9% compared to the control mixture. Moreover, the interaction between asphalt binder and melted plastic during a mixing process leads to improved bonding and cohesiveness between the asphalt binder and the modified aggregate [17]. Since failure in an asphalt mixture is mainly related to adhesive and adhesion properties, enhancing the asphalt binder and aggregate surface properties could significantly increase the interaction and adhesion properties of the mixture thus leading to high-performance mixture production [33]. The high bonding properties and stiffness of asphalt binder around the aggregate particles could consequently increase the friction force and lead to solid interlocking between the aggregate particles. In the current study, the tensile strength of the control and PLW mixtures increased by 11% and 4% accordingly under short-term ageing, and by 26% and 16% respectively under long-term ageing.

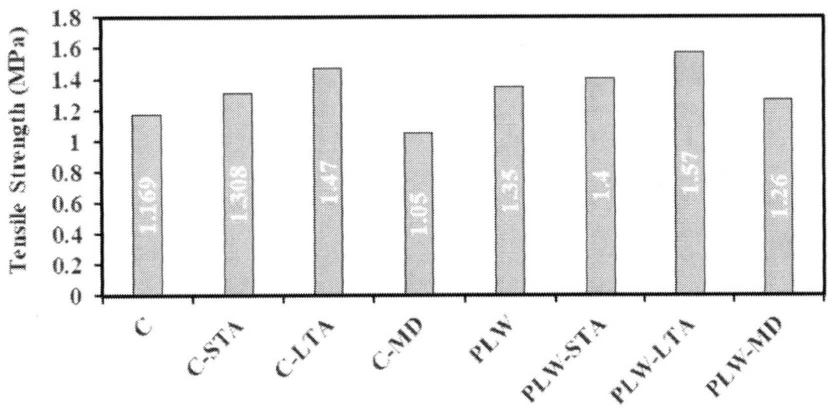

Figure 7. Tensile strength results of various asphalt mixtures.

Based on the fracture test results of moisture conditioned samples, the incorporation of plastic waste into conventional asphalt mixture was found effective in increasing its resistance to moisture damage, which resulted in superior tensile strength than that of the control mixture. Moreover, moisture damage has decreased the indirect tensile strength of the control and plastic waste modified mixtures by 19% and 10%, respectively. Based on visual analysis on the fractured samples, it was observed that the plastic waste modified samples exhibited brittle cracking behaviour as circled in Figure 8. The plastic waste modified asphalt mixture (C-STA) demonstrated a complete fracture, whereas the control mixture (C) showed a partial fracture as observed in Figure 8. Such behaviour is related to the mechanical behaviour of the asphalt mixture when subjected to external applied energy which then dissipates into elastic and plastic deformation. For a plastic mixture, most of the released energy is consumed during the fracture process [34,35]. A similar behaviour was presented by the long-term aged control mixture due to the increase in the stiffness of the asphalt mixture.

Figure 8. Visual observation of post-fracture test control and modified mixtures.

5. Conclusions

Based on the findings of this study, the following conclusions can be drawn:

- The PLW asphalt mixtures demonstrated higher fracture energy than that of the control mixture. Moreover, it is vital to mention that the effect of plastic waste incorporation in increasing the fracture energy at maximum load was higher compared to the total fracture energy.
- STA and LTA had improved the fracture energy and tensile strength of the control and plastic waste modified asphalt mixtures. In addition, the impact of STA in increasing the tensile strength and fracture energy of the control mixture was higher compared to the plastic waste modified mixture. On the contrary, LTA had increased the fracture energy of both the control and modified mixtures.
- The CT-index of the plastic waste modified mixture was greater than that of the control mixture, suggesting the superior resistance of the former to crack propagation.
- The results of indirect tensile strength test indicated the capability of plastic waste in increasing the resistance of conditioned and unconditioned samples to tensile failure.
- The plastic waste modified mixtures were less susceptible to moisture damage compared to the control mixture. This indicates that the incorporation of plastic waste into asphalt mixture could enhance the mixture's resistance to moisture damage.

6. References

[1] Jambeck J R, Geyer R, Wilcox C, Siegler T R, Perryman M, Andrady A, Narayan R and Law K L 2015 Plastic waste inputs from land into the ocean *Science (80)* **347** 768–71

[2] Malaysian Plastics Manufacturers Association 2019 An Advanced Plastics Recycling Industry for Malaysia 1–29
[3] Ahmadinia E, Zargar M, Karim M R, Abdelaziz M and Ahmadinia E 2012 Performance evaluation of utilization of waste Polyethylene Terephthalate (PET) in stone mastic asphalt *Constr. Build. Mater.*
[4] Movilla-quesada D, Raposeiras A C, Silva-klein L T, Lastra-González P and Castro-fresno D 2019 Use of plastic scrap in asphalt mixtures added by dry method as a partial substitute for bitumen *Waste Manag.* **87** 751–60
[5] Khurshid M B, Qureshi N A, Hussain A and Iqbal M J 2019 Enhancement of Hot Mix Asphalt (HMA) Properties Using Waste Polymers *Arab. J. Sci. Eng.* **44** 8239–48
[6] Modarres A and Hamedi H 2014 Effect of waste plastic bottles on the stiffness and fatigue properties of modified asphalt mixes *Mater. Des.* **61** 8–15
[7] Dalhat M A and Adesina A Y 2020 Utilization of micronized recycled polyethylene waste to improve the hydrophobicity of asphalt surfaces *Constr. Build. Mater.* **240** 117966
[8] Al-Adham K and Al-Abdul Wahhab H 2018 Influence of temperature on Jnr values of polymer modified asphalt binders *Int. J. Pavement Res. Technol.* **11** 603–10
[9] White G and Reid G 2018 Recycled waste plastic for extending and modifying asphalt binders *Symp. Pavement Surf. Charact.* 1–13
[10] Radeef H 2021 "MIXTURE DESIGN AND TEST PARAMETER EFFECT ON FRACTURE PERFORMANCE OF ASPHALT: A REVIEW" *ASEAN Eng. J.* **Preprint**
[11] Porto M, Caputo P, Loise V, Eskandarsefat S, Teltayev B and Rossi C O 2019 Bitumen and bitumen modification: A review on latest advances *Appl. Sci.* **9**
[12] Radeef H R 2015 Effect of Lab . Compaction Effort on Evaluation of Modified HMA Specimens *Civ. Environ. Res.* **7** 112–25
[13] Brasileiro L, Moreno-Navarro F, Tauste-Martínez R, Matos J and Rubio-Gámez M del C 2019 Reclaimed polymers as asphalt binder modifiers for more sustainable roads: A review *Sustain.* **11** 1–20
[14] García-Morales M, Partal P, Navarro F J and Gallegos C 2006 Effect of waste polymer addition on the rheology of modified bitumen *Fuel* **85** 936–43
[15] Padhan R K, Sreeram A and Mohanta C S 2019 Chemically recycled polyvinyl chloride as a bitumen modifier: synthesis, characterisation and performance evaluation *Road Mater. Pavement Des.*
[16] Mishra B and Gupta M 2018 Use of Plastic Waste in Bituminous Mixes by Wet and Dry Method *Proc. Inst. Civ. Eng. - Munic. Eng.* **173** 1–41
[17] Radeef H R, Abdul Hassan N, Abidin A R Z, Mahmud M Z H, Yusoffa N I M, Idham Mohd Satar M K and Warid M N M 2021 Enhanced Dry Process Method for Modified Asphalt Containing Plastic Waste *Front. Mater.* **8** 247
[18] Zhou F, Im S, Sun L and Scullion T 2017 Development of an IDEAL cracking test for asphalt mix design and QC/QA *Road Mater. Pavement Des.* **7** 405–27
[19] Yan C, Zhang Y and Bahia H U 2020 Comparison between SCB-IFIT, un-notched SCB-IFIT and IDEAL-CT for measuring cracking resistance of asphalt mixtures *Constr. Build. Mater.* **7** 119060
[20] Radeef H R, Abdul Hassan N, Zainal Abidin A R, Mahmud M Z H, Yaacob H, Mashros N and Mohamed A 2021 Effect of aging and moisture damage on the cracking resistance of rubberized asphalt mixture *Materials Today: Proceedings* **42** 2853–8
[21] Jabatan Kerja Raya Malaysia (JKR) 2008 *Standard Specification for Road Works,* (50580 Kuala Lumpur)
[22] Hesami S, Roshani H, Hamedi G H and Azarhoosh A 2013 Evaluate the mechanism of the effect of hydrated lime on moisture damage of warm mix asphalt *Constr. Build. Mater.*
[23] Sasidharan M, Torbaghan M E and Burrow M 2019 Using waste plastics in road construction *Univ. Birmingham* 1–19
[24] American Society for Testing Materials. (2014c). ASTM D2726 / D2726M-14. 2014 Standard Test Method for Bulk Specific Gravity and Density of Non-Absorptive Compacted

Bituminous Mixtures (West Conshohocken, PA ASTM Int)
[25] American Society for Testing Materials. ASTM D2041 / D2041M-11. ASTM International. 2011 Standard Test Method for Theoretical Maximum Specific Gravity and Density of Bituminous Paving Mixtures (West Conshohocken, PA)
[26] American Society for Testing Materials ASTM D1559-89 1989 Test Method for Resistance of Plastic Flow of Bituminous Mixtures Using Marshall *West Conshohocken, PA.*
[27] AASHTO R30 2002 Standard Practice for Mixture Conditioning of Hot Mix Asphalt (HMA) *Provisional Stand. Am. Assoc. State Highw. Transp. Off. Washington, DC, USA*
[28] ASTM D4867 2014 Standard Test Method for Effect of Moisture on Asphalt Concrete Paving Mixtures (West Conshohocken PA ASTM)
[29] Yin F, Taylor A J and Tran N 2020 *Performance Testing for Quality Control and Acceptance of Balanced Mix Design*
[30] American Society for Testing Materials. ASTM D6931-12 2012 Standard Test Method for Indirect Tensile (IDT) Strength of Bituminous Mixtures (West Conshohocken PA ASTM)
[31] Walubita L, Flórez B and Álvarez Lugo A 2012 Evaluation of hma cracking resistance as a function of laboratory aging based on the direct-tension test *Rev. EIA* 209–24
[32] Abdo A M A 2017 Investigation the effects of adding waste plastic on asphalt mixes performance *ARPN J. Eng. Appl. Sci.* **12** 4351–6
[33] Tiwari A V. and Rao Y R M 2019 Study of plastic waste mixed bituminous concrete using dry process for road construction *Sel. Sci. Pap. - J. Civ. Eng.* **13** 105–12
[34] Reich M, Esztergar E P, Ellison E G, Erdogan F, Gray T G F, Spence J and Wells C H 1979 Application of fracture mechanics methods in safety analysis of piping components in subcreep and creep behavior *Nucl. Eng. Des.*
[35] Radeef H R, Hassan N A, Abidin A R Z, Mahmud M Z H, Satar M K I M, Warid M N M, Saffar Z H Al, Shu X, Huang B and Vukosavljevic D 2021 Determining Fracture Energy in Asphalt Mixture: A Review *IOP Conference Series: Earth and Environmental Science* **682** 12069

Acknowledgement
This work was financially supported by the Ministry of Higher Education (Malaysia) under the Fundamental Research Grant Scheme (FRGS) FRGS/1/2020/TK01/UTM/02/5 and the UTM HIR grant (Q.J130000.2451.09G20) provided by Universiti Teknologi Malaysia (UTM).

IOP Conf. Series: Earth and Environmental Science **971** (2022) 012010 doi:10.1088/1755-1315/971/1/012010

Investigation physical and rheological properties of bio-asphalt containing waste cooking oil

A M Ali[1*]

[1] Department of Civil Engineering, Engineering College, Al-Muthanna University, Al-Muthanna, Iraq
*Corresponding author: atheer.muhammed@mu.edu.iq

Abstract. In recent years, several oil-based modifications have been employed to give clear benefits to the pavement sector, including the use of waste cooking oil (WCO). These activities are aimed at resolving waste management issues. The purpose of this research is to see if using waste cooking oil as an additive to the asphalt binder, which will be utilized later in the construction of flexible pavements, is feasible. This will be seen as an appropriate and lawful way of disposing of this kind of wastes. The current investigation examined the physical and rheological properties of aged and unaged binder modified with different percentages of WCO. In addition to the control sample, three different percentages of WCO content (1.5%, 3%, and 4.5% by weight of the binder) were blended with the PEN 60/70 control binder to evaluate for such related properties of the waste cooking oil-modified binder (WCOMB). Then, after aging the samples in a Rolling Thin Film Oven Test (RTFOT), the properties were verified through flash and fire points, penetration, softening point, viscosity, and Dynamic Shear Rheometer (DSR) tests. The findings proved that the penetration depth of the modified binder increased as WCO ratios increased, while the softening point decreased. The study was showed that the binder became soft in the presence of WCO. The results of the viscosity and DSR tests were satisfactory until the WCO content reached 3%. Based on the findings, the modified asphalt binder was workable and less rigid until it reached 3% of WCO. Moreover, recycling this waste material will be a suitable technique in waste materials recycling.

1. Introduction

At high temperatures, a bitumen binder binds aggregate particles (coarse and fine) together to form flexible pavement. The traditional bitumen binder is a by-product of the crude oil refining process. The cost of bitumen has lately grown because of improved environmental laws for new drilling, diminishing current resources, and modifications to the refining process that maximize gasoline amount while reducing bitumen residue [1]. As a result, it's more important than ever to look into potential alternative binder materials to entirely or partially replace petroleum-based bitumen for the sustainable development of flexible pavements.

Bitumen may be modified in a variety of ways. Modification in the road construction business refers to the addition of liquid or solid materials to bitumen or asphalt, such as fillers, chemical agents, wetting adhesion agents, polymers, waste oil, and viscosity-reducing additives. All additives should, in theory, improve the performance of newly built roads. Bitumen is used extensively in pavement construction, particularly for coating.

Bitumen should be used to increase the physical interlocking of bituminous mixes by acting as a satisfactory binder. However, due to aging, bulk storage, mixing, paving, transport, and on-site storage, its properties vary over time. The composition of bitumen has the biggest impact on aging. The interaction of oxidation and volatility changes physical and rheological properties, resulting in failures ranging from hardening (consistency change) to embrittlement. The loss of lower molecular weight volatiles is frequently accompanied by oxidation. Concentrations, temperature, and the surrounding environment all influence the rate of loss. Because these factors influence the loss rate, volatility is especially important when handling bitumen at high temperatures. As a result, pavement deterioration and significant distress, such as fatigue, rutting, and cracking, are caused by aging bitumen [2].

As the performance of asphalt binders improves, waste products are increasingly being employed as a partial replacement in the production of new asphalt binders. Generally, using waste products as asphalt substitutes or modifiers will reduce asphalt use while also improves pavement performance. Despite the fact that several studies have been conducted to investigate the creation of novel binders from biomass resources, only a few have looked into the potential of waste cooking oil as an asphalt modifier for use in asphalt pavements [3–6]. Saturates aromatic resins and asphaltenes are among the four fractions that make up asphalt binders [7]. Due to the low viscosity of WCO, it can be used in recycling aged asphalt, according to the rejuvenation compatibility theory [7]. The flash point of WCO was higher than 200°C as compared to other low viscosity materials, indicating that using WCO in hot mix asphalt mixes has high structural safety [7]. As a result, it was practically significant and would give WCO a wide range of applications in the asphalt pavement recycling sector. Besides the rapid tremendous growth of the human population[8], as a result of the enormous increase in food consumption, it is the main factor in generating large quantities of waste cooking oil from restaurants and homes [9]. During the cooking process, WCO is formed, which is a by-product of fresh oil [10]. The volatile and nonvolatile compounds will be formed as a result of the frying process, with the volatile compound evaporating into the atmosphere. At the same time, the WCO experiences a variety of chemical changes, and the non-volatile compounds in the oil change its physical and chemical properties, making it slightly different from fresh oil after it has been frying for a period of time [11]. Waste sunflower oil have proved its efficiency as a rejuvenating agent for asphalt self-healing [12–16]. Asli and Karim [2] conducted a study on WCO that could be used to rejuvenate the aged binders. By adding WCO to the aged binder, the researchers [2] discovered that it was mimicked the properties of the original binder, which were brittle and hard at a certain dosage. Finally, WCO's ability to renew the aged binder lowered the usage of the novel binder by restoring the aged binder to its former state, reducing the need for the novel binder. Chen et al.[5] studied the physical, chemical, and rheological properties of chemical treated WCO as an asphalt binder rejuvenator. During the physical assessment, the penetration value rose, the softening point dropped, and the viscosity decreased with the addition of WCO [17]. The viscosity of the asphalt mixture was reduced during processing, lowering the mixing and compaction temperatures. WCO's efficiency as a binder replacement was investigated by Wen et al.[1] at both high and low temperatures.This rheological result is consistent with the consequences of Maharaj et al.[18], who found that at low temperatures, fatigue cracking resistance was increased. Meanwhile, the addition of WCO had a negative impact on high-temperature output, as rutting deformation resistance decreased. WCO may be used to modify the bitumen binder by mixing varying amounts of oil content. A Previous study revealed that modifying the bitumen binder and blending recycled bituminous concrete with chemical treated WCO improved the performance attributes of the resultant mixes [1, 19, 20].

According to [21], bitumen binder obtained from the refining of crude oil petroleum typically lacks the properties required for use in road paving. Fresh bitumen has to be modified with different additives such as oils, fine minerals, carbonaceous compounds, and polymers if it is to be used as a binder in pavement construction. As a result, a substantial amount of research on enhancing asphalt binder properties should be conducted utilizing the method through the modification process [22]. Using additives in the bitumen modification process may be regarded one of the effective technical methods that should be used nowadays [23] to decrease deterioration and enhance the durability of a bituminous layer. The performance properties of the bituminous layer, such as resistance to permanent deformation, fatigue, and so on, should be enhanced.

IOP Conf. Series: Earth and Environmental Science **971** (2022) 012010 doi:10.1088/1755-1315/971/1/012010

Using pure WCO without chemical treatment as a rejuvenator in aged bitumen represents the novelty of this paper. WCO production is estimated to be in excess of 15 million tons per year across the world, with just a tiny fraction collected and recycled [24]. WCO was thrown in huge quantities into rivers and landfills, polluting the environment. Managing used cooking oil is a big concern due to disposal issues and the potential for contamination of water and land resources [25].

The major goals of this research are to see if pure (without chemical treatment) WCO can be used as an asphalt modifier, to see how different asphalt properties are altered by adding WCO, and to contribute to the solution of the problem of inappropriate and illegal disposing of WCO. The goal of this investigation was to see if WCO could be used to rejuvenate aging bitumen. The potential applications of WCO as a rejuvenator were investigated using traditional bitumen test techniques such as penetration, softening point, viscosity, flash and fire points, and the Dynamic Shear Rheometer (DSR) test. Thus, a penetration blending chart was used to show the appropriate percentage of WCO for the rejuvenated bitumen. As a result, statistical analysis was used to confirm the facts.

2. Materials and Preparation

2.1. Waste cooking oil

The WCO that was used multiple times had been collected from restaurants. It was purified from impurities and food residues during the purification process by using a piece of cotton cloth in the baker. It took roughly an hour to finish the filtering process. After the purification process, the remaining particles and impurities on the cotton cloth were separated, while the filtered oil passing through the cotton was collected to prepare the test samples. After that, WCO's fundamental properties were put to the test. The fundamental properties of WCO have been tested, and the results are presented in Table 1.

Table 1. The basic properties of (WCO).

Properties	Value
Flash Point (°C)	280
Viscosity (Pa.s)	0.065
Acid Value (ml/g)	0.59

2.2. Asphalt binder

In the current investigation, the PEN 60/70 binder was used as a control asphalt that was mixed with three different ratios (1.5%, 3%, and 4.5%) of WCO by binder weight. The control binder with 51°C as a softening point has met the standards as shown in Table 2. This type of binder represents the ideal grade to resist changes in the weather and is mostly used in flexible pavement construction.

Table 2. Basic properties of PEN 60/70 control binder.

Testing	Result	Standard
Penetration (0.1 mm)	65	ASTM D5/D5 [26].
Softening point (°C)	51	ASTM-D36 [27].
Viscosity (Pa.s)	0.41 (135°C) 0.2 (165°C)	ASTM-D4404 [28].
Ductility (cm)	104	ASTM-D113 [29].
G*/sinδ	1.23 Kpa (70°C)	ASTM D7175-15 [30].

2.3. Sample preparation
The PEN 60/70 bitumen binder which is semi-solid at room temperature was converted to a liquid by heating it at 130°C in an oven for easy mixing with the WCO in a container. The mixing temperature between the selected bitumen binder for this study and WCO was chosen at160°C, and the percentages of WCO to be used were 0%, 1.5%, 3%, and 4.5% (WCO). The modified bitumen binder was made by replacing WCO with a bitumen binder at various ratios. To achieve homogenous mixes, these ingredients were mixed for 1 hour in a high shear mixer at a speed of 1000 rpm [31]. The modified samples were placed in penetration cups that were labeled and kept separate for samples of aged and non-aged penetration, softening point, and viscosity testing. The same procedure was conducted on 0%, 1.5%, 3%, and 4.5% of WCO, respectively. After that, the samples were put to the test.

3. Experimental works

3.1. Physical Properties Tests
Physical testing was carried out in the lab to assess the physical properties of the WCO modified binder. Flash and fire points, penetration, softening point, and viscosity tests were among the physical testing performed.

3.1.1. Flash and Fire points test. The final fundamental test to do is the flash and fire points test (ASTM D92). In the presence of an open flame, the flashpoint is the temperature at which asphalt may be safely heated without generating a sudden flash. The fire point also refers to a tested asphalt sample's resistance to burning. As a consequence, the flash and fire points were employed to ensure that the asphalt could be safe to use at high temperatures.

3.1.2. Penetration and softening point tests. To determine the hardness of asphalt materials, a penetration test was performed due to ASTM-D5 [26]. The penetration test was carried out by filling the container with asphalt samples and storing it in a 25°C water bath for 90 minutes. The containers were then penetrated with a needle weighing 100 g. The penetration depth was measured in 0.1(mm).

The main goal of the ASTM-D36 softening point test is to establish the exact temperature at which a phase transition in the bituminous binder occurs using the ring and ball method [27]. Two horizontal bitumen discs, each supported by a steel ball, are heated at a regulated pace in a liquid bath, cast in a shouldered brass ring. These materials gradually and imperceptibly shift from brittle or exceedingly slow-flowing materials to softer and less viscous liquids as the temperature rises. The average temperature at which the two disks were encased in asphalt and dropped at a distance of 25 mm (1.0 in) was defined as the softening point. Ring and ball softening temperature is the name for this method [32].

3.1.3. Viscosity test. According to ASTM-D4404 [28], the Brookfield viscometer is employed to test the flowing resistance and internal friction of the asphalt binder in rotational viscosity. A cylindrical spindle no. 27 is utilized in a container filled with an 11g asphalt binder and spun at a speed of 20 rpm at a temperature of 135°C.

3.2. Rheological Properties Test
A DSR test was used to assess the rheological properties. The dynamic shear rheometer (DSR) is used to determine the viscosity and elastic behavior of asphalt binders at medium to high temperatures, according to ASTM D7175. The parameters of shear modulus (G*) and phase angle (δ) for samples were determined using a DSR test. While (δ) is a time lag between shear strain and shear stress, overall resistance to deformation under shear loading may be described as (G*). As a consequence of measuring G* and δ, this test identified the total complex shear modulus as well as the elastic and viscosity components [33].

4. Results and Discussion

4.1. Physical properties

4.1.1. Flash and fire points test. In terms of safety, this test is critical. The higher results of flash and fire points represent the greater the potential for using this type of asphalt in desert-like areas. As the WCO content increases, both the flash and fire points raise, as illustrated in Figure 1, with the fire point almost always being higher than the flash point. The difference in ignition temperatures of the two materials accounts for the increased flash and fire points in the WCOMB. This indicates that adding WCO as asphalt modifier can reduce the associated fire hazard.

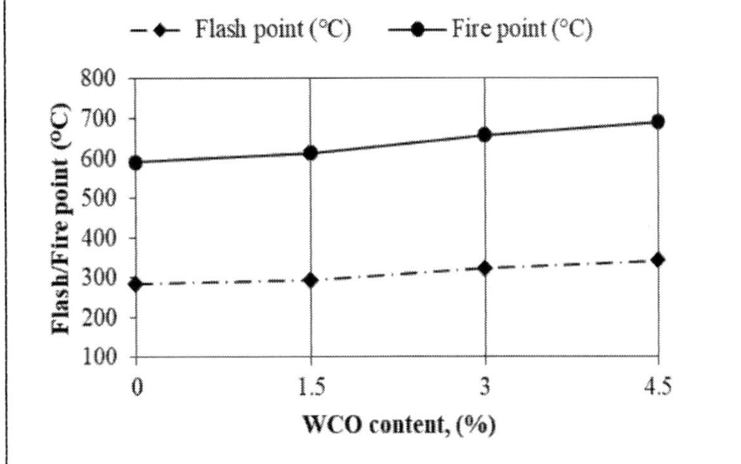

Figure 1. Flash and fire points results.

4.1.2. Penetration Test. The degree of softness and hardness of the binder can be determined by penetration. In Figure 2, the effects of WCO on the penetration value of unaged and aged binders are clearly shown. It is clear that as the percentage of WCO content was increased, the penetration values of different asphalt binders increased, indicating that adding WCO softens virgin binder by reducing its consistency. According to Sun et al. [34], a similar trend had been observed during studying the effect of adding bio-oil on the different physical and mechanical properties of the asphalt binder. Up to 3% of WCO, there was a linear increase in penetration value. When 4.5% of binder was replaced with WCO, the penetration value increased at a higher rate. It means that in that percentage, the modified binder became very soft. After aging the modified binder, the penetration value decreased. The change in penetration is exponential in the unaged case, but linear in the aged case. Moreover, the higher the waste cooking oil content, the lower the penetration aging ratio, resulting in a reduced degree of aging of the modified binder. As illustrated in Figure 2, the penetration values did not reach the high values of 200/300 suggested by Roberts et al. [33] for cold climates such as northern Canada, indicating that the penetration values for the entire range of the WCO used in this study would be suitable for moderate climates.

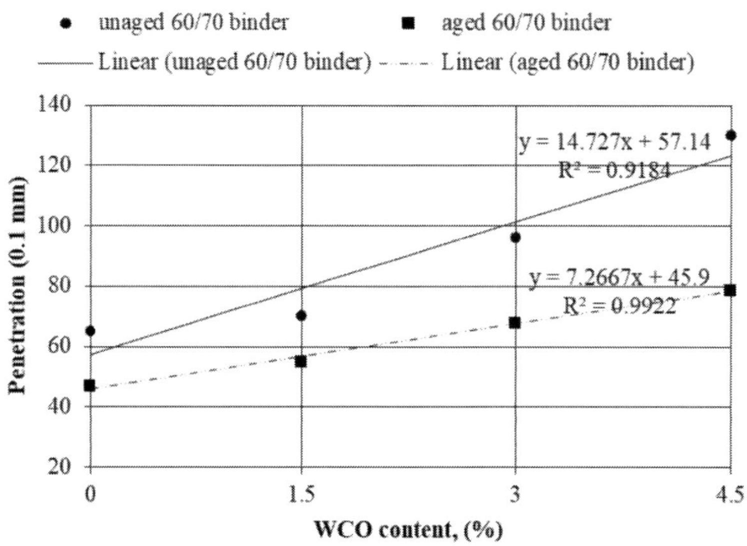

Figure 2. Penetration versus WCO content.

4.1.3. Softening point Test. As a complement to the penetration test, the softening point test may also be used to determine the sensitivity of an asphalt binder to temperature susceptibility. The experimental values for the softening point test of the WCO-modified binder are illustrated in Figure 3. The softening point results were decreasing with increasing WCO percentages. There is no big change in the softening temperature after replacing WCO in the binder till 3%. While there was a clear reduction with the 4.5% unaged WCO modified binder. On the other hand, the changing of the softening point temperatures of the aged binder was linear. As shown from Figure 3, the trend line steadily fell from 54.5°C to 42.3°C of the aged binder. At 1.5% of WCO, the softening point temperature of the aged binder was greater than the unaged binder. The asphalt with a higher softening point refers to the lower temperature susceptibility. Adding WCO to the asphalt enhanced the softness and decreased the softening point, which improved the temperature susceptibility, the same conclusion founded by Sun et al. [34]. In addition, the oil which is soft in nature constitutes the main reason behind this. In hot weather, the asphalt with a high softening point will be effective.

IOP Conf. Series: Earth and Environmental Science **971** (2022) 012010 doi:10.1088/1755-1315/971/1/012010

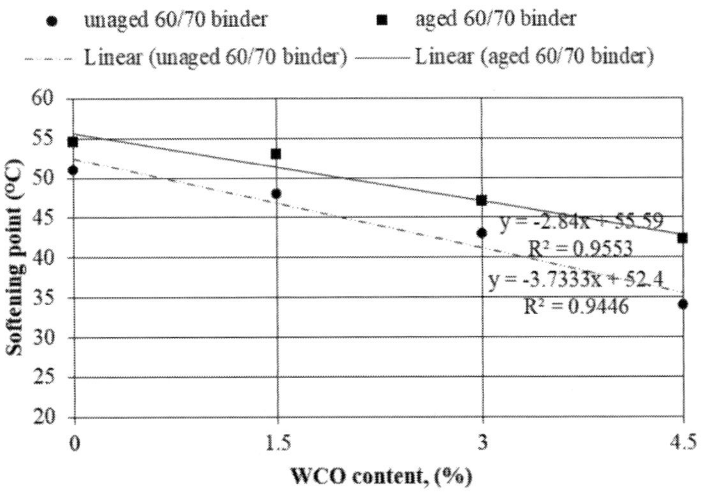

Figure 3. Softening point versus WCO content.

4.1.4. Viscosity Test. According to Al-Khateeb and Al-Akhras [35], the average mixing and laydown temperature for hot mix binder is 135°C. This study is performed and tested at elevated temperatures of 135°C and 165°C. Figure 4 represents the viscosity findings for unaged and aged modified binder at elevated temperatures. As expected by Sun et al. [34] and Chen et al. [8], as the viscosity of asphalt decreases, its fluidity increases. Thereafter, mixing, handling, and working with asphalt becomes much easier. Because of the lubricating effect of oil, the viscosity of asphalt binder decreases as the percentage of WCO increases. Because of the reduced viscosity of the oil sample, which is lower than the viscosity of the control asphalt, the percentage of WCO replaced with asphalt resulted in a drop in modified asphalt viscosity as compared to the control asphalt that was used. Furthermore, A higher WCO concentration will also allow for a greater volume of asphalt to be replaced, resulting in a new sample that is less viscous. As a result, adding WCO to asphalt can help achieve higher workability at lower working temperatures in pavement construction. Low viscosity is beneficial from an economic point of view because it reduces the mixing and compaction temperatures of the mixture, reducing expenses. An excessive amount of WCO in the binder, on the other hand, results in a low viscosity value, which might result in reduced serviceability performance due to the softer condition, as well as rutting problems. However, because of the necessity for greater mixing, laying and compaction temperatures, high viscosity values are undesirable since they result in excessive energy consumption and waste of production expenses [2]. As compared to the aged WCO-modified binder, the unaged asphalt binder modified with WCO had the lowest viscosity value at 135°C.

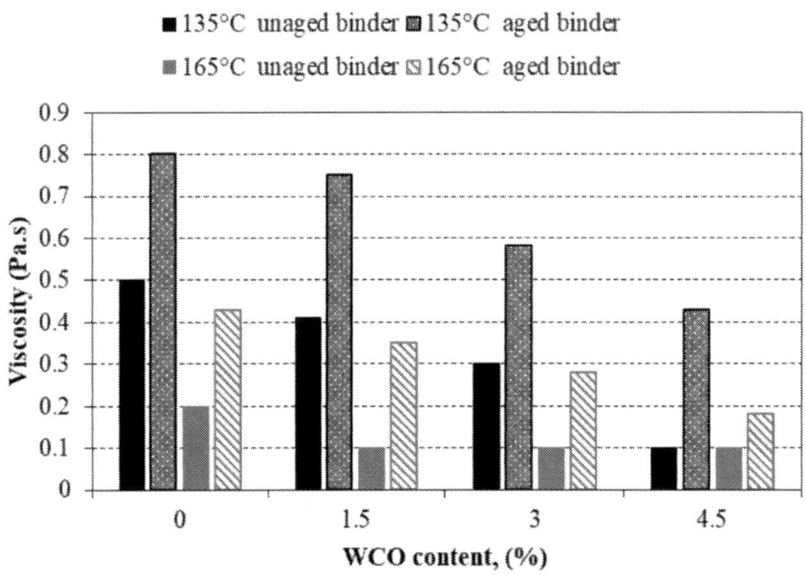

Figure 4. Viscosity versus WCO content.

4.2. Rheological Properties

4.2.1. Dynamic Shear Rheometer (DSR) Test. In this test, the behavior of fatigue cracking and rutting resistance in both original and modified asphalt was evaluated at moderate and high temperatures, respectively. Because of rutting and fatigue cracking, the pavement's early life is the most significant factor to take into consideration. The DSR test determines the specimen's complex shear modulus (G*) and phase angle (δ). When a sample is repeatedly sheared, G* represents its entire resistance to deformation. When the asphalt has a bigger δ, it is said to be more viscous. Both G* and δ predict the rutting and fatigue cracking properties of HMA. Figures 5 and 6 illustrate the behavior of rutting resistance properties in asphalt with different proportions of WCO added at different temperatures. When the G*/sin δ value is high and the deformation is low at high temperatures, the rutting resistance will be higher [20]. As the test temperature and WCO content are increased, the rutting resistance factor will gradually decrease. The unaged control binder had a rutting resistance value of 14.86 kPa at 46°C, followed by a modified binder with 1.5 % WCO at 12.22 kPa. While a modified binder with 4.5% WCO recorded the minimum rutting resistance which is 7 kPa. It can be shown that as the test temperature increased, the rutting resistance value was decreased incrementally. All trend lines met at 70°C, with the exception of 4.5% WCO binder, which failed to meet the rutting requirements due to binder resistance. The WCO is soluble because of its high melting point and this represents the reason of that. Moreover, the excessive amount of WCO was reduced viscosity in the sample, which reduced rutting resistance ability and asphalt binder performance. This is due to modified asphalt's poor adhesive performance [5].

The shear modulus was increased because of aging, while the phase angle was decreased. As a result, rutting resistance increased to 35.87 kPa at 46°C, followed by 27.11 kPa for a modified binder containing 1.5 % WCO. For the aged modified binder with 4.5% WCO, the minimum rutting resistance value was 17.33 kPa. The rutting resistance at different temperatures is illustrated in figure 6 for each aged asphalt sample modified with 0, 1.5, 3, and 4.5% WCO. The complex shear modulus and phase angle of the aged binder will increase as the temperature rises, while G* decreases, indicating that the binder will become less resistant to deformation. Figures 5 and 6 show that when WCO is replaced from 0% to 3%, rutting complies until 64°C for unaged and aged samples, but only until 58°C for 4.5 % WCO. It can be concluded from two figures 5 and 6 that adding additional WCO reduces the modified asphalt's rutting resistance. Sun et al. [34] came up with the same results. They discovered that adding a high amount of bio-oil results in a decrease in G* and an increase in δ. As a result, the G*/sin (δ) value will decrease overall. Chen et al. [20] also came up with similar findings.

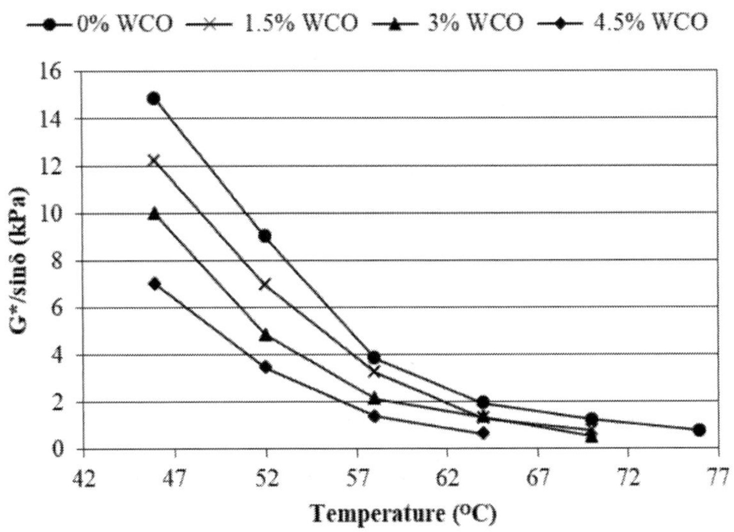

Figure 5. Rutting resistance for unaged asphalt.

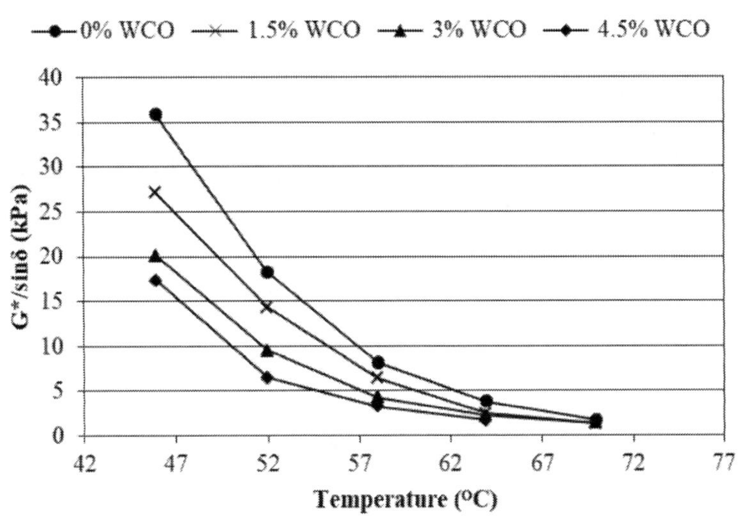

Figure 6. Rutting resistance for aged asphalt.

5. Conclusions

The current study has led to the following advantages of modifying the control asphalt binder with WCO:

- The modified binder became safer to use and work with due to the higher flash and fire points. Furthermore, when the viscosity decreased and fluidity increased, the modified binder became more workable and simpler to use, especially in low-temperature regions.
- Despite its enhanced penetration and temperature susceptibility for all utilized WCO concentrations, the modified asphalt may be acceptable for moderate climate countries.
- As the softness of the asphalt increases and the softening point decreases, the temperature susceptibility increases. The minimum softening point for the WCO range used for aged binder was 42°C, which is still adequate for moderate weather regions.
- With different WCO contents, the rheological properties of aged asphalts could also be restored to a normal level of VA. Overall, WCO proved to be an adaptable regenerative agent for improving the performance of aged asphalt. It established a new application method for WCO in the field of asphalt pavement recycling, allowing for waste cyclic utilization while also protecting the environment.
- Because WCO improves some of the original properties of asphalt, it can be used as a modifier for control asphalt. The quantity of effection can also be controlled by the percentage of WCO added to the asphalt.
- One of the most effective approaches to solve the problem of disposing of WCO is to use it in asphalt modification technologies.

6. References

[1] Wen H, Bhusal S and Wen B 2013 Laboratory evaluation of waste cooking oil-based bioasphalt as an alternative binder for hot mix asphalt *J. Mater. Civ Eng.* **25** 1432-1437
[2] Asli H and Karim M R 2011 Implementition Of Waste Cooking Oil As Rap Rejuvenator *Proceedings of the Eastern Asia Society for Transportation Studies* **8** 267
[3] Zofka A and Yut I 2012 Investigation of rheology and aging properties of asphalt binder modified with waste coffee grounds *Transp. Res. E-Circular.* 61-72.
[4] Seidel J C and Haddock J E 2012 Soy fatty acids as sustainable modifier for asphalt binders *Altern Bind Sustain Asph Pavements Washingt DC*
[5] Chen M, Leng B, Wu S and Sang Y 2014 Physical, chemical and rheological properties of waste edible vegetable oil rejuvenated asphalt binders *Constr. Build. Mater.* **66** 286-298
[6] Yang X, You Z, Dai Q and Mills-Beale J 2014 Mechanical performance of asphalt mixtures modified by bio-oils derived from waste wood resources *Constr. Build. Mater.* **51** 424-431
[7] Li H, Dong B, Zhao D, Guo P and Zhang J 2019 Physical, rheological and stability properties of desulfurized rubber asphalt and crumb rubber asphalt *Arab J. Sci. Eng.* **44** 5043-5056
[8] Chen Y, Xiao B, Chang J, Fu Y, Lv P and Wang X 2009 Synthesis of biodiesel from waste cooking oil using immobilized lipase in fixed bed reactor *Energy Convers. Manag.* **50** 668-673.
[9] Phan A N and Phan T M 2008 Biodiesel production from waste cooking oils *Fuel* **87** 3490-3496
[10] Choe E and Min D B 2007 Chemistry of deep-fat frying oils *J. Food Sci.* **72** R77-R86
[11] Cvengroš J and Cvengrošová Z 2004 Used frying oils and fats and their utilization in the production of methyl esters of higher fatty acids *Biomass and Bioenergy* **27** 173-181
[12] Tariq Hussein Al-Mansoori 2018 *Encapsulated healing agents for asphalt self-healing*
[13] Al-Mansoori T, Hussein J and Aziz H 2019 Asphalt damage reservation due to self-repair: A literature review *IOP Conference Series Materials Science and Engineering* **584** 12019
[14] Al-Mansoori T, Al-Adhadh A and Hussein J 2020 Influence of Temperature and Rest Period on Damage Repair of Aged Asphalt *Key Engineering Materials* **857** 138-144

[15] Al-Mansoori T, Garcia A, Norambuena-Contreras J and Artamendi I 2018 Self-healing properties of asphalt mixtures with embedded capsules *Advances in Materials and Pavement Performance Prediction* 551-555
[16] Al-Mansoori T, Micaelo R and Garcia A 2016 Characterization of polymer capsules used for self-healing asphalt pavements *Functional Pavement Design* 1167-1174
[17] Azahar W N A W, Jaya R P, Hainin M R, Bujang M and Ngadi N 2016 Chemical modification of waste cooking oil to improve the physical and rheological properties of asphalt binder *Constr. Build. Mater.* **126** 218-226
[18] Maharaj R, Ramjattan-Harry V and Mohamed N 2015 Rutting and fatigue cracking resistance of waste cooking oil modified trinidad asphaltic materials *Sci World J.*
[19] Sun D, Sun G and Du Y, et al 2017 Evaluation of optimized bio-asphalt containing high content waste cooking oil residues *Fuel* **202** 529-540
[20] Chen M, Xiao F, Putman B, Leng B and Wu S 2014 High temperature properties of rejuvenating recovered binder with rejuvenator, waste cooking and cotton seed oils *Constr. Build. Mater.* **59** 10-16
[21] Chebil S, Chaala A and Roy C 2000 Use of softwood bark charcoal as a modifier for road bitumen *Fuel* **79** 671-683
[22] Wan Azahar W N A, Bujang M, Jaya R P, Hainin M R, Ngadi N and Mohamad M 2016 Chemical identification of waste cooking oil as additive in bitumen *Key Engineering Materials* **700** 207-215
[23] Bahia H U, Hanson D I, Zeng M, Zhai H, Khatri M A and Anderson R M 2001 *Characterization of Modified Asphalt Binders in Superpave Mix Design*
[24] Castellanelli C and Mello C 2007 Analyzes of the used fried oil under environmental perspective and its possibilities for production of biodiesel *Dep. Prod. Eng. Fed. Univ. St Maria, Brazil* **1** 1-10.
[25] Math M C, Kumar S P and Chetty S V 2010 Technologies for biodiesel production from used cooking oil: A review *Energy Sustain Dev.* **14** 339-345
[26] ASTM D5 2013 Standard Test Method for Penetration of Bituminous Materials (West Conshohocken: ASTM Standard) pp 5-8
[27] ASTM 2014 Standard test method for softening point of bitumen (ring-and-ball apparatus) (West Conshohocken: ASTM Standard)
[28] D4402M–12 A 2012 Standard Test Method for Viscosity Determination of Asphalt at Elevated Temperatures Using a Rotational Viscometer American Society for Testing and Materials West Conshohocken: ASTM Standard)
[29] ASTM D113 2007 Standard Test Method for Ductility of Bituminous Materials West Conshohocken: ASTM Standard)
[30] ASTM D7175 2015 Stand Test Method Determ Rheol Prop Asph Bind Using a Dyn Shear Rheometer West Conshohocken: ASTM Standard)
[31] Wan Azahar W N A, Bujang M, Jaya R P, Hainin M R, Ngadi N and Al Bakri M M A 2016 Performance of waste cooking oil in asphalt binder modification *Key Engineering Materials* **700** 216-226
[32] Chen J S and Lin K Y 2005 Mechanism and behavior of bitumen strength reinforcement using fibers *J. Mater. Sci.* **40** 87-95
[33] Roberts F L, Kandhal P S, Brown E R, Lee D Y and Kennedy T W 1991 *Hot mix asphalt materials mixture design and construction*
[34] Sun Z, Yi J, Huang Y, Feng D and Guo C 2016 Properties of asphalt binder modified by bio-oil derived from waste cooking oil *Constr. Build. Mater.* **102** 496-504
[35] Al-Khateeb G G and Al-Akhras N M 2011 Properties of Portland cement-modified asphalt binder using Superpave tests *Constr. Build. Mater.* **25** 926-932

Utilization of Conch Shell Waste and Sugar Cane Dregs as Filler Substitutes in Concrete Asphalt Mixtures

Veranita[1*], Y Andi[1] and D Rahmat[1]

[1] Civil Engineering, Teuku Umar University, West Aceh, Indonesia
*Corresponding author: veranita@utu.ac.id

Abstract. Aceh province is an area that has very diverse natural resource potentials, such as rivers and extensive plantations. In the river sector, we can find many types of animals that live in water, one of which is a freshwater conch. In plantation areas we often encounter sugar cane plants. Both of these wastes have not been used properly and only become waste. Therefore, this study aim to use conch and sugarcane as a filler material for the asphalt mixture. The addition of filler conch and sugar cane was made in 3 variations, namely 60:40; 50:50, and 25:75. The number of test objects for the first stage of testing was 45 specimens and the second stage was 18 specimens. The results obtained are the optimum asphalt content for 3 variations, namely 6.5%; 6.25%, and 6.5%. The second stage of testing obtained durability values in the three variations, namely 98.72%, 94.83%, and 106.45%. From the three filler variations, it can be concluded that the 25:75 variation has a higher durability value and good stability value. The conclusion of the three variations of conch shell filler and sugarcane is suitable for heavy traffic because it still meets General Bina Marga 2018 specifications.

1. Introduction

Aceh province is an area that has very diverse natural resource potentials, such as a large area of coastline, rivers and plantations. In the river sector, there are many types of animals that live in water, one of which is conch. This conch is widely used as food for Acehnese people, while the shell is not used properly, it is only become a waste. This conch shell is almost identical to clam shell in terms of its mineral content. Therefore, it is believed that the conch shell can be used as a material filler in asphalt mixture. Another potential mineral filler is sugar cane dregs in which Aceh Province is also famous for its extensive plantation areas, this raw material is mainly used in making sugar or sugarcane juice, while the waste has not been properly utilized and only becomes waste.

Conch shells contain Calcium Oksida (CaO) lime compounds with a percentage of 66.70%. CaO compound is one of the main chemical compounds of cement constituents [1]. The conch shell used in this study is a conch shell that comes from Gunong Kulu, Indonesia. Conch shells took about 25 kg in the field and produced a filler of 6 kg that escaped the 200 mm sieve. Until now the utilization of waste conch shell still has not been utilized only wasted next to the stall and has not been processed into a useful product even into waste that interferes with the smell, sight and eat the place because it does not decompose or rot. The owner of Gunung Kulu restaurant says that, they found that shell waste per day is 70 kg/day.

Sugarcane dregs is the outcome of chemical changes from burning pure sugarcane [2]. Sugarcane dregs is obtained from a seller of sugarcane water on the side of the road in the city of Meulaboh, Indonesia. The sugarcane dregs that the researchers took in the field is about 8 sacks of 50 kg size, until now sugarcane dregs have not been properly utilized. This only make the materials becomes waste to the environment. A survey conducted by researchers that sellers of sugar cane produce 30 kg of sugarcane dregs waste/day. The conch and sugarcane dregs that come from restaurant and culinary waste are the inspiration for further research on the use of conch shell ash and sugarcane dregs as fillers in asphalt pavement mixtures.

The objective of this research is to determine the effect of adding sugarcane waste and conch shells as a substitute for filler in the asphalt concrete mixture (AC-WC). The use of variations of conch shell and sugarcane dregs ash used were in the ratio 60:40; 50:50 and 25:75. The comparative value of these three variations based on previous research [3]. Variation in filler content greatly influence of Marshalls properties [4].

2. Materials Properties and specimen preparation

The methods and stages of the research stages include procurement and examination of the physical properties of the material, planning mixtures, manufacturing and testing of the objects.

2.1. Materials properties

Aggregate and 60/70 pen asphalt used in this study came from PT. Wirataco, Nagan Raya. The gradation used is a continuous gradation. Before using the aggregate and asphalt, its physical properties are checked, in order to determine whether of two materials can be used. The fillers used are conch shell and sugarcane dregs. The conch shells used come from the Gunung Kulu restaurant, The sugarcane dregs used comes from a seller of sugarcane water at the edge of the Jembes Bridge. Testing of these specimens were carried out in the laboratory of the PUPR of West Aceh Regency, the variations of conch shell ash and sugarcane dregs used were 60:40; 50:50 and 25: 75. Making the clam shell is almost the same with making the conch shell, the clam shell filler should be heated at a temperature 110 ℃ [5]. How to make conch shell fillers, namely the conch shells that are taken are cleaned of the dirt that is sticking then the conch shells are burned on a bamboo arrangement that has been arranged with a fire burning above a temperature of 110℃. The conch shells are removed from the fireplace after they become soft and easily destroyed, this can be seen from the shell skin that turns white. Heating time ± 25 minutes. After being lifted from the bamboo fireplace then pounded until it becomes smooth, after pounding, the conch shells are sieved with sieve number 200. The conch shell ash that passes through sieve number 200 is used as a filler to replace cement.

2.2. Making Sugarcane Dregs Filler

The sugarcane dregs that was cleaned from the dirt that sticks on it. Then the sugarcane dregs were put into a frying pan, heated in the pan directly with a fire above 110℃, burning time is ± 10 minutes. After being removed from the pan then pounded until it becomes smooth, after crushing, the sugarcane dregs is filtered with sieve number 200 [6]. Sugarcane dregs that passed sieve No.200 (75μm) and this was used as a filler to replace cement [6].

2.3. Preparation of Specimen Test

Clean the aggregate and filler that will be used for the test specimen mixture then dry in the oven at 110℃ for 24 hours [7]. After that, separate the aggregate and filler into a container in the form of a tray / pan for easy collection. Clean the specimen mold then make the test object base according to the mold diameter. Before pouring, the inside of the mold is coated with oil so that the test object does not stick with the mold.

2.4. Specimen Mixing Preparations

Heat the required aggregate and filler by roasting at a temperature above 110℃. Also heat the asphalt until it reaches a temperature above 110℃ then weigh the required asphalt content from the composition of the mixture that has been obtained, after that pour the asphalt according to the required amount into the aggregate and filler. Stir the mixture until evenly distributed.

IOP Conf. Series: Earth and Environmental Science **971** (2022) 012011 doi:10.1088/1755-1315/971/1/012011

2.5. Compaction of the mixture

After the asphalt mixture is evenly mixed above a temperature of 110℃, move it to a mold that has been coated with filter paper, whose bottom has been coated with paper and stabbed at the edges of the mold and the middle of the mold that has been filled with the mixture. Remove the neck of the mold, smooth the surface of the mixture with a cement spoon so that it becomes slightly convex. Place the mold on the compactor and then blow it 75 times for each of the faces.

2.6. Test of Specimen

The compacted specimen is removed from the mold with a hydraulic jack (extruder) by placing the test object eject plate on the top of the mold and removing the mold base plate. Remove the specimen carefully and soak the specimen for about 1 day [8].

The composition of each filler conch shell and sugarcane dregs in the AC - WC mixture can be seen in Table 1 below.

Table 1. Composition of the mixture.

Mixture Composition	Asphalt content (%)				
	5	5.5	6	6.5	7
% Aggregate without filler (g)	1065.5	1060.3	1054.7	1049.1	1043.5
Mineral filler (g)	74.1	73.71	73.32	72.93	72.54
Aggregate weight (g)	1095.5	1089.8	1084	1078.2	1072.5
Asphalt weight (g)	60	66	72	78	84
Mix weight (g)	1200	1200	1200	1200	1200
Filler variation (g) 60:40 (CS:SC)	44.46	44.23	43.99	43.76	43.52
	29.64	29.48	29.33	29.17	29.02
Filler variation (g) 50:50 (CS:SC)	37.10	36.86	36.70	36.50	36.30
	37.10	36.86	36.70	36.50	36.30
Filler variation (g) 25:75 (CS:SC)	18.53	18.43	18.33	18.23	18.14
	55.57	55.28	54.99	54.70	54.41

2.7. Manufacture and testing of specimen

The manufacture of specimen test is carried out based on the results of continuous gradation checks in accordance with the technical specifications [6]. The aggregate gradient determines the pore size in the mixture [9].

The mixing of the specimen material is carried out in two stages. The initial stage is to obtain the optimum asphalt content with a total of 45 test specimens and the next stage is to test the stability and durability with 18 specimens. The specimen test for the first stage was made of three variations, the ratio of conch shell and sugarcane dregs from the three variations was the ratio (60:40, 50:50 and 25:75). This mixture is planned for heavy traffic with a number of collisions of 2 x 75 collisions. Recapitulation for the number of specimen test can be seen in Table 2.

Table 2. Design of specimen test for the first stage.

Remarks	Asphalt content (%)					Amount
	Pb-1	Pb-0,5	Pb	Pb+0,5	Pb+1	
V1	3	3	3	3	3	15
V2	3	3	3	3	3	15
V3	3	3	3	3	3	15
		Total				45

Note: Comparison of conch shells and sugarcane dregs:
V1 = 60: 40, V2 = 50:50, V3 = 25:75

The results of the first stage test were obtained optimum asphalt content (OAC) which was then used to carry out the second stage testing. The second stage of the test design can be seen in Table 3.

Table 3. Design of the second stage of specimens.

Testing	Total			Total
	V1	V2	V3	
Stability	3	3	3	9
Durability	3	3	3	9
	Total			18

3. Results of Discussion

3.1. Marshall Test Results

The research results presented are tables of examination results and test results as well as graphs that state the relationship between variations of Conch Shell Ash and Sugarcane Dregs Ash with Pen 60/70 asphalt mixture. This mixture is planned for heavy traffic with a number of blow was 75 for each of the faces.

The data generated from the Marshall test are values of stability, density, flow, Void in Mixture (VIM), Voids in Mineral Aggregate (VMA), Voids Filled by Asphalt (VFA), and Marshall Quotient (MQ). The OAC value was from compacting 45 specimens for middle asphalt content and carrying out the Marshall test. Comparison of each specimen for 3 different variations, namely, Conch Shell Ash and Sugarcane Dregs Ash filler variations 60:40; 50:50 and 25:75 for the AC-WC mixture.

The data obtained from the Marshall test are each plotted on a cross axis and form a scatter point. The x-axis represents the asphalt content and the y-axis for Marshall parameter data. The optimum asphalt content obtained in variation 1 is the ratio of conch shells and sugarcane dregs ash for the 60:40 variation of 6.5% (Figure 1), the second variation of 50:50 obtained OAC of 6.25% (Figure 2) and the third variation of 25:75 which is 6.5% (Figure 3). The calculation of the test recapitulation of the Marshall parameter for the OAC value with the variation of sugarcane and conch fillers can be seen in Tables 4, 5 and 6 below.

Table 4. Marshall test results for the first variation (60:40).

No	Characteristic	Asphalt content (%)					Requirement
		5	5.5	6	6.5	7	
1.	Density (g/cm^2)	2.35	2.37	2.37	2.32	2.29	>2
2.	VIM (%)	5.92	4.21	4.05	5.34	6.14	3-5
3.	*VMA (%)*	15.53	14.92	15.68	17.60	19.70	>15
4.	*VFB (%)*	61.90	71.77	74.19	70.36	68.18	>65
5.	Stability(kg)	885.5	981.2	1124.1	1171.1	1240.3	>800
6.	Flow (mm)	4.9	4.23	4.1	3.5	3.4	2-4
7.	Stiffness (kg/mm)	181.3	235.1	276.93	334.8	365.9	200-300

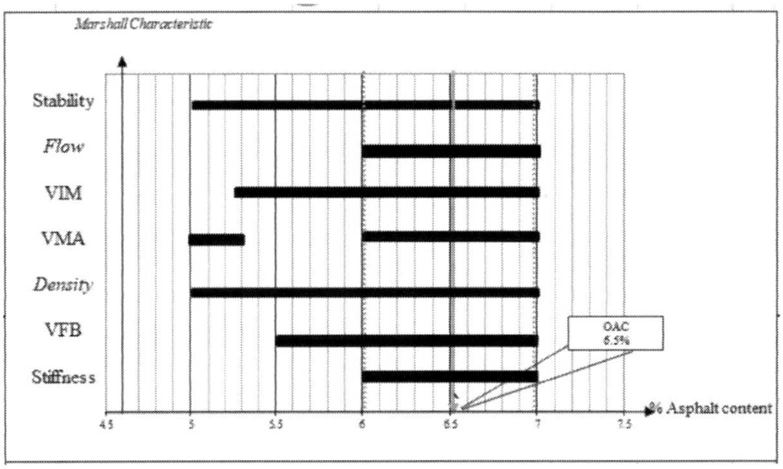

Figure 1. Graph value of asphalt optimum content (60:40).

Table 5. Marshall test results for the second variation (50:50).

No	Characteristic	Asphalt content (%)					Requirement
		5	5.5	6	6.5	7	
1.	Density (g/cm^2)	2.40	2.38	2.35	2.32	2.29	>2
2.	VIM (%)	3.92	4.02	4.52	5.20	6.33	3-5
3.	*VMA (%)*	13.74	14.74	16.10	17.59	19.53	>15
4.	*VFB (%)*	71.47	72.80	71.92	70.42	67.45	>65
5.	Stability(kg)	935.2	1043.6	1110	1146.6	1160.5	>800
6.	Flow (mm)	6.33	4.60	3.67	3.70	3.30	2-4
7.	Stiffness (kg/mm)	148.3	228.1	304.05	311.12	353.09	200-300

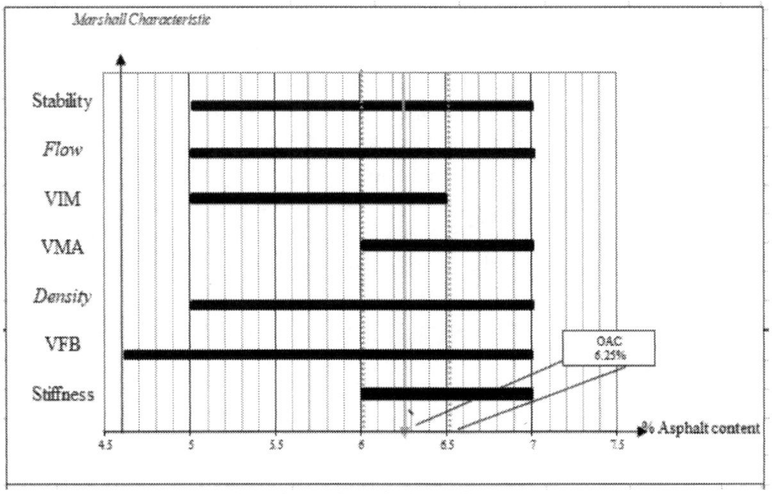

Figure 2. Graph value of asphalt optimum content (50:50).

Table 6. Marshall test results for the third variation (25:75)

No	Characteristic	Asphalt content (%)					Requirement
		5	5.5	6	6.5	7	
1.	Density (g/cm^2)	2.35	2.37	2.37	2.32	2.29	>2
2.	VIM (%)	5.92	4.21	4.05	5.34	6.14	3-5
3.	*VMA(%)*	15.53	14.92	15.68	17.60	19.70	>15
4.	*VFB (%)*	61.90	71.77	74.19	70.36	68.18	>65
5.	Stability(kg)	885.5	981.2	1124.1	1171	1240.3	>800
6.	Flow (mm)	4.9	4.23	4.1	3.5	3.4	2-4
7.	Stiffness (kg/mm)	181.3	235.1	276.9	334.8	365.9	200-300

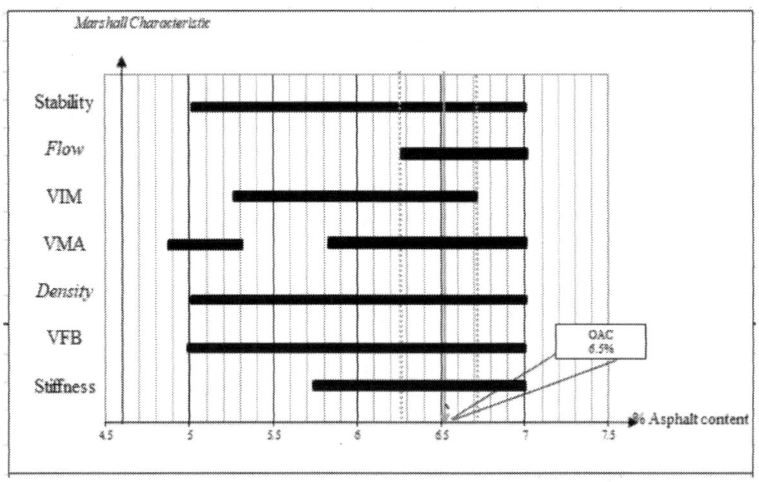

Figure 3. Graph value of asphalt optimum content (25:75).

3.2. Marshall test results on optimal asphalt levels

After obtaining the optimum bitumen content of each variation, the next step is to make the specimen test for stability testing of 30 minutes and 24 hours. The following is the recapitulation results for the Marshall test for 30 minutes and 24 hours of immersion, can be seen in Tables 7 and 8.

From the calculation results of each Marshall parameter shows that the stability values of all variations meet the requirements, namely greater than 800 kg. The highest stability value was obtained in the 50:50 variation of 1046 kg in the 30 minutes immersion while the 24 hour immersion obtained the highest stability at the 25:75 variation. This shows that the use of a combination of 50: 50% conch shell and sugarcane dregs will increase the stability value, so that it can provide better resistance to high temperatures and traffic loads.

The flow value shows the flexibility of a mixture. A high flow value indicates that the mixture is plastic so it is better able to keep up with the changes that occur due to load. Judging from tables 4 and 5, it shows that the flow values of both the 30 minute immersion and the 24 hour immersion both meet the specified requirements. At 30 and 24 hours of immersion, the highest flow values were obtained with a 60:40 filler ratio. In the field, high flow values have a bad impact, because heavy slow moving traffic and high temperatures result in high deformation. If this happens, the asphalt mixture will be damaged before the service life.

Of the three variations, all flow values meet the requirements. The higher the percentage of conch shells in a mixture will cause an increase in the flow value, too high the stability value will also result in stiffer and less flexible pavement.

The VMA value shows the pore percentage between the aggregate grains in the solid asphalt concrete. VMA will increase if the asphalt cover in the mixture is thicker. The highest VMA value is in the 50:50 variation. For the 60:40 variation and the 25:75 variation there is a tendency to decrease the VMA value but the three variations have values that are close to and enter the required specifications.

The VIM value is needed to determine the percentage of pore volume that remains after the asphalt mixture is solidified. The VIM values obtained from the three variations all meet the requirements. The highest VIM value is in the 50:50 variation, which is 5.51%. A high VIM value indicates that a comparable mixture of cue shell ash and bagasse ash will produce a better cavity value. This high VIM value indicates that the cavity in the specimen is large and the water tightness of the specimen is lacking.

The MQ value is an approach to the strength and flexibility of a pavement layer. If the mixture has a high MQ value, it means that the mixture is stiffer, in other words, it has low flexibility. The MQ values for each variation have all met the requirements. The highest MQ value was obtained at the 50:50 variation. The ratio of stability and flow is expressed as MQ and MQ is expressed as an empirical indicator of the stiffness of a mixture. The higher the stability, the higher the MQ value. This shows that MQ is a measure of a material's resistance to shear stress, deformation and rutting.

The value of density for the three variations has basically the same value. This density value depends on the density and content of the two fillers and very much depends on the effect of their hardening on the asphalt mixture. The density value for variation 60:40 and variation 25:75 is 2.35 g/cm^3. For the 50:50 variation, the density value is 2.30 g/cm^3. Density values for all variations fall into the required specifications.

Table 7. Recapitulation of the result of the 30 minute immersion test.

No	Characteristic	Requirement	60:40	50:50	25:75
			OAC 6.5%	OAC 6.25%	OAC 6.5%
1.	Stability (kg)	>800	954.2	1046	954.2
2.	Flow (mm)	2-4	3.4	3.27	3.4
3.	VMA (%)	>15	16.68	17.11	16.68
4.	Stiffness (kg/mm)	200-300	280.4	321.3	280.4
5.	VIM (%)	3-5	4.2	5.16	4.2
6.	VFB (%)	>65	75.1	69.8	75.1
7.	Density (g/cm^3)	>2	2.35	2.3	2.35

Table 8. Recapitulation of the result of the 24 hours immersion test.

No	Characteristic	Requirement	60:40	50:50	25:75
			OAC 6.5%	OAC 6.25%	OAC 6.5%
1.	Stability (kg)	>800	941.9	991.9	1015.7
2.	Flow (mm)	2-4	3.2	3.0	3.0
3.	VMA (%)	>15	16.67	17.41	16.7
4.	Stiffness (kg/mm)	200-300	298.1	332.6	342.8
5.	VIM (%)	3-5	4.1	5.5	4.1
6.	VFB (%)	>65	75.2	68.4	75.2
7.	Density (g/cm^3)	>2	2.35	2.32	2.35

3.3. Durability Value

The results of the optimum asphalt content obtained in the Marshall test phase I were then used for the manufacture of stage II specimens. In stage II, durability testing is carried out, namely to obtain the durability value of a mixture of the three variations. The results of calculating the durability value for the three variations can be seen in Table 9 below:

Table 9. Durability result.

Sample variation	Stability 30 minutes (kg)	Stability 24 hour (kg)	Durability (%)
60:40	954.21	941.95	98.72
50:50	1046	991.97	94.83
25:75	954.20	1015.71	106.44

The durability value required by the General Specifications [2]. The table shows that all variations meet the requirements. The durability value in the 25:75 variation shows that the mixture produces a higher bearing capacity level than the other two variations so that the asphalt concrete mixture is able to accept loads, friction and wear due to weather effects.

The greater durability value in the 25:75 variation indicates that this mixture has a slightly smaller cavity so that water does not easily enter the pores, so it is rather difficult for the asphalt blanket to oxidize with air and result in increased durability values.

4. Conclusion

The result of the examination of three different variation, namely conch shell ash and sugarcane dregs ash filler variations 60:40; 50:50 and 25:75 showed that the optimum asphalt content (OAC) value were 6.5%; 6.25% and 6.5%.

From all the Marshall parameters obtained, it shows that the addition of conch shell filler and sugarcane dregs filler in the three variations gives high strength and stiffness values. Besides that, it also provides a better level of durability. This mixture is suitable for heavy traffic.

5. References

[1] Katrina G 2014 Pemanfaatan Limbah Kulit Kerang Sebagai Substitusi Pasir dan Abu Ampas Tebu Sebagai Substitusi Semen Pada Campuran Beton Mutu K-225 *J. Teknik Sipil Dan Lingkungan* **2(3)** 308-313

[2] Sofyan M S 2018 Pengaruh Filler Abu Ampas Tebu Dengan Bahan Pengikat Aspal Pen 60/70 J. *Arsip Rekayasa Sipil Dan Perencanaan Universitas Syiah Kuala, Banda Aceh*

[3] Advanti Esentia 2014 *Pengaruh Penggantian Sebagian Filler Semen Dengan Kombiinasi 40%. Serbuk Batu Bata dan 60% Abu Cangkang Lokan Pada Campuran Asphalt Concrete Binder Course (AC-BC)* (Unib Scholar Repository, Fakultas Teknik, Universitas Bengkulu)

[4] Nunung W dan Fiko H 2019 Pengaruh Variasi Kadar Filler Abu Cangkang Kerang Terhadap Parameter Marshall di Lapisan Laston AC-WC *J. Teknika.* **14(1)** 127-136

[5] Veranita 2019 Pengaruh Pemberian Filler Abu Cangkang Lokan Terhadap Parameter Marshall Pada Campuran Aspal Beton J. *Teknik Sipil Dan Teknologi Konstruksi, Universitas Teuku Umar, Meulaboh* **5(1)** 73-75

[6] Putra M A 2014 PemanfaatanKombinasi Limbah Abu Ampas Tebu dan Abu Kulit Kerang Sebagai Substitusi Semen Pada Campuran Beton Mutu K225 Dengan NaCL Sebagai Rendaman *J. Teknik Sipil Dan Lingkungan* **2(3)** 413-417

[7] Bina Marga 2010 *Petunjuk Pelaksanaan Lapis Aspal Beton (Laston) Untuk Jalan Raya*, SNI No. 1737 1989 (Departemen PU, Jakarta) pp 35-40

[8] Fajar Himawan W 2012 Pengaruh Pemanfaatan Limbah Ampas Tebu Sebagai Pengganti Filler *J. Karya Teknik Sipil Universitas Diponegoro, Semarang* **2(2).**

[9] Bukhari dkk 2017 *Rekayasa Bahan dan Tebal Perkerasan* (Fakultas Teknik, Universitas Syiah Kuala Banda Aceh) pp 27-28

IOP Conf. Series: Earth and Environmental Science **971** (2022) 012012 doi:10.1088/1755-1315/971/1/012012

The Influence of Coal Fly Ash on the Mechanical Properties of Hot Mix Asphalt Mixture

A A M Radwan[1,2], M K I M Satar[1*], N A Hassan[1], K U Rogo[1,3]

[1] School of Civil Engineering, Faculty of Engineering, Universiti Teknologi Malaysia, 81310, Skudai, Johor, Malaysia
[2] Department of Civil Engineering, Faculty of Engineering, University of Tripoli, Tripoli, Libya
[3] Institute Department of Civil Engineering Technology, School of Engineering Technology, Nuhu Bamalli Polytechnic Zaria, P.M.B 1016, Zaria, Kaduna State, Nigeria
*Corresponding author: khairulidham@utm.my

Abstract. Coal is one of the most important sources of energy, providing for over 40% of global electricity generation. Coal fly ash (CFA) is the by-product of thermal generation of energy in coal-fired power plant. CFA has been widely employed in the construction of concrete; however, there are only a few cases in which asphalt pavements have employed coal fly ash. This paper aims to determine the performances of CFA as filler in hot mix asphalt (HMA) mixture. This study used four CFA contents as filler by weight of aggregate in the dry method, namely CFA-0, CFA-2, CFA-4, and CFA-6. The mixtures were tested for stability, flow, stiffness, moisture damage, and Cantabro loss test. The findings indicated that the stability values of asphalt mixtures containing CFA were higher than conventional mixture. In addition, incorporating CFA also improved resistance towards moisture damage and durability. As a result, it can be inferred that CFA can be employed as filler substitute in HMA mixtures.

1. Introduction

This day, one of the primary global concerns in waste management is the vast amount of waste material created on a daily basis. Concerns about the environment and a limiting number of disposal sites are prompting countries to find new ways to recycle waste [1]. The use of waste and alternative materials in concrete and asphalt pavements has been studied extensively to conserve natural resources. Fly ash, a by-product of the thermal generation of energy in coal-fired power plants, is among the most promising from these materials. The annual global production of coal fly ash is approximately a billion tonnes [1]. The possibility of using it in asphalt mixtures was investigated in the mid-nineteenth century as part of efforts to contribute to sustainable engineering, to obtain mixtures with satisfactory properties, minimising the negative effects of landfilling, and conserving natural resources [2]. For many years fly ash has been widely used in concrete studies because of its pozzolanic activity; however, it has only a limited number of applications in asphalt pavement. Since 1988, seven of Malaysia's power plants have used coal as a raw material to generate electricity [3]. Coal power stations create two forms of ash: coal fly ash (CFA) with an ash content of 80% and coal bottom ash (CBA) with an ash content of 20%. Malaysian electricity production generates a massive 6.8 million tons of CFA and 1.7 million tons of CBA [4]. The use of coal fly ash in mineral-asphalt mixtures has not been extensively investigated. Only a few research have shown that fly ash can improve the performance of asphalt concrete [1]. The studies by Mistry and Kumar [5] have showed that fly ash as filler can give reduced deformation and improved strength compared to hydrated lime as filler replacement in hot mix asphalt (HMA) [6]. The purpose of this research was to determine the effect of CFA on the mechanical properties of HMA. CFA was obtained from the coal thermal plant in Tanjung Bin in Johor. It was compared to the control mixture to evaluate the feasibility as filler in an optimum proportion.

2. Materials and Methods

2.1. Material

The 60/70 PEN asphalt binder was acquired from Kemaman Bitumen Company (KBC). The asphalt binder properties are shown in Table 1 and all properties met the requirement. The aggregate was collected from the inventory of Hanson Quarry in Kulai, Malaysia and met all of the requirements by Malaysia Public Work Department (2008) [7]. Table 2 show the aggregate properties. Hot mix asphalt (HMA) mixture was designed according to the AC14 aggregate gradation as listed in Table 3. The conventional fillers used in this research were fine aggregates and hydrated limestone. Coal fly ash (CFA) with Class-F category was collected from Tanjung Bin coal power thermal plant Johor, Malaysia. It contains a high percentage of SiO2 (65.9%), Al2O3 (19.7%), and a little amount of CaO. (3.29%) as shown in Table 4. The above-mentioned fillers were preheating before being sieved on a 0.075 mm sieve, with the material finer than 0.075 mm being utilised in this research Table 5.

Table 1. The properties of the asphalt binder.

Material	Characteristics	Unit	Test Method	Requirement	Value	Check
Asphalt binder 60\70 PEN	Density	gm/cm^3	ASTM D70	1.01-1.05	1.02	√
	Penetration at 25°C	mm	ASM D5	60 – 70	64.5	√
	Softening Point	°C	ASTM D36	48 – 56	51	√
	Ductility at 25°C	cm	ASTM D113	Min. 100	117	√
	Viscosity @ 135 °C	(mPa.s)	ASTM D4402	-	3500	√
	Viscosity @ 165 °C	(mPa.s)	ASTM D4402	-	700	√

Table 2. Aggregate properties.

Aggregate Testing	Standard method	Results	value
SG. Coarse	ASTM C 128	2.863	-
SG. Fine	ASTM C 127	2.681	-
Aggregate Impact Value (%)	BS EN 1097-2	13.37	< 30
Aggregate Crushing Value (%)	BS EN 1097-2	17.1	< 30
Los Angeles Abrasion (%)	AASHTO T96	17.4	< 45
Flakiness (%)	BS EN 933-3	18.4	< 25
Elongation (%)	BS 812	22.19	< 25

IOP Conf. Series: Earth and Environmental Science **971** (2022) 012012 doi:10.1088/1755-1315/971/1/012012

Table 3. Aggregate gradation for AC 14.

Sieve Size (mm)	Passing %
20	100
14	95
10	81
5	56
3.35	47
1.18	26
0.425	18
0.150	10
0.075	6
Pan	0

Table 4. The chemical component of CFA from (XRF).

Component	CFA %
Al_2O_3	19.7
SiO_2	65.9
Fe_2O_3	5.82
TiO_2	1.21
Na_2O	1.28
MgO	0.958
SO_3	0.322
K_2O	1.52
CaO	3.29

Table 5. Fine material utilised.

Samples	Hydrated Lime	Fine Aggregate	CFA
CFA-0	2	4	-
CFA-2	2	2	2
CFA-4	2	-	4
CFA-6	-	-	6

2=24g, 4=48g, 6=72g

2.2. Samples Preparation

The Marshall mix design method was used to design the HMA mixture. The aggregate and CFA were heated in the oven at 150 °C before mixing with asphalt binder. While 60/70 PEN asphalt binder was kept in the oven at 110 °C before mixing with the aggregate. The samples were placed in a pre-heated mould and compacted by applying 75 blows on each face according to ASTM: D6926 [8] having five different bitumen content between 4 and 6% by total weight of aggregate at 0.5% increments for control mixture as well as a mixtures containing CFA-2, CFA-4 and CFA-6 as filler. In this study, a 5.05% optimum asphalt content was utilized for all four mixtures to maintain consistency throughout the research for comparison purposes.

2.3. Tests

Marshall stability and flow tests were performed on ASTM D6927-compliant control and modified asphalt mixes. The stiffness value was then calculated using the stability and flow correlation. Then, using ASTM D4867 [9], the indirect tensile strength (ITS) and tensile strength ratio (TSR) were evaluated. Finally, a Cantabro loss test was performed in compliance with Texas Department of

Transportation (TxDOT) Tex-245-F in order to determine the mixture's resistance to disintegration induced by traffic load.

3. Result and Discussion

3.1 Marshall Properties

Table 6 summarises the Marshall properties results for the control and modified mixes. It can be observed that the stability value increases linearly and the flow decreases with the increase in the CFA content. The regular shape of FA particles (usually spherical) may act as rollers, allowing for less friction in the mastic during compaction, resulting in tighter packing [10]. A stiffer mixture was represented by a higher value of stability divided by a lower value of flow. The high values of the stability and the low value of the flow in the CFA mixtures all comply with the Malaysia Public Work Department (2008) specification [7].

In addition, the void in total mix (VTM) of the control mixture was 3.6% where the value was decreasing as the CFA content increased. These results can be explained by the addition of CFA to the mixtures increased the surface area, which decreased the air voids in the specimens. According to Likitlersuang & Chompoorat [11], the larger the surface area, the more asphalt is required to coat the aggregates. Thus, it will increase the void filled with asphalt (VFA) and decrease the void in total mix (VTM) in the mixture.

Table 6. Marshall properties of asphalt mixtures.

Marshall properties	Type of Mixture				Requirement
	0% CFA	2% CFA	4% CFA	6% CFA	
Stability (N)	15225	16848	17148	18243	> 8000
Flow (mm)	3.22	3.12	2.80	2.37	2.0 – 4.0
Stiffness (N/mm)	4728	5400	6124	7697	> 2000
VFA %	75.5	76.1	76.9	78	70 – 80
VTM %	3.60	3.54	3.23	3.1	3.0 – 5.0

3.2. Moisture Damage

The findings of indirect tensile strength (ITS) for all of the mixtures tested at temperatures of 25°C and 60°C are shown in Figure 1. Generally, it can be seen that the ITS values of the unconditioned samples are higher for the control and modified compared to those of wet conditioned. In the case of a wet condition, it shows that moisture had a minor impact on the tensile strength of the CFA mixtures. A higher TSR value indicates that asphalt mixtures are more resistant to moisture damage [11]. It is clear that all of the combinations evaluated in this study met or exceeded the AASHTO TSR minimum criterion (> 80%) as shown in Figure 2. CFA-6 mixture had the highest TSR value of all the mixtures, indicating that the inclusion of CFA in the asphalt mixture reduces moisture damage susceptibility.

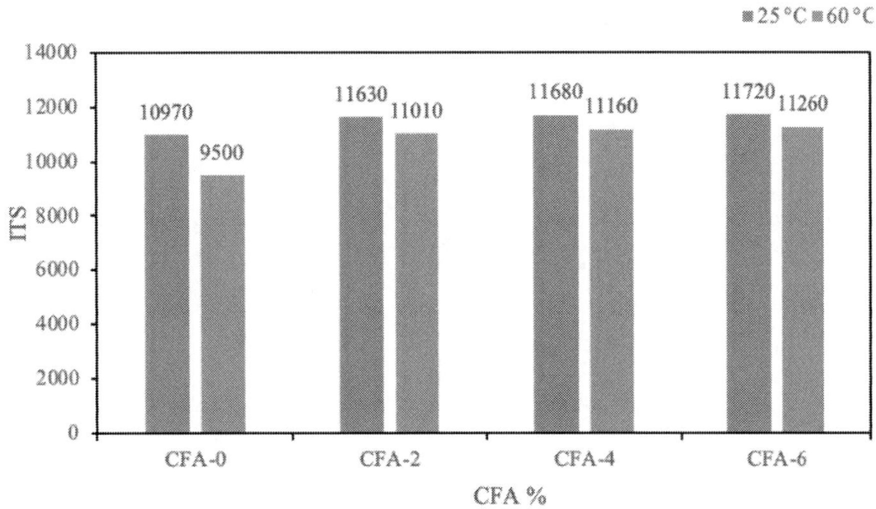

Figure 1. ITS of asphalt mixtures.

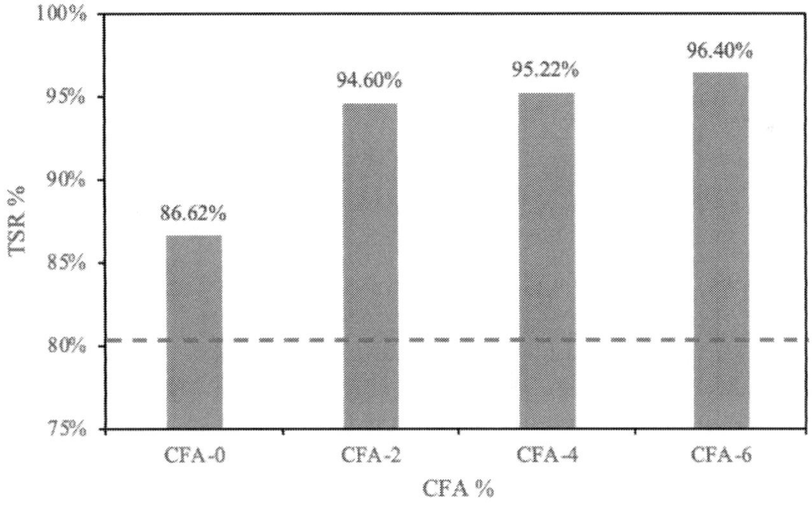

Figure 2. TSR for the asphalt mixtures

3.3. Cantabro Loss

Figure 3 depicts the mass loss test results for mixtures, showing that adding CFA could reduce mass loss as CFA content increases. As a result, the mixture showed more resistance to disintegration. This improvement was caused by the enhancement in adhesion and cohesion properties with the addition of CFA [12]. It's worth noting that the proportion of air voids in the specimen is strongly connected with Cantabro mass loss. The CFA, also known as bitumen extender, can provide a partial replacement of bitumen [13], reducing the air void of HMA. The addition of CFA reduces the percentage of air voids in the mixtures. As a result, mass loss was reduced.

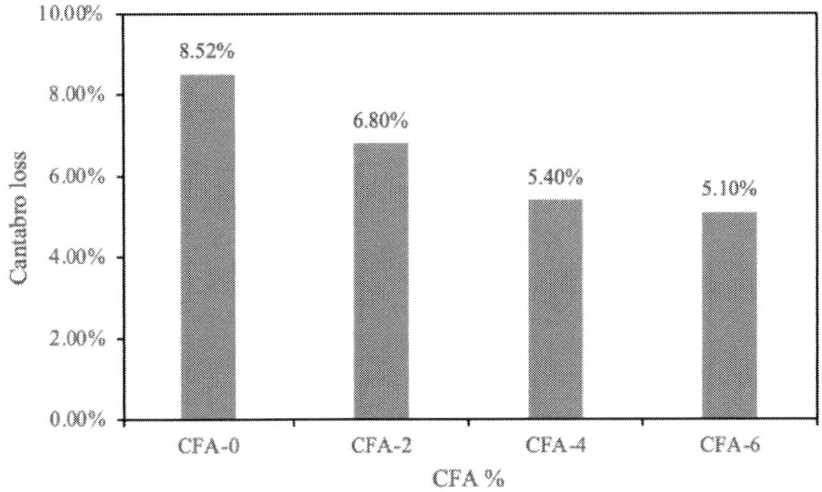

Figure 3. Cantabro loss for asphalt mixtures.

4. Conclusion
The main findings are summarised follows:

- Incorporating CFA in the asphalt by weight of aggregate increases the Marshall properties of the mixtures. It increases stiffness, thereby increasing the stability and decreased the flow
- The ITS and TSR values of modified mixtures with CFA have dramatically risen, showing that the CFA mixtures are the more resistant to moisture damage compared to the conventional mixture.
- Incorporating CFA into the mixture showed a reduction in the Cantabro loss, indicating an improvement in term of durability of the mixture.

5. References

[1] Woszuk A, Bandura L and Franus W 2019 Fly ash as low cost and environmentally friendly filler and its effect on the properties of mix asphalt, *J. Clean. Prod* **235** 493–502

[2] Mirkovi K, Toˇ N and Mladenovi G 2019 Effect of Different Types of Fly Ash on Properties of Asphalt Mixtures, *Advances in Civil Engineering* **2019**

[3] Aminaton M C S T 2016 Properties Of Coal Bottom Ash From Power Plants In Malaysia And Its Suitability As Geotechnical Engineering Material, *J. Teknol.* **5** 1–10

[4] Muhardi A, Marto A, Kassim A, Makhtar A, Wei L F and Lim Y S 2010 Engineering characteristics of Tanjung Bin coal ash, *Electron. J. Geotech. Eng.* **15** 1117–1129

[5] Mistry R and Roy T K 2016 Effect of using fly ash as alternative filler in hot mix asphalt, *Perspect. Sci.* **8** 307–309

[6] Bi Y and Jakarni F 2019 Evaluating properties of wood ash modified asphalt mixtures *IOP Conf. Ser. Mater. Sci. Eng.* **512** 012004

[7] JKR Standard specification for road works Section 4 2008 *Flexible Pavement* (Kuala Lumpur: Jabatan Kerja Raya Malaysia) pp S4-58-S4-69

[8] ASTM D6926-10, "Standard Practice for Preparation of Bituminous Specimens Using Marshall," (ASTM International) pp 1–6

[9] ASTM D4867/D4867M-9, "Standard Test Method for Effect of Moisture on Asphalt Concrete Paving Mixtures 1," (ASTM International) pp 1–5

[10] Mistry R and Kumar R T 2020 Predicting Marshall stability and flow of bituminous mix containing waste fillers by fuzzy logic *Revista De La Construccion* **19** 209-219
[11] Likitlersuang S and Chompoorat T 2016 Laboratory investigation of the performances of cement and fly ash modified asphalt concrete mixtures, *Int. J. Pavement Res. Technol.* **9** 337–344
[12] Tian Y 2021 Laboratory investigation on effects of solid waste filler on mechanical properties of porous asphalt mixture, *Constr. Build. Mater.* **279** 122436
[13] Sobolev K, Flores V I, Saha R, Wasiuddin (Wasi) N M and Saltibus N E 2014 The effect of fly ash on the rheological properties of bituminous materials, *Fuel* **116** 471–477

Acknowledgments
This work was supported/funded by the Universiti Teknologi Malaysia for UTMShine grant, Q.J130000.2451.09G26. The author also expressing appreciation to the Libyan Government for providing a scholarship through the Ministry of Higher Education and Scientific for supporting this research.

IOP Conf. Series: Earth and Environmental Science **971** (2022) 012013 doi:10.1088/1755-1315/971/1/012013

Alternative Bituminous Binder for Sustainable Flexible Pavement: A Review

R Yuniarti[1,2], E Ahyudanari[1*] and C A Prastyanto[1]

[1] Department of Civil Engineering; Faculty of Civil, Planning and Geo Engineering, Institut Teknologi Sepuluh Nopember
Kampus ITS Sukolilo, Surabaya 60111, East Java-Indonesia
[2] Department of Civil Engineering, Faculty of Engineering, University of Mataram
Jalan Majapahit 62 Mataram 83125, West Nusa Tenggara-Indonesia
*Corresponding author: ervina@ce.its.ac.id

Abstract. Bituminous binders used for flexible pavement materials are primarily obtained from refinement of petroleum. Since petroleum is non-renewable resource, there is an urgent need to produce alternative binders from bio-renewable resources. Standard alternative binders include pyrolyzed materials, jatropha curcas oil, waste cooking oil, pine resin, etc. This article presents an overview on the application of various types of alternative binders as a modifier for virgin asphalt, rejuvenator for aging asphalt, and recycling agent for reclaimed asphalt pavement. This review discusses the effect of using alternative binders on the physical and rheological properties of asphalt 60/70 and shows that alternative binders' in appropriate dosage can enhance the physical properties of virgin asphalt and restore the properties of the aged binders. It is also concluded that rutting resistance and fatigue performance are affected by an alternative binder. The possibility of using alternative bituminous binder from economics point of view is also discussed for consideration of its applications in the paving industry.

1. Introduction

Bituminous binders for flexible pavement materials are residue generated from the process of crude oil refining. Technology development in the process of refining tends to maximize the fuel quantity and minimize asphalt residue [1]. As petroleum is a non-renewable resource and the recoverable amount is predicted to become depleted, the supply of petroleum asphalt tends to decrease.

On the other hand, bituminous binders are in great demand for new construction and maintenance of existing asphalt pavements. With increasing consumption of bituminous binders and enhanced environmental concern to minimize natural resources, it is important to seek alternative binders to reduce petroleum asphalt [2].

The alternative binder is environmentally-friendly asphalt gained from renewable resources material [3]. Using alternative binders as modifiers or partial replacement of asphalt could reduce petroleum asphalt in pavements. Standard alternative binders include pyrolyzed materials, jatropha curcas oil, waste cooking oil, pine resin, etc. According to Huang et al. [4], alternative binders produced from plant matter and residue comprise chemical compositions slightly identic to conventional asphalt. Yang et al. [5] also stated that bio-oil components could fall into the four distinct petroleum asphalt compounds: saturates, aromatics, resin, and asphaltenes.

The use of bio-binder as modifier or rejuvenator of asphalt has been stated in several publications. Neubert [6] produced a superior asphalt cement includes petroleum asphalt, natural asphalt, vegetable oil, and an elastomer which has low viscosity during the application and high viscosity, toughness, and tenacity after curing on the highway.

Bailey and Phillips [7] claimed that the oils derived from sesame, sunflower, soybean, corn, palm, or peanut are suitable as asphalt rejuvenator. The amount of plant derived oil needed to rejuvenate an aged asphalt is in the range of 2-20% by mass of bitumen. It is also mentioned that by blending the conventional penetration grade bitumen with the oil, the viscosity of the binder can be modified and resulting in performance grade being obtained. Said rejuvenator can be applied both ex-situ and in-situ method for asphalt rejuvenation.

Freisthler [8] has patented the sealant composition involves soy and other vegetable products and another one is a mixture of soy products and asphalt. In addition, Nigen-Chaidron et al. [9] claimed rejuvenating agents for recycling which consist of 10-90% palm oil and 90-10% asphalt. The rejuvenator can be used in the process of hot asphalt pavement in-place and in-plant recycling. Hot-in place means the recycling process is operated at the road location, whereas hot in-plant is executed by removing road surface and transporting it to an asphalt mixing plant.

From the above elucidation, it is shown that a large number of alternative binders can be used as a partial replacement of bituminous adhesives derived from fossil fuels. This study evaluates various types of alternative binders and their effect on asphalt physical and rheological properties.

2. Methodology

The methodology followed in the review on an alternative bituminous binder for sustainable flexible pavement was carried out by collecting and selecting appropriate literature resources. The well-known databases such as Scopus, Science Direct, Web of Science and Google Scholar were used to collect documents. Recycling agent, rejuvenating agent, modifier, softening agent, reclaiming agent, bio-oil, and fluxing oil were applied as key words. The title and abstract from hundreds of documents then analyzed. The irrelevant and duplicated documents were excluded. Through reading, summarizing, and evaluating the collected documents, the fundamental ideas related to the topic and the finding in literature resources then compiled in this review.

This study provides an overview of using bio-binder as modifier or rejuvenating/recycling agent for virgin or aged and recycled asphalt in four steps. The first step briefly describes the production process of bio-binder from various resources. The second step discusses the effect of bio-binder on the physical properties of asphalt and the third step focuses on bio-binder on its rheological properties. Meanwhile, the final step highlights the feasibility of using alternative bituminous binders for future sustainable material.

3. Results and Discussion

3.1. Production Process of Bio-binder

One of the processes used to produce bio-oil as bio-binder is fast pyrolysis. Fast pyrolysis can be defined as the rapid decomposition of biomass without oxygen to produce char, liquid smoke (bio-oil), and gas. Another definition explains fast pyrolysis as a process at high-temperature in which organic matters are rapidly heated in a vacuum. From decomposition process, it produces vapors, aerosols, charcoal-like char, and a dark brown liquid [10]. A vast range of organic matters can be resources of pyrolyzed material such as palm kernel shells, palm trees, yard waste, waste wood, rice husk, switchgrass, wheat straw, corn stover, and sugarcane bagasse [11-21]. Due to various resources of bio-oil, the preparation process and characteristics of products are different. However, the compositions of the bio-oil are mainly identical and similar to petroleum asphalt, it including saturated, aromatic, asphaltene, and colloid [22].

The other bio-binders are produced from the ripe or mature seeds of the plant after sun drying and following a sequence of seed processing operations. The process of seed oil includes dehulling, seed coat removal, winnowing, sorting, cleaning, grinding, and preheating [23]. Grinding before extraction ensures that oil-bearing implanted in fibrous structures is ruptured to release the oil. Preheating is important process to ascertain that the moisture content of the oilseed in appropriate range or not too moist for pressing [24].

On the industrial scale, expellers are used to extract seed oils mechanically. Mechanical expression encompasses pressure application to force oil out of an oil-bearing material [25]. The percentage of oil yield obtained can be calculated by dividing the oil mass by the abundance of oilseed. It depends on many factors such as varieties of oilseed, environmental conditions around the oil-bearing resource plant, oilseed storage and handling, pretreatment procedure, and the extraction methods. Jatropha curcas oil, soybean oil, sunflower oil, *Calophyllum inophyllum* L. oil are among bio-binders derived from this extraction process [26-31].

A bio-binder that can also be used for pavement material is waste cooking oil [32,33]. To remove undesired particulates from food left in waste cooking oil, separation through filtration with specific materials can be applied. This process used technological approaches such as membrane treatment, conventional, and activate filtration [34]. Membrane is elucidated as a semipermeable barrier, allowing the passage of components and the retention of others. In addition, conventional filters involve cellulose fibres pad and metal screens. These pads of cellulose absorb a substantial amount of the filtrated oil, and only particles with dimensions larger than 1 μm can be retained.

On the other hand, stainless steel filter screen is more expensive than pad of cellulose, and its efficiency filtration is restricted to particle size larger than 80 μm. Activated filtration is employed to increase filtration efficiency. Adsorbent or neutralizing agent provides this system. Activate filtration removes not only particulates but also removes oil-soluble components from the cooking oil by chemical reactions.

Furthermore, crude pine resin from a pine tree can be employed as an asphalt modifier for the pavement [35]. To collect the pine resin, a minor damage such as a hole is made to the tree. The hole is far enough into the tree trunk to penetrate the vacuoles so that resin can leak from the tree. Simple filtration can be applied to remove impurities such as flower, leaf, and branch flakes before mixing pine resin with asphalt.

3.2. Effect of Bio-binder on Physical Properties of Asphalt

The physical properties of asphalt including penetration, softening point, ductility, mass loss, and flash point. Penetration test identifies the consistency and resistance to deformation of asphalt at the temperature of 25°C. Softening point determines the thermal susceptibility at a higher temperature, at which asphalt changes from semi-solid to semi-liquid. Ductility test assesses the ability of asphalt for sticking, whereas the mass loss test provides a quantitative measure of the volatiles lost on heating due to temperature. Flashpoint is measured for safety production process at the asphalt mixing plant [36].

Due to age, asphalt pavement deteriorates and becomes more stiff, hard and brittle. The aged asphalt can be obtained from the mixture of reclaimed asphalt pavement by extraction and recovery process [37]. Extracted asphalt binder must be mixed with recycling agents to restore its physical and chemical properties.

Standard laboratory aging procedures can also prepare the aged asphalt. The virgin asphalt was conditioned by rolling thin film oven test (RTFOT) at the temperature of 163°C for 75 minutes and pressurized aging vessel (PAV) test at 90°C for 20 hours. RTFOT simulates short-term aging of asphalt mixtures during mixing and paving, whereas PAV simulates long-term aging of the mixture during service life of road pavement [38].

Several researchers stated that bio-binder was used as a modifier for virgin asphalt and a rejuvenating or recycling agent for the aged asphalt and reclaimed asphalt pavement [39-41]. El-Shorbagy et al. have investigated the possibility of using waste cooking oils as rejuvenator for the aged asphalt binder extracted and recovered from reclaimed asphalt pavement. The percentages of waste cooking oils used were 2%, 3%, and 3.5 and 4% by the total weight of asphalt binder. The addition of rejuvenators increased penetration value in parallel as increased the addition of waste cooking oil. In contrast, the softening point decreased with the addition of this rejuvenating agent. To reach a penetration value of 60/70 target grade, a 3% rejuvenator was needed. Meanwhile, about 3.5% of rejuvenator was required to gain the target value of softening point [42].

Lie et al. [44] studied artificially aged asphalt by rolling thin film oven at various aging times of 5; 7; 9; 11; 13; and 15 hours at 163°C. It is reported that as aging time increases, a higher amount of waste cooking oil is needed to gain the penetration and ductility value of the virgin asphalt. The softening point requirement could be fulfilled with > 2% of waste cooking oil at more than 13 hours aging time.

The influence of waste vegetable oil on the physical properties of virgin binder was also examined by Al-Omari et al. [45]. It is concluded that the more significant percentages of waste vegetable oil added, the higher value of penetration, ductility, and flashpoint. Meanwhile, softening point value decreased as an increased amount of waste vegetable oil indicated that asphalt has higher temperature susceptibility.

The influence of waste vegetable oil on the physical properties of asphalt was also investigated by Xinxin et al. [43]. Virgin binder conditioned aged in rolling thin film oven and pressurized aging vessel was used and mixed with 5%, 10%, 15%, and 20% of waste vegetable oil. It is found that penetration value increased and softening value decreased as an additional amount of this rejuvenator. The research concluded that 13.4% of waste vegetable oil has similar physical properties as a virgin binder.

Furthermore, Yuniarti [46] investigated the physical properties of pine resin and gum rosin modified asphalt. The percentages of pine resin used were 2%, 3%, 4%, and 5%, whereas 4% of gum rosin was applied for all mixture of pine asphalt. From this study, it is concluded that the more the amount of pine resin, the higher penetration value and the lower the temperature of softening point. Softening point value of asphalt modified with pine resin and gum rosin is higher and penetration value is lower compare to the physical properties of asphalt grade of 60/70. According to Perum Perhutani [47], softening point of gum rosin in the range of 78°C-82°C. Therefore, applied 4% of gum rosin increasing the hardness of pine asphalt.

The test results about the influence of bio-binder on the physical properties of asphalt are shown in Table 1.

Table 1. Effect of bio-binders on physical properties of asphalt.

Bio-binder	Percentage	Effect of bio-binder					Results
		Penetration	Softening point	Ductility	Flashpoint	Mass loss	
Waste cooking oil [42]	0; 2; 3; 3.5; 4	⇧	⇩	N/A	N/A	N/A	Penetration value increases, and softening point decrease compared to extracted binder without recycling agent.
Waste vegetable oil [43]	0; 5; 10; 15; 20	⇧	⇩	⇧	⇩	⇧	Using 13.4% of waste vegetable oil has resulted in similar physical properties as the virgin binder.

Waste cooking oil [44]	0; 1; 2; 3; 4; 5	↑	↓	↑	N/A	N/A	Penetration value increases, softening point decrease and 4% of waste cooking oil is needed to reach minimum ductility value of 100 cm for aging time of 15 hours.
Waste vegetable Oil [45]	0; 1; 2; 4; 6; 8	↑	↓	↑	↑	N/A	Penetration value, ductility, and flash point rise compare to the virgin binder.
Pine resin and gumrosin [46]	0; 2; 3; 4; 5	↑	↓	↑	↓	↑	Penetration value, ductility, and mass loss increase as increase the amount of pine resin.

N/A = not available

Comparison for several types of bio-binder is shown in Figure 1-5. In these figures, the test results conducted by Lie et al. [42] are presented according to the aging time of 15 hours.

According to the Public Work Ministry in Indonesia, the physical properties of rejuvenated or modified asphalt are discussed based on the specification of 60/70 penetration grade [48], even though the other countries may have a different requirement.

Figure 1 illustrates the changes in penetration, which can exhibite the consistency of asphalt. It is investigated in Figure 1 that penetration values of asphalt without bio-binder addition fail to meet the requirement as studied by El-Shorbagy et al. [42], Xinxin et al. [43], and Li et al. [44] due to aging naturally or artificially. It is showed that as increasing bio-binder proportions, penetration value of asphalt are all risen. The percentage of 3-4% waste cooking oil dosage is suggested to rejuvenate the penetration value of an extracted asphalt to an equal 60/70 target binder [42].

According to Li et al. [44], about 3% dosage is proposed to recover the penetration value of artificially aged asphalt. However, a higher proportion of waste vegetable oil is recommended by Xinxin et al. [43]. It is about 10% dosage needed to reach the penetration value of the target penetration grade asphalt. Compare to the finding of El-Shorbagy et al. [42] and Li et al. [44], there is around 3 times larger dosage to produce the value that meets penetration requirement of asphalt 60/70. This is because the neat asphalt investigated by Xinxin et al. [43] is harder than that of El-Shorbagy et al. [42] and Li et al. [44]. Blending virgin asphalt with waste vegetable oil increases penetration value which implies this bio-binder can lower the hardness of asphalt.

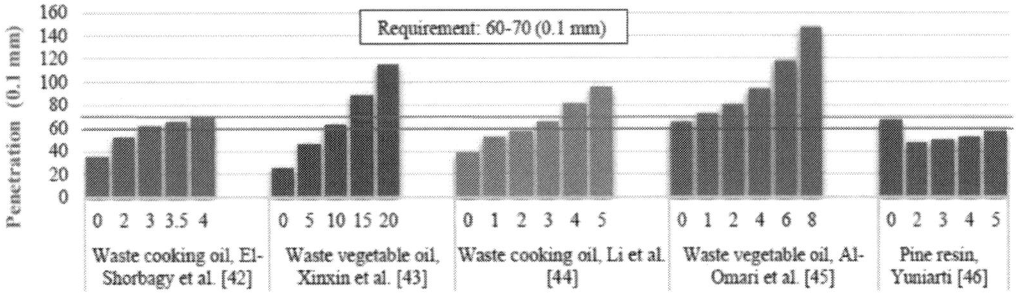

Figure 1. Relation between bio-binder and penetration value.

The influence of bio-binder on softening point is presented in Figure 2. Compared to softening point of asphalt without bio-binder addition, aged asphalt has a higher value of softening point as studied by El-Shorbagy et al. [42], Xinxin et al. [43], and Li et al. [44]. Moreover, it is seen in Figure 2 that the value of softening point decrease with the increase of bio-binder percentages, which represent that the addition of bio-binder can ease aged asphalts. Figure 2 also exhibites that softening point of virgin asphalt fall down with the improving amount of waste vegetable oils as studied by Al-Omari et al. [45]. Therefore, it can be concluded that the phenomenon effects of bio-binder on penetration and softening point are the same.

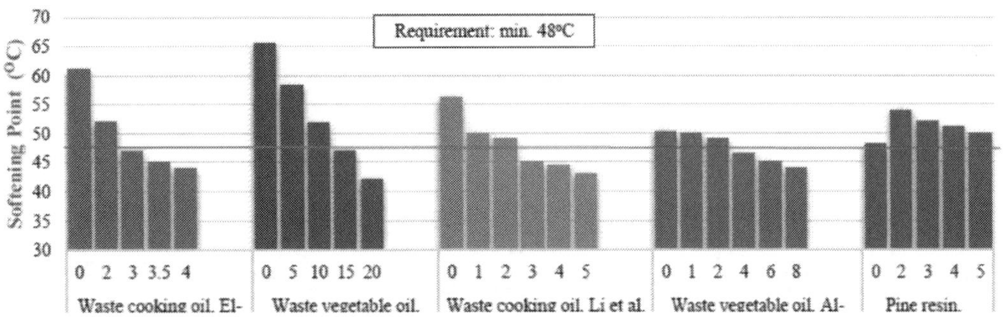

Figure 2. Relation between bio-binder and softening point value.

Figure 3 shows the influences of bio-binder contents on ductility, which indicates the flexibility of asphalt. It is found in Figure 3 that ductility value of asphalt rises with the addition of bio-binder content. However, as Xinxin et al. [43] studied, up to 20% waste vegetable oil addition gave ductility value of aged asphalt below the minimum acceptable value of 100 cm. In contrast, about 4% dosage is needed to achieve ductility value of the target binder for aged asphalt studied by Li et al. [44]. According to Al-Omari et al. [45], blending virgin asphalt with waste vegetable oil in the range of 1-8% increased its ductility. As seen in Figure 3, ductility of asphalt tends to increase due to rising percentages of pine resin. Thus, it can be said that pine resin-modified asphalt likely more adhesive than asphalt 60/70 [46].

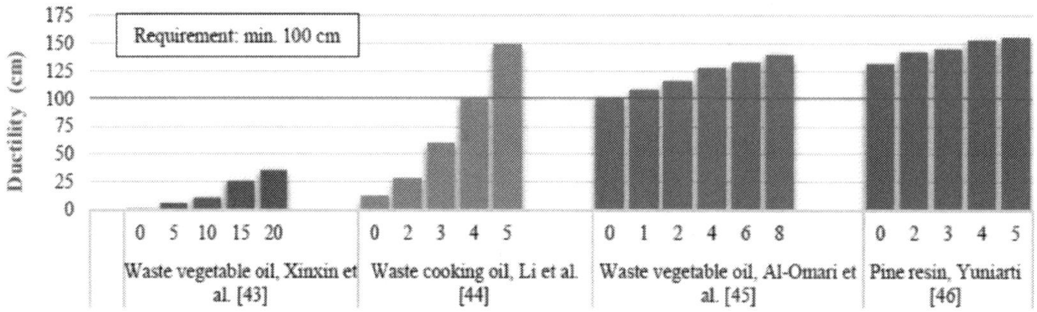

Figure 3. Relation between bio-binder and ductility value.

The infulence of bio-binder percentage on the flashpoint was investigated in research conducted by Xinxin et al. [43], Al-Omari et al. [45], and Yuniarti [46]. From three studies that examined the flashpoint value, results were opposition in conclusion. By mixing bio-binder with asphalt, Xinxin et al. [43] and Yuniarti [46] highlighted a decrease of flash point temperature with the increase of waste vegetable oil and pine resin percentages. Meanwhile, Al-Omari et al. [45] revealed a possible rise in the flashpoint of the modified asphalt.

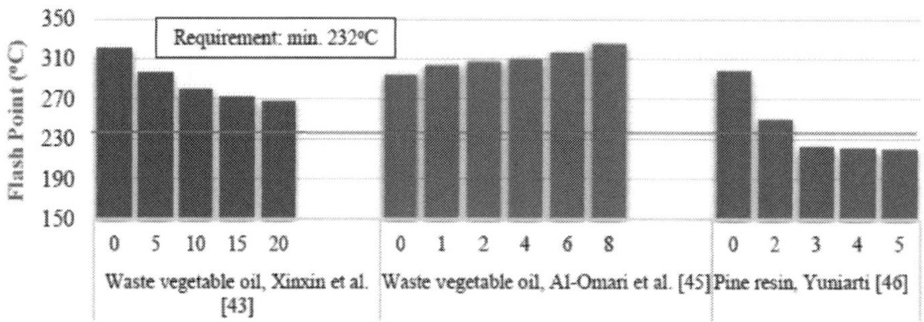

Figure 4. Relation between bio-binder and the flashpoint value.

Figure 5 shows the influences of bio-binder dosages on the loss mass as measured by the volatility of asphalt. As shown in Figure 5, the loss on the heating increase with the increment proportion of waste vegetable oil and pine resin. Xinxin et al. [43] and Yuniarti [46] stated that all the dosing of bio-binder met the maximum mass loss of 0.8%. Therefore, it can be predicted that the hardening of asphalt during mixing and service life of pavement has fulfilled the acceptable value.

Overall, the physical properties of modified asphalt depend on the virgin and the aged asphalt characteristics, also the type of modifier or rejuvenator and its amount.

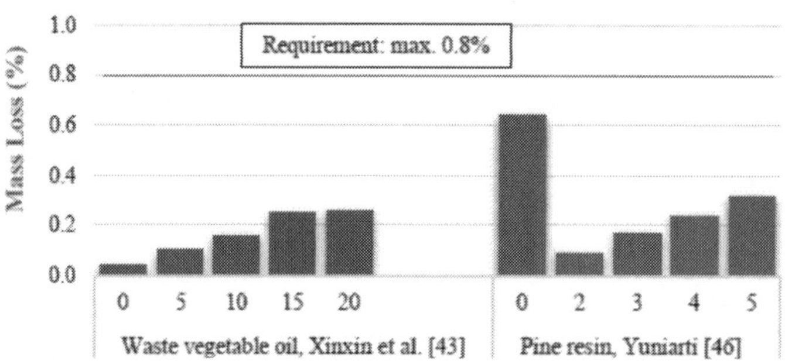

Figure 5. Relation between bio-binder and mass loss value.

3.3. Effect of Bio-binder on Rheological Properties of Asphalt

The rheological properties of asphalt are defined mainly by the dynamic shear rheometer (DSR) test. The outcomes of the DSR test are the complex shear modulus (G*) and phase angle (δ) of specimens. The complex shear modulus (G*) is related to rutting, which determines the binder's total resistance to deformation from shear loading [49]. Meanwhile, the phase angle (δ) is connected to fatigue cracking performance which determines lag time between shear strain and shear stress. Rutting occurs initially at the early stage of road pavement, whereas fatigue cracking accumulates during life period of the pavement.

Shorbagi et al. [42] found that 3.5% blending of the waste cooking oil to the aged asphalt from reclaimed asphalt pavement was the optimum percentage to almost achieve the identical rutting parameter of the 60/70 penetration grade control asphalt. With the use of waste cooking oil, rejuvenated asphalt showed lower values of the fatigue parameter compare to control asphalt. Thus it was indicated that the rejuvenated asphalt has better fatigue cracking resistance.

Xinxin et al. [43] concluded that adding more waste vegetable oil decreases the viscosity of rejuvenated asphalt and increases its workability. It also mentioned that as the percentage of waste vegetable oil increases, the resistance to rutting is reduced, but he fatigue life is extended.

The same results were obtained by Li et al. [44]. It can be asserted that G* value decrease as the waste cooking oil percentage increase. The lower complex modulus indicated a lessened rutting resistance ability, however, the phase angle increase with larger waste cooking oil percentage. The higher value of phase angle means the aged asphalt has better fatigue cracking resistance. The findings show that the waste cooking oil could restore the rheological properties of aged asphalt.

Determining the performance of asphalt rutting and fatigue life based on complex modulus and phase angle also studied by Al-Omari et al. [45]. It is reported that blending of waste cooking oil to virgin asphalt 60/70 penetration grade achieves in a decrease of G* value and an increase of δ parameter. Accordingly, the use of waste vegetable oil reduced the deformation resistance of asphalt, especially for the percentage of more than 4%, while improving the resistance from fatigue cracking. This is because the viscosity of the mixed asphalt decreases and becomes soft so that improve its ability against cracking. The influence of bio-binder on asphalt rheological properties are presented in Table 2.

Table 2. Effect of bio-binder on rheological properties of asphalt.

Bio-binder	Percentage	Results
Waste cooking oil	0; 2; 3; 3.5; 4 [42]	Mixing 3.5% waste cooking oil to aged bitumen made the resistance to rutting almost similar to the 60/70 control asphalt.
	0; 1; 2; 3; 4; 5 [43]	Whilst the percentage of waste cooking oil increase, resistance to rutting decreases but life of fatigue increases.
Waste vegetable oil	0; 5; 10; 15; 20 [44]	The use of waste vegetable oil decreases rejuvenated asphalt viscosity, increases its workability, falls down the resistance to rutting but improves the life of fatigue.
	0; 1; 2; 4; 6; 8 [45]	Due to addition of waste vegetable oil, resistance of asphalt to rutting decreases but resistance to fatigue cracking increases especially for the percentage of waste vegetable oil more than 4%.

3.4 Alternative Bituminous Binder for Future Sustainable Material

Based on some researchers' conclusion, the utilization of bio-oils produced from pyrolysis as pavement material is hoping. However, their application can be limited, and there is a lack of report that illustrates the applicability of using bio-oils as an asphalt modifier on a larger scale. This is because of techno-economic barriers associated with pyrolysis production.

During the production process, the thermal decomposition of biomass needs energy. According to Rasul and Jahirul [50], the external energy consumption for pyrolysis consist of three components. Dewatering the wet biomass to dry biomass is encompassed in the first component. The second is needed to heat dried biomass to the pyrolysis temperature. The last one is consumed to decompose the biomass during the reaction of pyrolysis. Due to the power needed and high production cost, pyrolysis products cannot compete economically with fossil fuel products [51]. Besides reactor, the cost of pyrolysis involve collecting of biomass, cutting into appropriate particle size, product handling and storage, etc.

Bio-oils cannot also be mixed directly with asphalt because of the water and a large amounts of volatile content. The water occupies about 30%-40% of bio-oil from the source materials or dehydration reaction during the pyrolysis process [52]. Bio-oils require upgrading procedures such as heating at a temperature around 95°C to remove water. Due to the possibility of moisture intrusion, the rheological properties of pavement influenced by water should be examined carefully before using the developed bio-binders.

Meanwhile, bio-oil application from plant seed is limited due to relatively low oil yields, time-consuming and labor-intensive. To obtain a higher oil yields, chemical extraction process can be employed by using n-hexane as a solvent. However, the constraint for the chemical extraction is the cost of solvent [53]. Therefore, it is needed to develop a bio-binder that is inexpensive and simple to apply. Waste cooking oil satisfy these requirements since it can be mixed directly with the virgin or aged asphalt or after removing its impurities [54]. Besides bio-oil from plant, alternative binders can also derive from blending of asphalt and polymer. Using polymer as modifier of asphalt is needed to explore broadly to overcome the problem related to non-biodegradable property of polimer which cause damage to the environment. Due to its huge amount and potentially induce environmental hazard, it is also necessary to investigate the use of waste engine oil as asphalt softening agent.

As stated previously, pine resin can also be used directly after a simple filtration process to remove impurities. Another advantage of using pine resin is its availability. Indonesian Forest State Corporation mentioned that in 2012 the production of pine resin was 101,266 tons come from 163,150 hectares of pine forest [55]. The potential for increase pine resin production is enormous due to a massive land area eligible for reforestation in Indonesia. Rubber latex also has a good prospect to be used as asphalt modifier because of inexpensive and ready to use. Therefore, there is a great potential to conduct research and developments on the applicability of using crude pine resin and rubber latex as modifier of asphalt binder. The collaboration research between academic and industry will be crucial to get more efficient and effective results that can enhance the quality of existing asphalt products.

4. Conclusion

Based on the study, it is concluded that there are enormous of alternative binder that can be used as modifier or rejuvenator since it has similar chemical properties with petroleum asphalt. Physical properties of virgin, aged, and reclaimed asphalt are influenced by the type and amount of bio-binder applied. Modified or rejuvenated asphalt becomes soften as penetration and ductility increase while softening point decreases with additional bio-binder percentages. The requirement of asphalt 60/70 is satisfied in the range of 1-10% bio-binder. It also highlights that the rutting resistance decreased, and the addition of bio-binder can significantly increase fatigue resistance. From an economics point of view, the possibility of using certain economical bio-binder should be considered for conducting research and development in this field. Because the long life performance of pavement is strongly affected by the presence of water, it is suggested to examine the moisture susceptibility of asphalt mixture containing alternative binder for future research.

5. References

[1] Wen H, Bushal S and Wen B, 2013 Laboratory evaluation of waste cooking oil-based bio asphalt as an alternative binder for hot mix asphalt *J. Mater. Civ. Eng.* **25** 1432-1437

[2] You Z, Beale J M, Fini H E, Goh S W, and Colbert B, 2011 Evaluation of low-temperature binder properties of warm mix asphalt, extracted and recovery RAP and RAS, and bio asphalt *J. Mater. Civ. Eng.* **23** 1569-1574

[3] Raouf M A and Williams R C, 2010 Temperature susceptibility of non-petroleum binders derived from bio-oils: *The 7th Asia Pacific Conference on Transportation and the Environment* Semarang Indonesia

[4] Huang S C, Salomon D, and Haddock J E, 2012 Alternative binders for sustainable asphalt pavements: work-shop introduction *Transportation Research Circular E-C165* Transportation Research Board Washington DC

[5] Yang X, Mills-Beale J and You Z, 2016 Chemical characterization and oxidative aging of bio-asphalt and its compatibility with petroleum asphalt *J. Clean. Prod.* **142(4)** 1837-1847

[6] Neubert T C, 1991 *Asphalt containing gilsonite, reactive oil and elastomer,* Patent Number 5023282, United States Patent & Trademark Office

[7] Bailey H and Phillips P, 2010 *Asphalt rejuvenation* United Kingdom Patent Application No. GB 2462371 A (Cambridge United Kingdom)

[8] Freisthler M, 2006 *Bituminous composition* Patent Number 7008670 B1, United States Patent & Trademark Office

[9] Nigen-Chaidron S and Porot L, 2010 *Rejuvenating agent and process for the recycling of asphalt, Patent Number* US 7811372 B2 United States Patent & Trademark Office

[10] Mohan D, Pittman C U and Steele P H, 2006 Pyrolysis of wood/biomass for bio-oil: A critical review *Energy Fuels* **20(3)** 848-889

[11] Kim S J, Jung S H, Kim, J S 2010 Fast pyrolysis of palm kernel shells: Influence of operation parameters on the bio-oil yield and the yield of phenol and phenolic compounds *Bioresource Tech.* **101** 9294-9300

[12] Abnisa F, Arami-Niya A, Wan Daud W M A, Sahu J N and Noor I M 2013 Utilization of oil palm tree residues to produce bio-oil and bio-char via pyrolysis *Energy Convers. Manage.* **76** 1073-1082

[13] Williams R C, Satrio J, Rover M, Brown R C and Teng S 2009 Utilization of fractionated bio-oil in asphalt, *88th Annual Meeting of the Transportation Research Board* Washington *DC*

[14] Hill D R and Jennings A A 2011 *Bioasphalt from urban yard waste carbonization a student study, Case Western Reserve* (University Ohio Department of Transportation)

[15] Yang X, You Z, Dai Q and Mills-Beale J 2014 Mechanical performance of asphalt mixtures modified by bio-oils derived from waste wood resources *Constr. Build. Mater.* **51** 424-431

[16] Zheng J-L and Kong Y-P 2010 Spray combustion properties of fast pyrolysis bio-oil produced from rice husk, *Energy Conv. Manage.* **51(1)** 182-188

[17] Raouf M A and Williams C R 2010 General rheological properties of fractionated switchgrass bio-oil as a pavement material *Road Mater. Pave. Design*, **11** 325-353

[18] Ren Y, Zhang L, Duan W, Han Z, Guo J, Heydenrych M D, Zhang A, Nie K, Tan T and Liu L 2020 Performance of bitumen coating sheet using biomass pyrolysis oil *J. Air & Waste Manage. Ass.* **70(2)** 219-227

[19] Lei Y, Wang H, Chen X, Yang X, You Z, Dong S and Gao J 2018 Shear property, high-temperature rheological performance and low-temperature flexibility of asphalt mastics modified with bio-oil *Constr. Build. Mater.* **174** 30-37

[20] Gonçalves G C, Pereira N C and Veit M T 2016 Production of bio-oil and activated carbon from sugarcane bagasse and molasses *Biomass and Bioenergy* **85** 178-186

[21] Ahmed N, Zeeshan M, Iqbal N, Farooq M Z, Shah S A 2018 Investigation on bio-oil yield and quality with scrap tire addition in sugarcane bagasse pyrolysis *J. Clean. Prod.* **196** 927-934

[22] Yang X and You Z 2015 High temperature performance evaluation of bio-oil modified asphalt binders using the DSR and MSCR tests *Constr. Build. Mater.* **76** 380-387

[23] Yusuf A K, Mamza P A P, Ahmed A S and Agunwa U 2015 Extraction and characterization of castor seed oil from wild *Ricinus communis Int. J. Science, Environ. Tech.* **4(5)** 1392-1404
[24] Patel V R, Durmancas G G, Viswanath L C K, Maples R and Subong B J J 2016 Castor oil: properties, uses and optimization of processing parameters in commercial production *Lipid Insights* **9** 1-12
[25] Arisanu A O 2013 Mechanical continuous oil expression from oil seeds: oil yield and press capacity. *International Conference Computational Mechanics and Virtual Engineering COMEC 2013* Brasov Romania
[26] Zeng M, Pan H, Zhao Y and Tian W 2016 Evaluation of asphalt binder containing castor oil-based bioasphalt using conventional tests *Constr. Build. Mater.* **126** 537-543
[27] Ahmad K A, Abdullah M E, Hassan N A, Usman N and Ambak K 2017 Investigating the feasibility of using jatropha curcas oil (JCO) as bio based rejuvenator in reclaimed asphalt pavement (RAP) *MATEC Web of Conferences* EDP Sciences
[28] Portugal A C X, Lucena L C D F L, Lucena A E D F L and Costa D 2017 Rheological performance of soybean in asphalt binder modification *Road Mater. Pave. Design* **19** 768-782
[29] Elkashef M, Podolsky J, Williams R C and Cochran E 2017 Preliminary examination of soybean oil derived material as a potential rejuvenator through superpave criteria and asphalt bitumen rheology *Constr. Build. Mater.* **149** 826-836
[30] Somé S C, Pavoine A and Chailleux E 2016 Evaluation of the potential use of waste sunflower and rapeseed oils-modified natural bitumen as binders for asphalt pavement design *Int. J. Pave. Res. Tech.* **9** 368-375
[31] Yuniarti R 2014 Pengaruh minyak biji nyamplung pada bio-flux oil sebagai modifier asbuton butiran terhadap kinerja asbuton campuran panas *J. Teknik Sipil ITB*, **21(3)** 251-258 (in Bahasa Indonesia)
[32] Asli H, Ahmadinia E, Zargar M, Karim M R 2012 Investigation on physical properties of waste cooking oil-rejuvenated bitumen binder *Constr. Build. Mater.* **37** 398-405
[33] Sun Z, Yi J, Huang Y, Feng D and Guo C 2016 Properties of asphalt binder modified by bio-oil derived from waste cooking oil *Constr. Build. Mater.* **102** 496-504
[34] Mannu A, Garroni S, Porras J I, Mele A 2020 Available technologies and materials for waste cooking oil recycling *Processes* **8** 366
[35] Yuniarti R 2015 Performance of bio-flux oil as modifier of buton granular asphalt in asphalt concrete-wearing course *Journal of JSCE* **3(1)** 33-44
[36] Zahoor M, Nizamuddin S, Madapusi S and Giustozzi F 2021 Sustainable asphalt rejuvenation using waste cooking oil: A comprehensive review *J. Clean. Prod.* **278** 123304
[37] Zaumanis M, Mallick R B and Frank R 2014 Determining optimum rejuvenator dose for asphalt recycling based on superpave performance grade specifications *Constr. Build. Mater.* **69** 159-166
[38] Ingrassia L P, Lu X, Ferrotti G and Canestrari F 2019 Chemical and rheological investigation on the short-and long-term aging properties of bio-binders for road pavements *Constr. Build. Mater.* **217** 518-529
[39] Gong M H, Yang J, Zhang J Y, Zhu H R and Tong T Z 2016 Physical-chemical properties of aged asphalt rejuvenated by bio-oil derived from biodiesel residue *Constr. Build. Mater.* **10** 35-45
[40] Joni H H, Al-Rubaee R H A and Al-zerkani M A 2019 Rejuvenation of aged asphalt binder extracted from reclaimed asphalt pavement using waste vegetable and engine oils *Case Studies Constr. Mater.* **11** e00279
[41] Kaseer F, Martin A E, Arámbula-Mercado E 2019 Use of recycling agents in asphalt mixtures with high recycled materials contents in the united states: A literature review *Constr. Build. Mater.* **211** 974-987
[42] El-Shorbagy A M, El-Badawy S M and Gabr A R 2019 Investigation of waste oils as rejuvenators of aged bitumen for sustainable pavement *Constr. Build. Mater.* **220** 228-237

[43] Li H, Dong B, Wang W, Zhao G, Guo P and Ma Q 2019 Effect of waste engine oil and waste cooking oil on performance improvement of aged asphalt *Appl. Sci.* **9** 1767
[44] Xinxin C, Xuejuan C, Boming T, Yuanyuan W and Xiaolong L 2018 Investigation on possibility of waste vegetable oil rejuvenating aged asphalt *Appl. Sci.* **8** 765
[45] Al-Omari A A, Khedaywi T S and Khasawneh M A 2018 Laboratory characterization of asphalt binders modified with waste vegetable oil using superpave specifications *Int. J. Pave. Res. Tech.* **11** 68-76
[46] Yuniarti R 2019 Resistance to degradation of porous asphalt mixture using pine resin as asphalt modifier *JJCE* **13(1)** 113-123
[47] Perhutani P 2007 *Gum Rosin Turpentine Factory* Perum Perhutani Unit II East Java
[48] Directorate General of Bina Marga 2018 *General Specification* (Ministry of Public Works, Jakarta)
[49] Azahar W N A W, Bujang M, Jaya R P, Hainin M R, Ngadi N, Abdullah M M A B 2016 Performance of waste cooking oil in asphalt binder modification *Key Eng. Mater.* **700** 216-226
[50] Rasul M G and Jahirul M I, 2012 Recent developments in biomass pyrolysis for bio-fuel production: Its potential for commercial applications, Recent researches in environmental and geological sciences, *IOP Conference Paper* 256-265
[51] Jahirul M I, Rasul M G, Chowdhury A A and Ashwath N 2012 Biofuels Production through Biomass Pyrolysis-A Technological Review *Energies* **5** 4952-5001
[52] Gupta S, Mondal P, Borugadda V B and Dalai A K 2021 Advances in upgradation of pyrolysis bio-oil and biochar towards improvement in bio-refinery economics: A comprehensive review *Environ. Tech. & Innovation* **21** 101276
[53] Bhuiya M M K, Rasul M G, Khan M M K, Ashwath, N, Azad A K and Mofijur M 2015 Optimization of oil extraction process from Australian native beauty leaf seed (*Calophyllum innophyllum*). 7th International Conference on Applied Energy- ICAE2015 *Energy Procedia* **75** 56-61
[54] Azahar W N A W, Jaya R P, Hainin M R, Bujang M and Ngadi N 2017 Mechanical performance of asphaltic concrete incorporating untreated and treated waste cooking oil *Constr. Build. Mater.* **150** 653-663
[55] Perhutani P 2012 *Consolidation business process towards excellent Perhutani* (Annual Report, Jakarta)

Acknowledgment
Authors would like to thank Institut Teknologi Sepuluh Nopember and University of Mataram for all kinds of supports.

Optimizing the Physical Properties of Waste Denim Fiber-Modified Rubberized Bitumen Through Response Surface Methodology

A M Al-Sabaeei[1*], M Napiah[1], M Al Salaheen[1], R M Badri[2], S Noura[3], M I Khan[1], T M Al-Bahr[4] and K Alzubi[1]

[1]Department of Civil & Environmental Engineering, Universiti Teknologi PETRONAS, 32610 Bandar Seri Iskandar, Perak, Malaysia.
[2]Depertment of Civil Engineering, Dijlah University College, 00964, Baghdad, Iraq.
[3]Department of Civil Engineering, Universiti Putra Malaysia, 43400, Seri Kambangan, Selangor, Malaysia.
[4]School of Civil Engineering, Faculty of Engineering, Universiti Teknologi Malaysia, 81310, Skudai, Johor Bahru, Malaysia.
*Corresponding author: abdulnaser_17005477@utp.edu.my; abdulnaser_mohd@tu.edu.ye

Abstract. In this study, waste denim fiber (WDF) was used as a modifier to enhance the properties of bitumen incorporating tire pyrolysis oil (TPO). The designing of the experiment, analysis data, modeling and multi-objective optimization performed using response surface methodology (RSM). Base bitumen was modified with 0%, 5%, 10% and 15% TPO and 0%, 2%, 4% and 6% of WDF using high shear mixer. The interaction effects of TPO and WDF on the penetration, softening point and temperature susceptibility of bitumen were evaluated. Results showed that the addition of TPO and/or WDF has a significant effect on the consistency and temperature susceptibility of base bitumen. RSM analysis exhibited high degrees of correlation (R^2) of 0.982, 0.973 and 0.947 for the dependent variables' penetration, softening point and temperature susceptibility, respectively. That indicates there is an adequate agreement between models developed by RSM and experimental results. The numerical multi-objective optimization showed that optimum properties can be achieved with 8.4% of TPO and 6% of WDF.

1. Introduction

The modified bitumen has become popular during last few decades as an alternative for base bitumen to improve the performance and durability of asphalt mixture. Thermoplastic elastomers are used as a one of the most common polymers in bitumen modification to improve the resistance of asphalt mixtures to main distresses by increasing their temperature sensitivity resistance [1, 2]. Rubber in form of crumb rubber (CR) is one of these elastomers has been widely studied as a bitumen modifier due to the economical saving, environmental concerns and acceptable performance obtained that encouraged the use of waste rubber in bitumen [1, 3]. In contrast, the incorporation of CR in the bitumen modification has several disadvantages such as low storage stability as a result the non-dissolve rubber particles, poor workability due to of high stiffness of rubberized bitumen and high energy consumption and CO_2 emissions during the modification and construction processes as a result to the high operating temperatures required which could reach 180 °C [4]. Tire pyrolysis oil (TPO) has been proposed to overcome the disadvantages of conventional CR as a modifier for rubberized bitumen technology. Fini et al. [5] evaluated the possibility of use the TPO as a bitumen modifier. It was found that TPO enhances the physical, rheological and aging properties of bitumen in low and intermediate temperatures. However, a reduction was noticed in the viscosity and high-temperature rheological properties due to the lower molecular weight and higher aromatics content in the TPO. Similar findings on the effects of TPO on the physical and rheological properties of bitumen have been reported [3, 4, 6]. Recent study has also showed that the incorporation of TPO with other modifiers such as polymers and nanomaterials exhibited an improvement in the physical and rheological performance of TPO-modified bitumen particularly at high temperature conditions [7]. Therefore, further evaluation on the incorporation of various polymers and/or fibers as modifiers to improve the performance of TPO-modified bitumen has been expressed.

Lately, there is a tendency for utilizing a combination modifiers, particularly those have a hybrid influence on the environment and performance of bitumen toward more sustainable modified bitumen [8, 9]. Waste denim fiber (WDF) has been proposed as an alternative waste fiber to improve the concrete and building materials performance and contribute to solve the environmental issue due to the disposal of denim waste in a not efficient manner [10-12]. Waste denim fiber was introduced by Al-Sabaeei et al. [8] for the first time as a bitumen modifier. It was found that WDF improved the viscoelastic and rutting performance of bitumen binder at high temperatures conditions. This finding observed to be consistent with another study showed that the use of WDF exhibited a remarkable enhancement in the elasticity and rutting performance of bitumen [13]. Such improvement was attributed to the high molecular weight and tensile strength of WDF, in addition to the lower water content and the adequate content of cellulose that contribute to desired adhesion, durability and performance of bitumen and asphalt mixtures [14]. Therefore, WDF is one of the strongest additives has been investigated in this study for possible improvement the properties of TPO-modified bitumen for durable and sustainable composite bitumen binder.

Recently, response surface methodology (RSM) has been widely utilized in various experimental research including pavement materials in order to design the experiments, develop predictive modells and perform multi-objective optimization [15]. Also recently, the utilization of RSM in pavement materials has been increased due to the excellent capability of RSM to establish the correlation among one or more independent variables and dependent variable (response) at a few experimental runs [16, 17]. For example, RSM has been successfully applied for modelling and optimization of the nano-silica composite asphalt mixtures performance by Bala et al. [15]. Another study has shown the excellence of using RSM in the modelling and optimizing the performance and cost of liquid rubber modified bitumen and RSM also showed to be adequate for producing analytical functions among the design parameters with providing an accurate future prediction [3]. Therefore, this study aims to use RSM to analyse, develope predictive models and perform multi-objective optimization for the physical and consistency properties of WDF/TPO composite bitumen.

2. Materials and methods

2.1. Materials

Conventional bitumen has been used in this study is 60-70 penetration-grade, that was provided by PETRONAS Refinery, Malaysia. Conventional and advanced properties of unmodified bitumen were checked to ensure fulfilling the Malaysian specifications. The TPO was obtained as the sludge of the process of conversion of rubber waste into bio-gas in the bio-fuel manufactures. TPO was provided by Tyre Oil (M) Sdn. Bhd, Malaysia. The TPO percentages used in this study are 0, 5, 10 and 15% by total weight of blend. The WDF used was obtained from the waste of Jeans which was cut to small pieces (2000 μm as an average) using a Knife mill for 10 cycles of 15 s for each cycle. The characteristics of WDF showed a density of 1.55 g/cm^3, pH value of 6.8 and moisture content of 4.9%. WDF used in this study at 0, 2, 4 and 6 % by the weight of blends. Figure 1 shows the appearance of TPO and WDF.

IOP Conf. Series: Earth and Environmental Science **971** (2022) 012014 doi:10.1088/1755-1315/971/1/012014

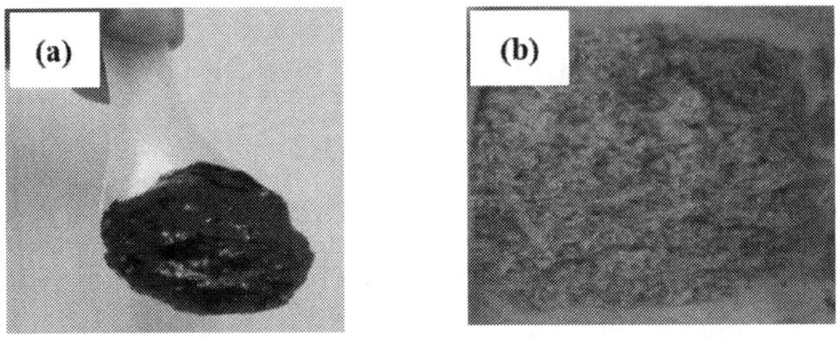

Figure 1. (a) Tire pyrolysis oil (TPO) (b) Waste denim fiber (WDF).

2.2. Methods

2.2.1. Preparing of WDF-modified rubberized bitumen samples. Base bitumen was kept at 160 °C in an oven to be sufficiently fluid for mixing with TPO and/or WDF. A certain amount of TPO and/or WDF by total weight of the blend, which were recommended in accordance with RSM experimental design, were added to virgin bitumen and manually blended for about two minutes. The homogenous blends of TPO/WDF composite bitumen were achieved at mixing temperature of 160 °C for 90 minutes at shear rate of 4000 rpm. The mixing parameters were selected in this study based on the trial and error method which comes up with consistent blends. Finally, prepared blends were carefully stored in tight containers for being use for further investigations.

2.2.2. Penetration test. The penetration test was carried out at 25 °C following the ASTM D5-13 standard. In order to investigate the consistency of base and TPO and/or WDF composite bitumen binders at the intermediate temperature, the penetration test was performed for 5 s under 100g load. Resulted penetration depths were recorded as the penetration values of the specimens of binders.

2.2.3. Softening point test. To determine the effects of TPO and/or WDF on the consistency of the bitumen at high-temperature, the softening point temperatures were determined for all base and TPO/WDF modified binders prepared in this study accordance to ASTM D36 standard. The softening point obtained as the temperature at which binder cannot longer stand and support a steel ball with a weight of 3.5 g at the constant rate of temperature (5 °C/min).

2.2.4. Penetration index. The temperature susceptibility of base and TPO and/or WDF modified bitumen was assessed according to the penetration index (PI) obtained based on the penetration and softening point results. Equation (1) was used for calculating the PI values of base and modified binders.

$$\boldsymbol{PI} = \frac{\mathbf{1952 - 500 * \mathit{LogPen} - 20\ \mathit{SP}}}{\mathbf{50 * \mathit{LOGPen} - \mathit{SP} - 120}} \qquad (1)$$

where pen is the penetration value and SP is the softening point in °C.

2.2.5. Design of experiment and method of analysis using RSM. In order to conduct the response surface analysis, the Design Expert software was used. The user-defined design (UDD) method was applied to design the experiment, statistical analysis of data and modelling and establishing the correlation among the independent variables (TPO and WDF contents) and responses (penetration, softening point and penetration index). UDD has been selected due to its flexibility to be used for different nature of data such as for data has been collected. In addition, the higher number of experimental runs can be resulted compared to other design approach which enhance the understanding and compression the effects of various concentrations of materials separately and combined together on the responses. The independent variables considered in this study are: TPO at four levels of 0%, 5%, 10% and 15% and WDF at 0, 2, 4 and 6% by weight of blends, while the dependent variables are penetration, softening points and PI. The experimental design layout is presented in Table 1 with 16 experiments were conducted to evaluate each response.

Table 1. Experimental design matrix.

Run	Independent variable		Response		
	A: TPO (%)	B: WDF (%)	Penetration (dmm)	Softening point (°C)	Penetration index
1	0	0	60	49	-1.03
2	0	2	40	55	-0.54
3	0	4	37	56	-0.49
4	0	6	32	57	-0.58
5	5	0	50	47	-1.98
6	5	2	29	57	-0.78
7	5	4	24	58	-0.93
8	5	6	17	59	-1.33
9	10	0	57	44	-2.55
10	10	2	51	48	-1.67
11	10	4	38	51	-1.54
12	10	6	27	53	-1.74
13	15	0	60	43	-2.74
14	15	2	55	47	-1.77
15	15	4	51	50	-1.16
16	15	6	44	49	-1.73

The suitable higher-order polynomial regression models, that can represent the complex relationship among the independent and dependent variables and establish the optimum conditions, were used in this study to fit the data obtained from the laboratory experiments after verifying that the linear regression is inappropriate to represent the data. The second order function suggested by [18] was applied as shown below:

$$y = \beta_0 + \sum_{i=1}^{k} \beta_i X_i + \sum_{i=1}^{k} \beta_{ii} X_i^2 + \sum_{i<j} \sum \beta_{ij} X_i X_j + \varepsilon, \quad (2)$$

where y is the predicted dependent variable, β is the y-intercept, and k is the number of variables used in the analysis. X_i and X_j are the coded values of variables, where i and j are the coefficient for the linear and quadratic equations, respectively. ε denotes the error.

3. Results and discussions

3.1. Statistical analysis results

In order to have sufficient understanding on the relationships of TPO and/or WDF with physical properties of bitumen as responses, regression analysis has been conducted. Then, fitted cubic or quadratic models have been developed to represent the penetration, softening point and penetration index of TPO/WDF composite bitumen. Models were selected in accordance with the RSM suggestions to avoid the aliased models by the Design-Expert software. Equations 3-5 present the developed models can be used for prediction the consistency and temperature sensitivity of base and composite modified binders. Developed models can be used within the limits of materials used in this study such as concentration of TPO between 0% and 15% and WDF 0-6% with considering the bitumen modification parameters applied in this study.

$$Penetration = 33.56 + 25 * A - 12.06 * B + 2.39 * A * B + 12.09 * A^2 + 3.09 * B^2 + 5.74 * A^2 * B - 5.91 * A * B^2 - 16.59 * A^3 - 4.5 * B^3 \quad (3)$$

$$Softening\ point = 53.94 - 10.44 * A + 3.81 * B - 0.43 * A * B - 1.55 * A^2 - 2.95 * B^2 - 1.77 * A^2 * B + 0.59 * A * B^2 + 6.61 * A^3 + 1.55 * B^3 \quad (4)$$

$$Penetration\ index = -1.23 - 0.63 * A + 0.35 * B + 0.17 * A * B + 0.35 * A^2 - 0.67 * B^2 \quad (5)$$

Tables 2-4 exhibit the analysis of variance (ANOVA) results for the developed models of penetration, softening point and penetration index, respectively. It can be noticed that, for all models F-value show high values which indicate there is only 0.01% choice that the high F-value could occur due to noise. It can also be seen that the P values of all developed models and most of their terms lower than 0.05, which indicate the statistically significant of the selected models to present the experimental data within the confidence interval of 95%. The degree of fitness between the independent and dependent variables can be checked by the coefficient of correlation (R^2). As shown in tables below, all responses show R^2 value of 0.94, which indicate there is almost 94% of the change in the responses is due to the intervention of TPO and/or WDF as independent variables in this study. The adj. R^2 of all developed models also show a difference less than 0.2 relative to pred. R^2 which was recommended by [18] to ensure the capability of models to predict the responses with adequate accuracy. The high values of R^2, adj. R^2 and pred. R^2 indicate that the RSM approach was selected for establish the correlation and develop models among the TPO and/or WDF and physical properties of bitumen was suitable and resulted with low errors. The low values of standard deviation and coefficient of variation for all developed models also reveals the lower variability between the predicted physical properties from the developed models and that obtained from the experimental work.

Table 2. ANOVA result for penetration response.

Source	Sum of squares	df	Mean square	F Value	p-value prob > F	
Model	2710.14	9	301.13	36.24	0.0002	Significant
A: TPO	390.24	1	390.24	46.96	0.0005	
B: WDF	90.85	1	90.85	10.93	0.0163	
AB	28.09	1	28.09	3.38	0.1156	$R^2 = 0.982$
A^2	462.25	1	462.25	55.63	0.0003	Adj. R^2 =0.955
B^2	30.25	1	30.25	3.64	0.1050	Pred. R^2 =
A^2B	57.80	1	57.80	6.96	0.0387	Std. Dev. = 2.88
AB^2	61.25	1	61.25	7.37	0.0349	C.V. = 6.86%
A^3	174.05	1	174.05	20.94	0.0038	
B^3	12.80	1	12.80	1.54	0.2609	

Table 3. ANOVA results for softening point.

Source	Sum of squares	df	Mean square	F Value	p-value prob > F	
Model	379.40	9	42.16	24.01	0.0005	Significant
A: TPO	68.02	1	68.02	38.74	0.0008	
B: WDF	9.08	1	9.08	5.17	0.0634	
AB	0.90	1	0.90	0.51	0.5004	R^2 = 0.973
A^2	7.56	1	7.56	4.31	0.0833	Adj. R^2 =0.933
B^2	27.56	1	27.56	15.70	0.0074	Pred. R^2 = 0.824
A^2B	5.51	1	5.51	3.14	0.1268	Std. Dev. = 1.33
AB^2	0.61	1	0.61	0.35	0.5763	C.V. = 2.58%
A^3	27.61	1	27.61	15.73	0.0074	
B^3	1.51	1	1.51	0.86	0.3892	

Table 4. ANOVA results for penetration index.

Source	Sum of squares	df	Mean square	F Value	p-value prob > F	
Model	6.56	5	1.31	35.35	< 0.0001	Significant
A: TPO	3.50	1	3.50	94.35	< 0.0001	R^2 = 0.947
B: WDF	1.10	1	1.10	29.56	0.0003	Adj. R^2 =0.920
AB	0.14	1	0.14	3.77	0.0809	Pred. R^2 = 0.872
A^2	0.38	1	0.38	10.35	0.0092	Std. Dev. = 0.19
B^2	1.44	1	1.44	38.73	< 0.0001	C.V. = 13.65%

3.2. Response surface contour plots

The relationships of experimental (actual) values obtained from the laboratory versus the predict values from the developed models of penetration, softening point and PI are shown in Figure 2. It can be observed that almost all data points for the three depended variables (responses) are disseminated very close to the equality lines, which revealing the adequate agreement between the predicted values and experimental results indicating the satisfactory accuracy of developed models. The plots aid in understanding the behaviour of the developed models.

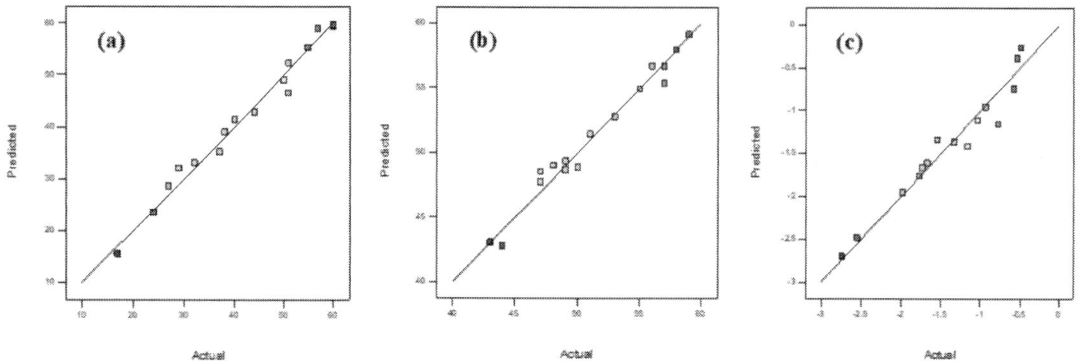

Figure. 2 Actual versus predicted plots of base and modified bitumen (a) penetration (b) softening point (c) penetration index.

Figures 3 and 4 show the 2D and 3D surface plots of the penetration, softening point and penetration index models at various TPO and/or WDF contents. Overall, it can be observed that the addition TPO increases the penetration value and decreases the softening point and penetration index values. Such finding could be due to the higher light components in TPO relative to base bitumen that results in more viscous behaviour and lower stiff binder. On the other hand, addition WDF to base and TPO-modified bitumen binders resulted in a significant reduction in the penetration values and increasing in the softening points and penetration index. This could be ascribed to the higher molecular weight of WDF that enhanced the stiffness of base and TPO-modified bitumen. In details, the curvature nature of the contour lines indicates the reasonable correlation among the variables and the red zone in the 2D and 3D plots reveal the highest values of penetration, softening point and PI. It can also be noted that up to 5% TPO there is an improvement in the physical properties of modified bitumen including the penetration value, which could be due to the compatibility of TPO with base bitumen at lower content. In terms of bitumen modified with high content of TPO (more than 5%), a negative effect was shown on the physical properties of bitumen. That could be ascribed to the higher light components in TPO compared to base bitumen. But the addition of WDF restored the properties that have been affected due to the higher concentration of TPO. The findings from this study in terms of the effects of TPO and WDF on the behaviour of bitumen is agreed with what was reported in literature [4, 8].

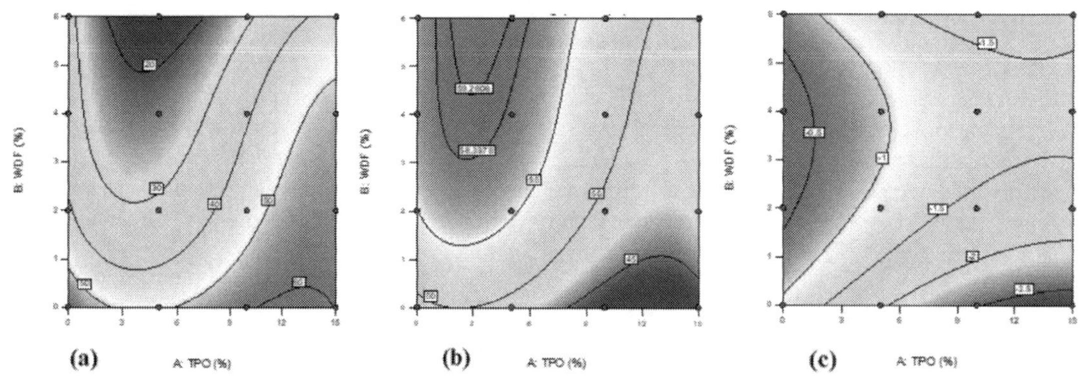

Figure. 3 2D contour plots of base and modified bitumen (a) penetration (b) softening point (c) penetration index.

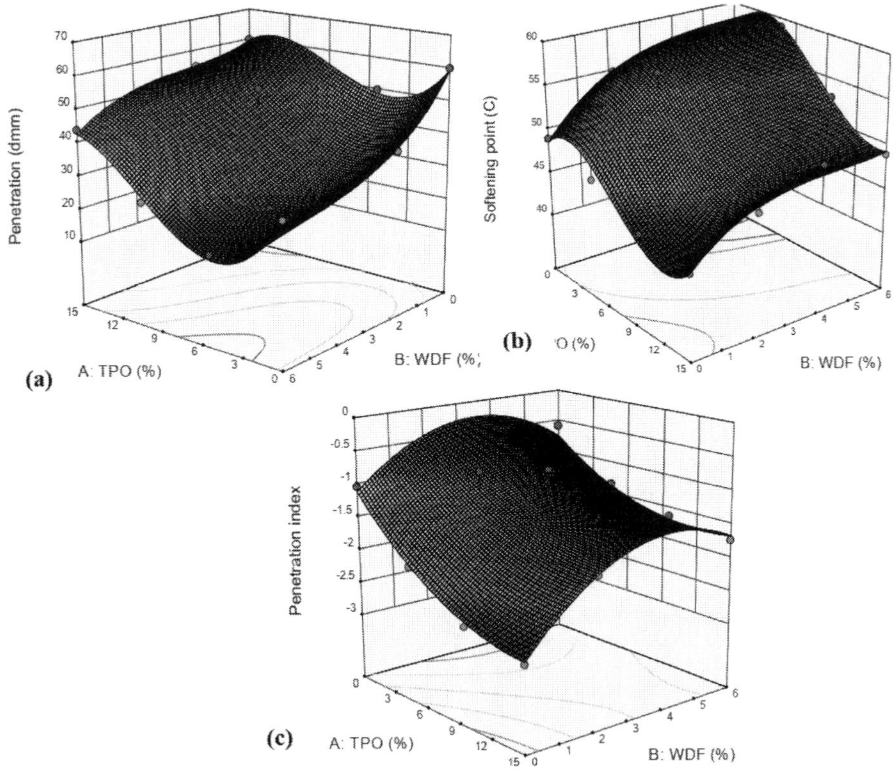

Figure. 4 3D surface plots of base and modified bitumen (a) penetration (b) softening point (c) penetration index.

3.3. Multi-objective optimization of responses

In order to achieve the optimal TPO and WDF contents that result in best physical properties (lower penetration and higher softening and PI values), multi-objective optimization has been conducted. The criteria of the optimization were set as following: TPO maximize, WDF in the range of 0% - 6%, penetration value minimize, softening point maximize and PI in the range of -2 to +2. The optimal situation was found at 8.4% TPO and 6% WDF at a desirability of 0.75. The experimental validation has been performed to the optimal binders and the difference between the responses obtained from the RSM prediction and validation experiment is less than 5%. That indicates the adequacy of developed models to be used for prediction the physical properties of TPO and/or WDF modified bitumen with the limitations applied in this study.

4. Conclusion

In this research, the RSM approach was used to design, model and optimize the separate and combined effects of TPO and WDF on the empirical properties of base bitumen binder. Based on the experimental and statistical analysis and modelling carried out the following points can be concluded:

- The addition TPO exhibited an improvement in the physical properties of base bitumen up to 5%, while at TPO content higher than 5% reduction of physical properties has been observed.
- The addition WDF into base and TPO-modified bitumen binders showed a significant improvement in penetration, softening point and temperature susceptibility with the increasing of WDF.
- RSM statistical analysis showed that the predicted physical properties of TPO and/or WDF modified bitumen from the developed models are agreed with the experimental characteristics obtained from the laboratory work.

- The correlation coefficients (R^2) showed values greater than 0.94 for all developed predictive models, which indicate that selected models are appropriate for fitting the experimental data toward more accurate prediction.
- Based on the multi-objective optimization, the optimal physical properties of modified bitumen were achieved at the 6% WDF and 8.4% TPO.

5. References

[1] Presti D L, Izquierdo M and del Barco Carrión A J 2018 Towards storage-stable high-content recycled tyre rubber modified bitumen *Constr. Build. Mater.* **172** 106-111

[2] Al-Sabaeei A, Yussof N I M, Napiah M and Sutanto M 2019 A review of using natural rubber in the modification of bitumen and asphalt mixtures used for road construction *J. Tek.* **81** 81-88

[3] Del Barco Carrion A J, Subhy A, Rodriguez M A I and Presti D L 2020 Optimisation of liquid rubber modified bitumen for road pavements and roofing applications *Constr. Build. Mater.* **249** 118630

[4] Al-Sabaeei A, Napiah M, Sutanto M, Alaloul W and Bala N 2021 Effects of Tire Pyrolysis Oil (TPO) on the Rheological Properties of Bitumen in *IOP Conference Series: Materials Science and Engineering* IOP Publishing

[5] Fini E H , Hosseinnezhad S, Oldham D J and Sharma B K 2016 Investigating the effectiveness of liquid rubber as a modifier for asphalt binder *Road Mater. and Pavement Design* **17** 825-840

[6] Wu X, S Wang and R Dong 2016 Lightly pyrolyzed tire rubber used as potential asphalt alternative *Constr. Build. Mater.* **112** 623-628

[7] Al-Sabaeei A M, Napiah M B, Sutanto M H, Habib N Z, Bala N, Kumalasari I and Ghaleb A 2021 Application of nano silica particles to improve high-temperature rheological performance of tyre pyrolysis oil-modified bitumen *Road Mater. and Pavement Design* p. 1-19

[8] Al-Sabaeei A, Napiah M, Sutanto M and Alaloul W 2019 Effects of Waste Denim Fibre (WDF) on the physical and rheological properties of bitumen in *IOP Conference Series: Mater. Sci. and Eng.* IOP Publishing

[9] Günay E 2018 *Natural and Artificial Fiber-Reinforced Composites as Renewable Sources* (London: IntechOpen)

[10] Paul R 2015 *Denim: manufacture, finishing and applications* (Elsevier)

[11] Peña-Pichardo P, Martínez-Barrera G, Martínez-López M, Ureña-Núñez F and dos Reis J M L 2018 Recovery of cotton fibers from waste Blue-Jeans and its use in polyester concrete *Constr. Build. Mater.* **177** 409-416

[12] Temmink R, Baghaei B and Skrifvars M 2018 Development of biocomposites from denim waste and thermoset bio-resins for structural applications *Composites Part A: Applied Science and Manufacturing* **106** 59-69

[13] Al-Sabaeei A M, Napiah M B, Sutanto M H, Alaloul W S, Zoorob S E and Usman, A 2020 Influence of nanosilica particles on the high-temperature performance of waste denim fibre-modified bitumen *Int. J. Pavement Eng.* 1-14

[14] Al-Sabaeei A, Napiah M, Sutanto M and Alaloul W 2018 An overview on using the waste denim fiber as a bitumen modifier for sustainable road construction *Inter. J. of Civil Eng. and Tech.* **13** 1194-1200

[15] Bala N, Napiah M and Kamaruddin I 2020 Nanosilica composite asphalt mixtures performance-based design and optimisation using response surface methodology *Int. J. Pavement Eng.* **21** 29-40

[16] Al-Sabaeei A, Napiah M, Sutanto M, Alaloul W and Ghaleb A A S 2020 Prediction of Rheological Properties of Bio-asphalt Binders Through Response Surface Methodology in *IOP Conference Series: Earth and Environmental Sci.* IOP Publishing

[17] Khairuddin F H, Alamawi M Y, Yusoff N I M, Badri K H, Ceylan H and Tawil S N M 2019 Physicochemical and thermal analyses of polyurethane modified bitumen incorporated with Cecabase and Rediset: optimization using response surface methodology *Fuel* **254** 115662

[18] Montgomery D C 2017 *Design and analysis of experiments* (John wiley & sons)

IOP Conf. Series: Earth and Environmental Science **971** (2022) 012015 doi:10.1088/1755-1315/971/1/012015

Effect of Processing Parameters on Properties of Bitumen Emulsions

S A Ghafar[1,2], M N M Warid[1*], N A Hassan[1] and Firas Anowr[1]

[1] School of Civil Engineering, Universiti Teknologi Malaysia, 81310 Johor Bahru, Johor, Malaysia
[2] Department of Civil Engineering, Faculty of Engineering, Balochistan University of Information Technology, Engineering, and Management Sciences, 87300 Quetta, Pakistan
*Corresponding author: naqiuddin@utm.my

Abstract. Bitumen emulsions are special type of paving binder materials used in road construction industry due to its environmentally friendly approach, cost efficiency and less energy consumption compared to other conventional bitumen binders. Despite so many benefits, the use of bitumen emulsions is still limited in many regions due to difficulties faced in its emulsification process. Controlling several process parameters are needed during formulation while achieving the sufficient storage stability and viscosity. Therefore, this study was conducted to investigate the effect of different processing parameters on the properties of bitumen emulsions in terms of storage stability and viscosity. The 80/100 grade bitumen was emulsified through colloid mill, considering different pH values and emulsifier concentrations as processing parameters during emulsification process. The results indicate that the storage stability has improved significantly with increased concentration of emulsifier whereas high acidic system in terms of pH level has also substantially enhanced storage stability at lower level of emulsifier concentration. In addition, the viscosity of bitumen emulsions has also been affected by the emulsifier concentration and higher amount of emulsifier resulted in higher values of viscosity at lower pH level. The data achieved through this investigation can be further used to study the effect of other formulation parameters on properties of bitumen emulsions.

1. Introduction

Bitumen emulsions are prioritized over a conventional asphalt binders to be used in a road maintenance applications due to its cost efficiency, environmental advantages and lower energy consumption [1,2]. The production of bitumen emulsion is a technical task, since several processing parameters need to be controlled and optimized during the emulsification process [3]. The parameters include bitumen/water ratio, emulsifiers concentration, temperature of both soap solution and disperse phase (bitumen), pH value, speed of mixing shear; these processing parameters directly or indirectly influence the physical and rheological properties of bitumen emulsions [4]. The basic physical and rheological properties of bitumen emulsions are its storage stability, viscosity, mean droplet size distribution and breakage behaviour [5]. The properties of bitumen emulsions dictate its suitability of particular types of emulsion in the field applications [6]. In addition, the interfacial tension on bitumen/water interfaces caused by emulsifier activity is affected by acidic or alkaline levels in terms of pH value. Since the interfacial tension among bitumen/water droplets decreases with the increased value of pH. The emulsifier has high emulsifying ability at high pH value and makes the dilation modulus high at particular value of pH

IOP Conf. Series: Earth and Environmental Science **971** (2022) 012015 doi:10.1088/1755-1315/971/1/012015

around 2-3, particularly for cationic type of emulsions which results in a more favourable storage stability [7,8].

Despite affecting stability of emulsions, the pH value also affects the rheology of emulsions. The viscosity of bitumen emulsions increases with increased amount of acid and alkaline under extra shear provided [9]. The viscosity of bitumen emulsion is also influenced by mixing speed and time, droplet size & distribution and emulsifier concentration [10]. However, the flow characteristics such as viscosity of emulsion plays significant role in applications of bitumen emulsions since the emulsions viscosity neither should be very low to flow from aggregates nor be more viscous for mixing and workability purposes [11].

The physical properties of bitumen emulsions are also affected by its droplet size and distribution. The size of emulsion droplets increases with increased volume of bitumen fraction, however, the droplets size decreases with increased of rotor speed, since small droplets result in higher viscosity [12]. The influence of mechanical parameters on storage stability also extends to the emulsifier type and penetration grade of bitumen which has significant effect on storage stability of bitumen emulsions [13]. Since it becomes necessary to study the processing parameters during the emulsification of bitumen and its effect on the properties of bitumen emulsions. The influence of different pH values and emulsifier concentration on properties of bitumen emulsions such as stability and viscosity are investigated in this study.

2. Experimental procedure

2.1. Materials

A base bitumen with a penetration grade of 80/100 was used in the emulsification and the physical properties are listed in table 1. The cationic slow & medium set emulsifier Redicote N39L was used in the emulsification of bitumen, used as received. It is designed to produce slow and medium set emulsions based on its concentration. The acid, hydrochloric (HCL) was employed to adjust the pH of soap solution and pH was measured by litmus paper. Since the deionized water was used for preparing soap solution. The colloid mill with shearing speed of 2840rpm was used to formulate the emulsions as shown in the figure 1.

Table 1. Physical properties of bitumen binder.

Penetration @ 25°C,100g, 5 Sec	Softening Point °C	Flash Point °C	Ductility @ 25°C 5 cm / min	Specific Gravity @25°C
80-100	43-54	232	100	1.01-1.06

2.2. Preparation of bitumen emulsions

The soap solution produced at temperature 60± 3°C by introducing different dosage such as 0.5%, 1% and 1.5 % of emulsifier by weight of aqueous phase at different pH 2, 4 and 6 values. The soap solution was poured followed by heated base bitumen up to 150± 3°C in a Colloid Mill, preparing oil in water 60:40 emulsions and mixed for 5 minutes as shown in the Figure 1 (a) & (b). The samples of bitumen emulsions were collected at the outlet of colloid mill and stored in sealed containers and then left to cool at room temperature. The emulsions were then tested for surface charge through particle charge test as per ASTM D7402 as shown in Figure 2 (a) [14]. The values of different parameters considered in the emulsification process are given in table 2. The emulsification process was repeated three times for each sample in order to avoid error in the results.

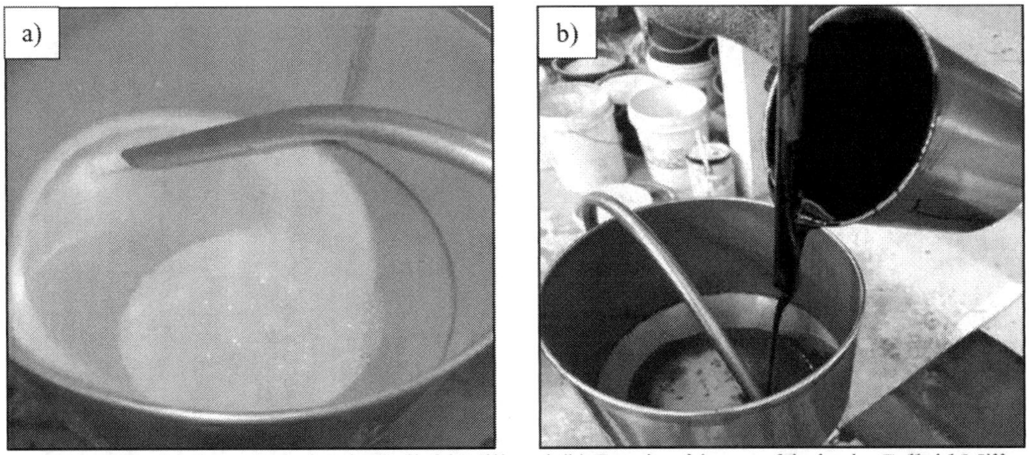

Figure 1. (a) Soap Solution in Colloid Mill and (b) Pouring bitumen binder in Colloid Mill.

Table 2. Selected Parameters during the emulsification process.

Emulsifier (%)	pH values	Mixing Time(mins)
0.5	2	5
1	4	5
1.5	6	5
0.5	2	5
1	4	5
1.5	6	5
0.5	2	5
1	4	5
1.5	6	5

2.3. Measurement of storage stability

The one-day storage stability of prepared samples was measured as per standard method of ASTM D6930 [12]. It is determined by the difference in percent of bitumen residue by evaporation of bitumen emulsions samples collected from the top and bottom from undisturbed samples placed in the cylinder for 24 hours shown in Figure 2 (b). The 50g samples from top and bottom each were collected in beakers and placed in an oven (temperature 163± 3°C) for 2 hours. Then the samples were stirred and kept for another 1 hour in the oven in order to get bitumen residues. The result for storage stability should be ≤ 1 and least value gives favourable storage stability.

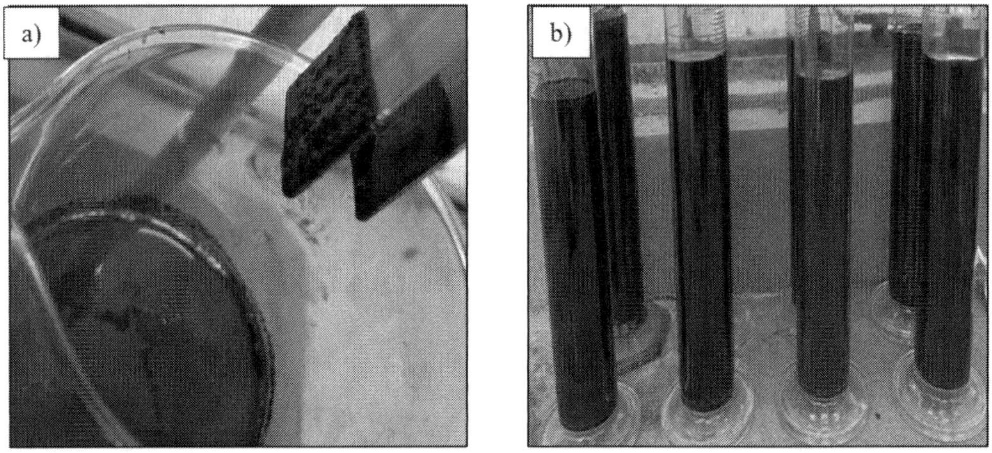

Figure 2. (a) Particle charge of emulsions; (b) Samples stored in cylinders for 1day storage stability.

2.4. Measurement of viscosity

The flow characteristic of bitumen emulsions affects its utility during the application. The bitumen emulsion should be viscous enough to be applied uniformly and resist flow from the crown of road pavement. The Saybolt Fural viscometer was used to measure the viscosity (SFS) of prepared emulsions at 25°C as per standard method of ASTM D7496 [15]. SFS is the efflux time in seconds of 60 mL of bitumen emulsions flowing through a calibrated Furol orifice at specified temperature as shown in Figure 3 (a) & (b). The SFS was measure at 0.5%, 1% and 1.5% emulsifier concentration to examine its influence.

Figure 3. (a) Saybolt Furol Viscometer; (b) Bitumen emulsions through Furol orifice.

3. Results and Discussion

3.1. Effect of pH values on storage stability of bitumen emulsions
The pH creates an environment for emulsifier to exhibit better emulsifying ability in bitumen emulsion system. Since this particular value of pH needs to be optimized during emulsification process [9]. The emulsifier becomes soluble in water at high acidic levels, thus resulting in polarizing the positively heads of cationic molecules [3]. In addition, the high pH value increases the dilation modulus of the emulsifier and result in better stability of emulsions. However, it is shown (see Figure 4.) that the values for storage stability increases with increased pH values linearly. The emulsions become unstable above value of pH 4 since the emulsifier is unable to emulsify the bitumen at less acidic system. The pH values 2-3 give significant results in terms of storage stability where the values for storage stability 0-0.4 are more favourable. The size of emulsifier charge on bitumen droplet is strongly pH dependent, thus strong pH exhibit high repulsion of droplets and result in a uniform suspension of droplets [11].

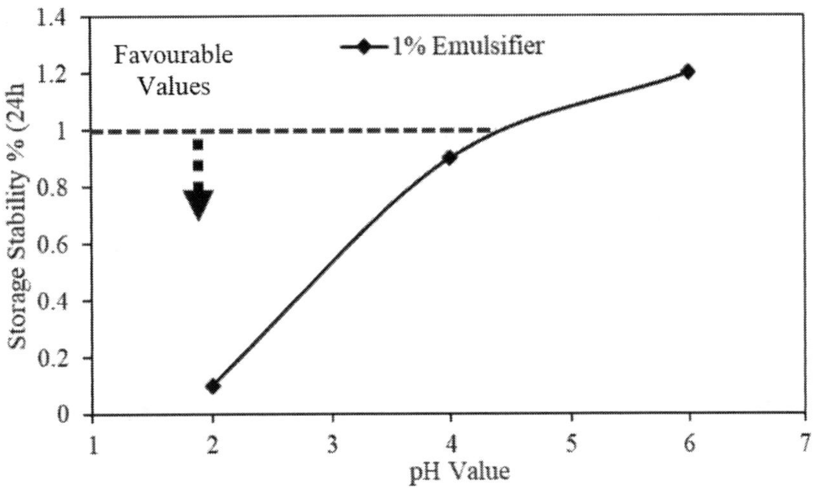

Figure 4. Effect of pH values on storage stability of bitumen emulsions.

3.2. Effect of emulsifier on storage stability of bitumen emulsions
The emulsifier reduces the interfacial tension on bitumen/water interfaces, resulting in a stable droplets suspension of bitumen and water. The emulsifier molecules are adsorbed on liquid-liquid interface on emulsion droplets and exhibit significant repulsion and suspension in the system [4]. Thus the storage stability can be defined as the resistance to coalescence of disperse droplets in emulsion system [7]. Since least values for storage stability ≤ 1 are favourable in bitumen emulsion stability. The results show (see Figure 5) that the values for storage stability decrease with increased percentage of emulsifier in the soap solution. However, the emulsifier dosage from 1% to 1.5% increment doesn't show any significant change in the results of storage stability since the storage stability for both concentrations remained unchanged. Despite, providing higher concentration of emulsifier up to 1.5%, it is feasible to provide emulsifier percentage of only 1%, resulting in saved quantity of emulsifier in terms of economy.

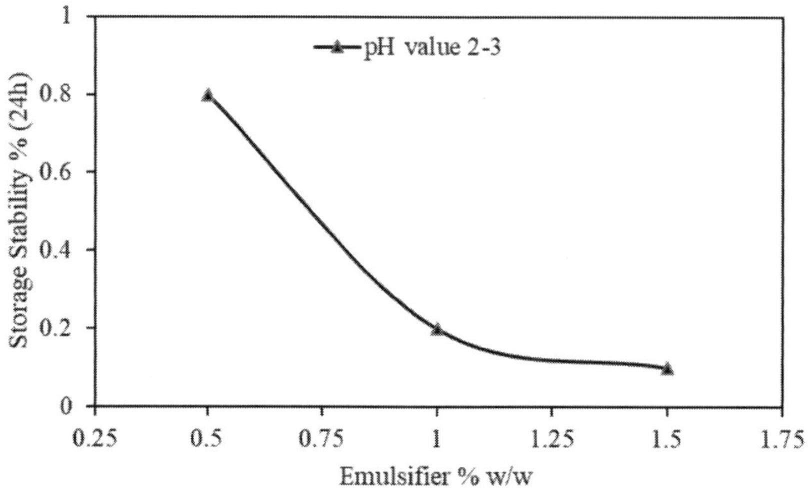

Figure 5. Effect of emulsifier % w/w on storage stability of bitumen emulsions.

3.3. Effect of emulsifier on viscosity of bitumen emulsions

The rheology of bitumen emulsions is identified for the purpose of its particular applications, although it doesn't affect the end performance, but emulsions with high viscosity are suitable for gape graded and low viscosity for dense graded mixture [10]. However, the viscosity of bitumen emulsion is generally influenced by several parameters such as emulsifier concentration, droplet size and distribution, shearing speed and viscosity of dispersion phase [4,5]. In addition, the emulsions with lower viscosity and small droplets size are more stable [16]. On contrary the stability of bitumen emulsions is very high in the existence of favorable amount of emulsifier concentration and least pH [17]. This can also be seen (see Figure 6) that the viscosity values increase with increased amount of emulsifier. Since higher amount of emulsifier produces strong charge on the surfaces of emulsion droplets, which also result in smaller size of droplet sizes. This leads to higher viscosity of bitumen emulsions [18].

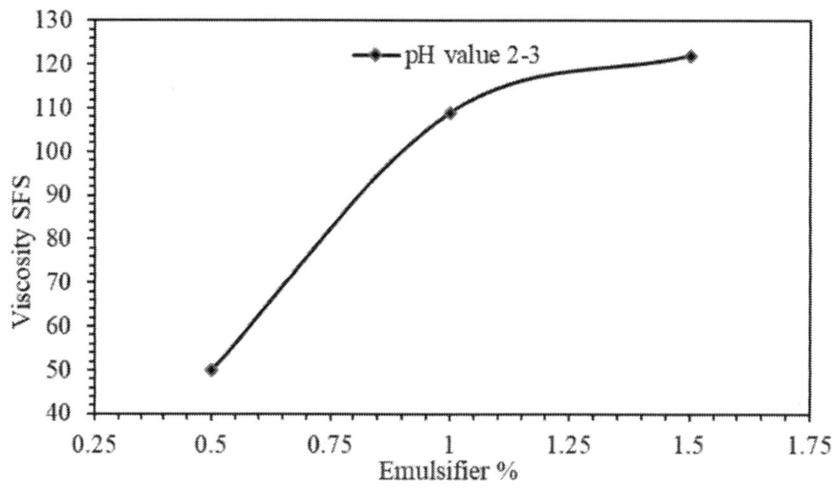

Figure 6. Effect of emulsifier % w/w on viscosity (SFS) of bitumen emulsions.

3.4. Effect of optimum pH value and emulsifier concentration on properties of bitumen emulsions

The parameters such as pH value and emulsifier concentration mutually play significant role in improving storage stability of formulated bitumen emulsions, since these both parameters are the function of Zeta potential which demonstrates the power of bitumen emulsion colloidal system in the presence of electrical charges [11]. However, these both variables in formulation are optimized as per required applications. The promising results of storage stability from (see figure 3) at pH value 4 or lower were at the existence of 1% emulsifier. While the results of storage stability at 1-1.5% emulsifier concentration in (see figure 4) were almost similar at pH value 2-3. Despite using higher amount of emulsifier, the optimum concentration should be used keeping the emulsification process of bitumen emulsions economical since emulsifier is an expensive material [4]. Hence, the bitumen emulsions can be formulated at optimum emulsifier concentration of 1% and pH values 2-3 resulting in a maximum storage stability. In addition, the interfacial viscosity of bitumen emulsion is associated with different formulation parameters, pH value and emulsifier concentration are two of them. However, the viscosity is optioned based on application specifications, since bitumen emulsions of high viscosity are more stable, this statement can be validated from Figure 3, 4 and 5 where bitumen emulsion with favorable stability are more viscous. The significant values of stability and viscosity of bitumen emulsions under the parameters such as pH value and emulsifier concentration are enlisted in Table 3.

Table 3. Optimized Values of bitumen emulsions.

Parameters		Properties	
pH Value	Emulsifier (%)	Stability (≤1)	Viscosity (SFS)
2-3	1	0.2-0.4	109

4. Conclusions

The objective of this study was to evaluate the influence of processing parameters on the characteristics of emulsified bitumen. The results indicate that the processing variables such as pH value and concentration of emulsifier have significant effect on the properties of bitumen emulsions such as storage stability and viscosity. The dependance shows that the storage stability improved at least pH value 2-3 producing high acidic level. However, the higher value of pH has negative impact on stability since stability values decreased with increased value of pH.

The results for emulsifier investigation show that the stability increases with increased amount of emulsifier since emulsifier works in the presence of strong pH values of soap solution. Thus, the emulsifier concentration of 1% at pH value 2-3 show significant results in terms of storage stability of 0.1%.

Although, the viscosity of bitumen emulsions is also influenced by mixing time and bitumen-water concentration, but this study only considered the effect of emulsifier concentration at pH of 2-3. However, the results show that the viscosity increases with increased emulsifier concentration since higher dosage of emulsifier results small droplet size and distribution which exhibit strong colloidal system of higher viscosity. Moreover, the choice of viscosity is based on application specifications of bitumen emulsions. This study also proposes to consider other mixing times for emulsification in future research in order to relate it with physical and rheological properties of bitumen emulsions.

5. References

[1] Hou S, Chen C, Zhang J, Shen H and Gu F 2018 Thermal and mechanical evaluations of asphalt emulsions and mixtures for microsurfacing *Constr. Build. Mater.* **191** 1221–9
[2] He R, Liang Y, Gao L, Chen H, Yang B, Guo F and Geng J 2020 Preparation and performance assessment of asphalt emulsion modified by the fabricated SBS latex *Adv. Civ. Eng.* **2020**
[3] Gingras J P, Tanguy P A, Mariotti S and Chaverot P 2005 Effect of process parameters on bitumen emulsions *Chem. Eng. Process. Process Intensif.* **44** 979–86

[4] Ronald M and Luis F P 2016 Asphalt emulsions formulation: State-of-the-art and dependency of formulation on emulsions properties *Constr. Build. Mater.* **123** 162–73
[5] Querol N, Barreneche C and Cabeza L F 2019 Asphalt emulsion formulation: State of the art of formulation, properties and results of HIPR emulsions *Constr. Build. Mater.* **212** 19–26
[6] Romero N, Cárdenas A, Henríquez M and Rivas H 2002 Viscoelastic properties and stability of highly concentrated bitumen in water emulsions *Colloids Surfaces A Physicochem. Eng. Asp.* **204** 271–84
[7] Cui D and Pang J 2017 The effect of ph on the properties of a cationic bitumen emulsifier *Tenside, Surfactants, Deterg.* **54** 386–92
[8] Xiao J J and Jiang W 2013 Study the influence of soap solution's pH value on the modified asphalt emulsion performance *Appl. Mech. Mater.* **253–255** 312–6
[9] You L, Dai Q, You Z, Zhou X and Washko S 2020 Stability and rheology of asphalt-emulsion under varying acidic and alkaline levels *J. Clean. Prod.* **256** 120417
[10] Querol N, Barreneche C and Cabeza L F 2017 Storage stability of bimodal emulsions vs. monomodal emulsions *Appl. Sci.* **7**
[11] Salomon D 2006 Asphalt emulsion technology
[12] Li M, Liu C, Liang C, Liu C and Li J 2014 Study of Bimodal Drop Size Distributions of Emulsion *J. Dispers. Sci. Technol.* **35** 397–402
[13] Querol N, Barreneche C and Cabeza L F 2017 Method for controlling mean droplet size in the manufacture of phase inversion bituminous emulsions *Colloids Surfaces A Physicochem. Eng. Asp.* **527** 49–54
[14] ASTM D7402-09(2017) 2017 Standard Practice for Identifying Cationic Emulsified Asphalts *ASTM Int. West Conshohocken, PA-2017*
[15] ASTM D7496−18 2020 Standard Test Method for Viscosity of Emulsified Asphalt by Saybolt Furol Viscometer *ASTM Int.* 18–20
[16] Querol N, Barreneche C and Cabeza L F 2017 Viscosity properties of bimodal bitumen emulsions: New approach *Afinidad* **74** 241–6
[17] Yuliestyan A, García-Morales M, Moreno E, Carrera V and Partal P 2017 Assessment of modified lignin cationic emulsifier for bitumen emulsions used in road paving *Mater. Des.* **131** 242–51
[18] Rodríguez-Valverde M A, Cabrerizo-Vílchez M A, Páez-Dueñas A and Hidalgo-Álvarez R 2003 Stability of highly charged particles: Bitumen-in-water dispersions *Colloids Surfaces A Physicochem. Eng. Asp.* **222** 233–51

Acknowledgements
This work was supported/ funded by Ministry of Higher Education under Fundamental Research Grant Scheme (FRGS/1/2019/TK01/UTM/02/1).

Evaluating the Quality of Service of Bus Performance in Johor Bahru

M F A Bakar[1], S Norhisham[1, 4*], N H Z Ibrahim A Abudeyab[1], N M Saad[2, 4], N N I M Azlan[1], N S M Shkuri [3, 4] and A M Mohamad[5]

[1]College of Engineering, Universiti Tenaga Nasional, 43000 Kajang, Selangor, Malaysia
[2]College of Computing & Informatics, Universiti Tenaga Nasional, 43000 Kajang, Selangor, Malaysia
[3]College of Energy Economics & Social Sciences, Universiti Tenaga Nasional, 43000 Kajang, Selangor, Malaysia
[4]**Institute of Energy Infrastructures (IEI),** Universiti Tenaga Nasional, 43000 Kajang, Selangor, Malaysia
[5]School of Law and Centre for Testing, Measurement and Appraisal (CeTMA), Universiti Utara Malaysia, 06010 Bukit Kayu Hitam, Kedah
*Corresponding author: shuhairy@uniten.edu.my

Abstract: The increase in living standard would also be a major factor for rapid urban development, which is caused by the increase in the daily traffic volume. Therefore, traffic congestion can occur daily, especially during peak hours in the morning and late afternoon to cover the huge number of people movement. Moreover, the quality of bus services should be assessed on operational aspects provided by urban operators. Therefore, this paper is to determine the quality of service (QOS) of bus performance in Johor Bahru, Johor. The study was conducted in April 2019. Service quality data were taken at Terminal JB Sentral, Johor Bahru. The *Transit Capacity and Quality of Service Manual* (TCRPM) was adopted as the specific guidelines and key indicators to determine and analyse the QOS of bus performance in the selected area. The study focused on four specific areas, such as service hour, passenger load factor, on-time performance and service frequency. The study concluded that the passenger load factor was classified as QOS of Class B since passengers could still choose their seats during the journey. On the other hand, on-time performance was classified as QOS of Class E, as most passengers made one round trip per weekday with a transfer experience of not-on-time vehicles every day. The overall QOS for bus services in Johor Bharu was classified as QOS of Class D, which was considered as moderate because several issues on QOS should be addressed and rectified by stakeholders. Some improvement on these three attributes should be made to increase the QOS. The study outcome should be used as an indicator for operators to measure their bus services. Since Johor Bahru is a major city in Malaysia, authorities and the government can use the outcome as inputs to measure the effectiveness of public transportation in urban areas.

1. Introduction

The continuous urbanisation process has caused rapid development and expansion of the population. At the same time, the demand of daily movement from one place to another is also increasing [8]. The increase in living standard would also be a major factor for the rapid development, which caused an increase in the road traffic amount. [9]. As the growing population in urban area increase tremendously, the high amount of road traffic volume would cause a huge traffic congestion in city and suburb areas [4]. Therefore, traffic congestion would occur daily, especially during peak hours in the morning and late afternoon to cover a huge amount of people movement [15]. In developing countries like Malaysia, people are also dependent on private cars to commute from one place to another. Data released by the Malaysian Government showed that public transport users in the Klang Valley area had significantly decreased from more than 30% in 1980 to just approximately 15 % in 2007 [7]. Therefore, by launching a few initiatives by the government though *Agensi Pengangkutan Awam Darat* (APAD) in 2012, improvement was shown in public transportation. Users in the Klang Valley showed that there were approximately more than one million journeys in the area daily. More than 25 % of the daily journeys had used public transportation for their main urban transportation [16, 8]. Therefore, it is important to identify QOS for bus services in the country, especially in urban areas [2].

The government has started to improve the bus service in urban areas by centralising its operation. The concession was handled to specific operators because they can control the QOS [4]. Globalisation and technology had led towards industrial revolution (IR) 4.0, in which every angle of the daily life is connected with Internet of Things (IOT) [34]. The government should assist key players in bus industries to develop and implement a new technology in bus services to improve its quality of for bus performance, such as actual time bus services [34], designing a significant feeder transit [32], optimisation of bus schedules [31], and systematic departure and arrival of buses [28]. By adapting this technology, trustworthiness and passenger and user perceptions towards bus services can also be increased as operators can deliver services of good quality [24]. Recently, a lot of issued were addressed by passengers and users towards bus services, which were punctuality, driver behaviour and services coverage [8]. Trustworthiness amongst passengers and users could also lead to other negative perceptions [25]. On the other hand, passengers demand for less waiting time by increasing the number of service frequency [27]. Therefore, it is the service provider's obligation to identify and address all customer perceptions and demands, and thus should try their best to rectify all weakness, accordingly [3, 30]. Operators should also implement level of services (LOS) assessment which act as a standard measurement to determine an acceptable QOS level for each attribute [26].

Johor Darul Takzim is the fifth largest state in Malaysia and Johor Bahru is the capital of Johor with a population of more than one million people living in the city [18]. Johor Bahru is located in southern Malaysia Peninsular. Johor Bahru is under the Johor Bahru City Council administration. Many bus operator companies exist in Johor Bahru, specifically at JB Sentral and Johor Bahru. JB Sentral is also connected with Bangunan Sultan Iskandar Custom Immigration and Quarantine Complex (CIQ), as the main entrance for Singaporeans to enter Malaysia. Therefore, the city plays an important role as the industrial, logistic and commercial centre for Malaysia [18, 11]. Some researchers had addressed that certain transport policies in this country did not focus on current issues and problem faced by bus operators because the implementation of public transport connection between interstate and local bus services is insufficient [18]. Currently, less than 25% of Malaysians are using public transportation [19]. A study found that passengers had emphasized on similar problems for almost a decade, such as punctuality of bus drivers, lack of schedule information and there is no continuous supervision by the authorities [5, 27]. The authorities and government should improve the bus services coordination, rectify the organisation of bus services, improve terminal facility [14, 15] and specific standard for QOS for bus services [18]. On the other note, passenger perception plays a significant impact on the QOS for bus performance because passengers act as customers demand for the good services [33].

Moreover, the bus services should be assessed on operational aspects and the understanding of passenger perceptions and expectations and users towards the bus services [23]. Passengers and users in this country are facing a continuous problem daily, such as lack of hours of operation (services hours), limited of area covered by bus services and excess number of passengers during peak hours on weekdays (passenger load) in developed areas, such as Klang Valley and Johor Bahru [1]. Therefore, customers would express good perception if the service given reach their expectation and reached a minimum acceptable standard of service [20, 21]. Previous studies also justified that overall satisfaction and behavioural intentions would be an important factor to analyse QOS of bus services on qualitative aspects [24]. Other attributes such as comfort, connivence, safety and security should be measured as well, and it would be a great influence on passenger satisfaction [13, 22]. A better understanding of the bus services is required so as to rectify the QOS and increasing of passenger and user used bus services [15]. Quantitative measurement on would include hours of services, service frequency, passenger load factors and on-time performance, as this attribute should be measured according to the transportation engineering knowledge [12]. Therefore, this paper will present a study that evaluates the QOS of bus performance in Johor Bahru, Johor.

2. Methodology

This section will explain all method and procedure applied to evaluate the QOS of bus performance in Johor Bahru, Johor. The attribute data were collected in August 2019. The study areas of bus services covered in this study is shown in Figure 1. JB Sentral, Johor Bahru were selected study areas as this terminal was considered as a transportation hub terminal and located at downtown JB Sentral, Johor. Several operators were operating local bus services in this area, namely Causeway Link, Maju Bus, City Bus, and S&S International. There are 29 routes departing for various locations in Johor Bahru. All buses in these routes have used a several types of buses based on demand on designated routes. These routes were Causeway Links (111, 505, 7B, 5B, 10B, 777B, KSL, Senai Airport, BET3, MV2, CT1, JPO1, LM1, TD1, & P101), Maju Bus (208, 227, P102, & Larkin), City Bus (15, 39, 123, 133, 188, & 33) and S&S International (1, 2, 7, & Larkin/Skudai). This study was only limited to local bus services which comprise bus services in Johor Bahru area.

IOP Conf. Series: Earth and Environmental Science **971** (2022) 012016 doi:10.1088/1755-1315/971/1/012016

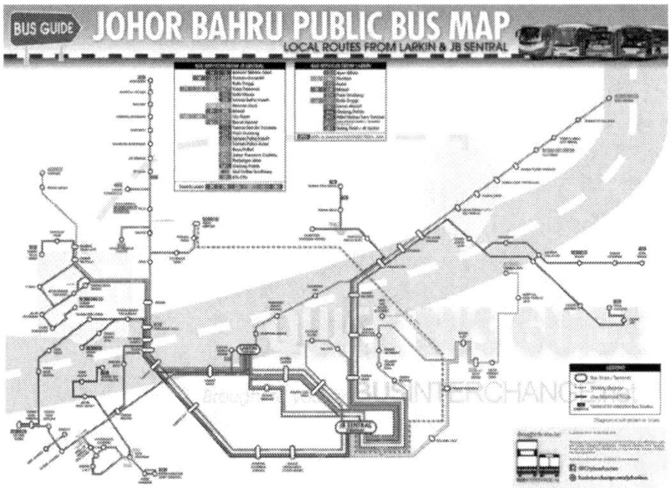

Figure 1. Routes of bus services in Johor Bahru [6].

The *Transit Capacity and Quality of Service Manual* (TCQSM) was selected as the standards and reference so as to evaluate the QOS for bus performance in this study. Several attributes were focused as this study factors, such as hours of services, service frequency, passenger load and on-time performance. The results for each attribute were classified based on QOS grades as stated in the guidelines. On the other hand, the average result of QOS for each attribute was calculated by representing the QOS of Class'A', Class 'B', Class 'C', Class 'D', Class 'E' and Class'F' by the score of '6', '5', '4', '3', '2', and '1', respectively. Lastly, and the average of QOS score was calculated, and the results of average QOS were classified based on the QOS score results as a summarisation of QOS of bus performance. The first attribute in this study was measuring the hour of service. Hour of services is also known as operation hours for each service from the first trip in the morning until the last trip of service in the late evening. The schedule was obtained from operators' websites or at the terminal. The procedure to collect the hour of service attributes as to record the first trip and last trips of bus services. Then, the hour of services or operation was calculated for each route by subtracting the last trips from the last trip of services. Since, hour of services was measured at the terminals, the hour of services for all routes could be calculated directly. Therefore, the QOS for hour of services can be compared based hours of services, as shown in Table 1.

Table 1. Quality of services (QOS) for hours of services (Sources: *Transit Capacity and Quality of Service Manual* [29]).

Quality of Services	A	B	C	D	E	F
Hours of Service (h)	19 – 24	17 – 18	16 – 14	15 – 12	11 – 4	0 – 3

Secondly, on-time performance percentage was to determine the punctuality of bus departure based on existing schedule provided by operators. The data would be collected at departure lanes at JB Sentral. Since, measurement of on-time performance could be conducted at the terminals the departure time for all routes was collected for a whole day period. The procedure to collect the followed, firstly acquired the schedule for each route. Secondly, the real time departure for the whole day period would be collected as to identify the different departure time between peak hour and off-peak hours period. Thirdly, the time difference between actual departure and schedule departure provided by operator and authority for each route was identified. Next, time different for each route was classified based on four types, namely on-time, late departure, early departure and no show. Then, the percentage of on time departure for each route overall departure for each route was calculated. Based on calculated on-time performance percentage, the QOS of on time performance for each route could be classified. Table 2 shows the category of QOS for on-time performance percentage. On-time performance was also measured directly at the terminal, and thus the on-time performance for all routes could be measured for all studied routes.

Table 2. Quality of services (QOS) for on-time performance percentage (Sources: *Transit Capacity and Quality of Service Manual* [29]).

Service Quality	A	B	C	D	E
On Time Performance (%)	95 - 100%	90 - 94%	80 - 89%	70 - 79%	< 70%

Next, the passenger load factor was determined based on the number of passenger ride and the number of seats provided. The passenger load data would be taken for roundtrip. Due timeframe limitation and number of numerators, only six different routes were randomly chosen from a total of 29 routes in Johor Bahru as the numerator should taek a bus ride for return trip for each route for the passenger load data collection. All selected routes were represented by different destinations in Johor Bahru, and it took almost three days to collect the passenger load data. Data of passenger load were collected on weekdays instead of weekends since the number of passengers on weekdays was expected to be higher than weekends. The procedure to collect the passenger load factor as followed, first step was select the specific routes and list the details on the selected routes, such as route code, departure time, and number of seats. Secondly, the number of preliminary passengers loads before bus departures at starting point were counted and recorded. Thirdly, the numerators took a ride on selected bus for return trip. Next, number of in-and-out passengers was recorded for each point of bus stopped until the selected bus reached the starting points on its return trips. Lastly, this procedure was repeated for other studied routes. As analysis, the average of passenger for the roundtrip journey would be calculated and divided by the total of seat provided in the bus. As per journey, the average of passenger load was counted. The passenger load data were taken for both return journeys. Based on the analysis, the quality of service for different routes in Johor Bahru for passenger load factor was identified. Passenger load data were only collected on weekdays. The Passenger load factor was calculated based on the number of passengers on the bus and divided by the number of seats provided in the bus. Table 3 shows quality of services based on load passenger factor.

Table 3. Quality of services (QOS) for passenger load factor (Sources: *Transit Capacity and Quality of Service Manual* [29]).

Quality of Service	A	B	C	D	E	F
Load Passenger (Passenger/seat)	0.00 – 0.50	0.51 – 0.75	0.76 – 1.00	1.01 – 1.25	1.26 – 1.50	>1.50

Last but not least, service frequency attributes were to determine the frequency for each route daily. In this study, the service frequency for all routes departed from JB Sentral for one whole day was taken. The number of frequencies also was obtained from bus schedules at bus terminals. The procedure to collect necessary information as followed, firstly the schedule for each route for one whole day was acquired. Secondly, the departure interval for each route was measured. Next, the average of departure time interval of each route was deliberated. Lastly, the quality of services would be classified based on the average of departure intervals for each route. Average of departure interval reflected the services frequency in this study. The QOS for service frequency would be classified based on the average of departure interval is shown in Table 4. The QOS for services frequency attribute was calculated directly for the one whole day period.

Table 4. Quality of services (QOS) for service frequency (Sources: *Transit Capacity and Quality of Service Manual* [29]).

Quality of Service	A	B	C	D	E	F
Average Departure Interval (min)	<10	10 – 14	15 – 20	21 – 30	31 – 60	>60

3. Results

This section explained the study outcomes. Four specific attributes were collected in this study, as explained in previous sections. Table 5 shows a summary for hour of services for all routes departed from JB Sentral. The result showed that the average of QOS for hours of service at JB Sentral was classified as QOS Class D. Table 5 showed there were six routes which achieved QOS of Class B, which were route 10B, 5B, 208, 227, 15 & 123. The routes operated more than 16 hours of service. The first trip departedwas recorded was as early as 5.15 am and the last trip departed wasrecorded at 11.30 pm. Besides, it was found that all operators at JB Sentral had different hours of service and it depended on the service coverage and demand from passengers and users.

Table 5. The quality of service for hours of services for all routes.

Route	First Trip	Last Trip	Hours of Services (hours (h): minute (m))	Quality of Service (QOS)
KSL	9.30 AM	11.00 PM	13h 30m	D
10B	5.15 AM	11.30 PM	17h 15m	B
505	6.45 AM	9.50 PM	14h 20m	C
5B	6.00 AM	10.00 PM	16h 0m	B
777B	7.40 AM	10.15 PM	14h 35m	C
7B	7.20 AM	10.15 PM	14h 55m	C
AA1	5.40 AM	10.00 PM	16h 20m	C
BET3	6.30 AM	8.45 PM	14h 15m	C
MV2	9.00 AM	10.15 PM	13h 15m	D
CT1	7.15 AM	6.30 PM	11h 15m	E
JPO1	9.30 AM	7.20 PM	9h 50m	E
LM1	8.00 AM	6.30 PM	10h 30m	E
111	7.00 AM	10.30 PM	15h 30m	D
TD1	8.45 AM	10.15 PM	13h 30m	D
P101	8.30 AM	12.30 PM	4h 0m	E
208	5.30 AM	10.15 PM	16h 45m	B
227	5.30 AM	10.00 PM	16h 30m	B
P102	8.45 AM	8.00 PM	11h 15m	E
Larkin	6.00 AM	9.30 PM	15h 30m	D
15	5.20 AM	10.20 PM	17h 0m	B

123	5.40 AM	10.20 PM	16h 40m	B
133	6.00 AM	9.00 PM	15h 0m	D
188	6.40 AM	10.45 PM	16h 5m	C
39	5.50 AM	9.20 PM	15h 30m	D
331	6.45 AM	10.30 PM	15h 45m	C
1	8.15 AM	10.00 PM	12h 45m	D
2	5.30 AM	7.00 PM	13h 30m	D
7	7.30 AM	8.20 PM	12h 50m	D
Larkin/Skudai	10.00 AM	7.00 PM	9h 0m	E
			Average	**D**

The method to measure these attributes was explained in detail in the Methodology section. On-time performance results for all routes are summarised in Table 6. The total of 652 of departures from this terminal was recorded in 24-hour period. The percentage of the total of on time departure percentage as compared to the total actual departure is 66.17%. Moreover, it showed that only routes10B and Larkin/Skudai had achieved QOS of Class A and both represent 91.43% and 92.86% of on-time departure percentage, respectively. On the other hand, the overall of bus departure only managed to depart less than 70% on-time departure according to the schedule provided by operators at JB Sentral.

Table 6. On-time performance percentage for all routes.

Routes	Total Actual Departure	No. of On Time Departure	On Time Departure Percentage (%)	Quality of Service (QOS)
KSL	35	22	62.86	E
10B	35	32	91.43	A
505	14	10	71.43	D
5B	12	9	75.00	D
777B	21	15	71.43	D
7B	29	22	75.86	D
AA1	20	11	55.00	E
BET3	13	8	61.54	E
MV2	25	14	56.00	E
CT1	7	6	85.71	C
JPO1	5	2	40.00	E
LM1	7	5	71.43	D
111	6	2	33.33	E
TD1	6	5	83.33	C
P101	4	2	50.00	E
208	26	19	73.08	D
227	56	22	39.29	E
P102	10	7	70.00	D
Larkin	62	44	70.97	D
15	42	26	61.90	E
123	44	33	75.00	D
133	8	5	62.50	E
188	19	13	68.42	E
39	31	20	64.52	E

331	5	2	40.00	E
1	20	15	75.00	D
2	6	4	66.67	E
7	70	52	74.29	D
Larkin/Skudai	14	13	92.86	A
			Average	**E**

To measure the passenger load factors, the total number of passenger in designated trip was divided by total of seats in designated trip. The collection was collected based on the return trip. Table 7 shows the summarisation of passenger load factors for Johor Bahru. The result showed that the average of passenger load for Johor Bahru was 0.56 and was considered as QOS of Class B. MV2 routes recorded the lowest passenger load factor which was as 0.22, followed by Route 5B and Route 505, which was represented as 0.55 of passenger load factors. Data were collected during peak hours of a weekday in order to get the maximum number of passengers.

Table 7. Summary of load factors for Johor Bahru routes.

Route	7B	10B	MV2	KSL	5B	505	Average
Passenger Load Factors	0.57	0.80	0.22	0.78	0.50	0.50	0.56
Quality of Service	B	B	A	B	A	A	**B**

Service frequency was also included in this study. Table 8 shows the overall frequency interval (in minutes interval) for each route. The quality of bus services was identified based on fixed route service frequency LOS. The overall QOS for service frequency in Johor Bahru was classified as QOS of Class E. Only one route managed to reach QOS of Class B, which was Route 7 and it represented an average of departure interval of 11 minutes. Besides, three routes also reached the QOS of Class C (Route KSL, Route 227 & RouteLarkin) which was represented as 20, 19 and 18 of service frequency respectively. Based on the observation, it is necessary to increase the frequency during peak hours since the bus driver intended to departure as earlier as 20 minutes of schedule due to the excessive passengers on the bus.

Table 8. The quality of service for service frequency for all routes.

Route	480	290	275	270	240	220	215	210	200	180	165	155	150	145	140	135	130	125	120	110	105	100	95	90	85	80	75	70	65	60	55	50	45	40	35	30	25	20	15	10	5	Average Interval	LOS
KSL																								1								1		3		4		17	5		5	20	C
10B																													1		1	2	4	4	2	7	3	1	5		1	33	E
505																			1			1	1			1	1	2		2		1	1	2								70	F
5B									1					1		1				1	1						1						1	2	1	1						87	F
777B																						1		1				1		2		2	2	3	1	3		2		2		54	E
7B																							1					1	1	2	1		1	1	2	4	2	3	3	4	2	32	E
AA1																														14			1			1		1	1			53	E
BET3			1														1									1		1	1	1		1	1	1		1				1		78	F
MV2																									1					1	1	5	5			6	1	1	3			39	F
CT1						1					1																2	2										1				113	F
JPO1		1											1				1																						1			148	F
LM1										1			1								1			2																		103	F
111				1				1														1				1																158	F
TD1					1					2									1					1																		162	F
P101										1																										2						78	F
208													1						2		1							1		1						4	3	7	1	3		42	F
227																																	3			8			44			19	E
P102																		1						4					1	1			1					1				75	F
Larkin																														2		1	1	2	2	3	2	10	10	17	11	18	C
15																														2				3	1	12	3	7	10	1	2	25	D
123																																				15	2	5	19	2		23	D
133				1						1							1		2											1									1			129	F
188									2			1					1				1				1							1	2			3	1	2		2	1	67	F
39																										1			1	1			4	3		8	1	1	9		1	31	E
331	1					1	1																													1						236	F
1															1							2					1	1						2		2	1	2	1	6		41	E
2																																2					2		1			33	E
7																																	1			1	1	4	18	30	13	11	B
Larkin/ Skudai																	1							1				1				1			1	4				4		41	E
																																										Average	E

Table 9 shows the quality of service score versus quality of service attribute. Overall quality of service could be determined based on quality of service based on attributes. Table 9 shows that hour of service, service frequency, passenger and on time performance have QOS of Class D, Class E, Class B and Class E, respectively. The overall QOSscore for each attributes was 9. The overall QOS for bus performance was classified as Class D. The QOS for each attribute measured in study was also affected by external factors such as the traffic volume, quality of bus and bus facility provided by the government. Since Johor Bahru was the second largest city in Malaysia and JB Sentral was also the main entrance to Singapore, an extreme number of passengers and users were expected to use bus service in weekdays. High traffic volume in the downtown of Bandaraya Johor Bahru also affected the quality of services, especially at peak hours in the early morning and late afternoon. A total of 652 of departures from this terminal were recorded for a 24-hour period. Authorities should upgrade certain areas in this terminal facility so as to serve the enormous number of departure every day. Besides, heavy traffic volume during peak hours was highly influenced by the frequency of services and departure and arrival of bus in this terminal. On the other hand, heavy traffic volume were reducing the headway of bus services and at the same time forced the operator to add the additional trip for high demand routes so as to ensure passengers and users have the optimum waiting time and followed the schedule, accordingly. Each attribute of bus performance in Johor Bahru can be improved in future by operators, local authority and state authority based on the outcomes form this study as such to provide a better public bus services in Johor Bahru city.

Table 9. QOS score vs. QOS attributes.

Attributes	QOS	QOS Score	Mean QOS	Overall QOS
Services Hours	D	2	2.00	**D**
On time performance	E	1		
Passenger Load	B	4		
Service Frequency	E	1		
Total		9		

4. Conclusion

This study is focused on evaluating the QOS for bus services in Johor Bahru, Johor. These studies were aimed to determine the quality of service for bus performance in JB Sentral, Johor Bahru. Johor Bahru is considered as the most developed city in southern Peninsular of Malaysia. This study was conducted by using procedure and approach that was proposed by *Transit Capacity and Quality of Service of Manual* (TCPRM). As a result, there were 29 different routes in Johor Bahru area and nearby districts. These routes were operated by various operators in the terminal. This terminal was also connected with other public transportation hubs. As mentioned in Methodology section, four attributes were focused in this study.

Only one attribute was classified as QOS of Class B. Besides, the passenger load and on-time performance were classified as QOS of Class E. The overall quality of bus services in Johor Bahru was classified as QOS of Class D services.Certain attributes should be improved by operators so as to improve the quality of bus services. Authorities and the government should make several improvements on the terminal facilities and information systems as this terminal serves 29 routes daily. Besides, the inconsistency on frequency of service was found in this study as all 29 routes managed by different operators. Lack of information on schedule should also be improved by operators and the authority as a big city like Johor Bahru consists of huge passenger and user demands daily. The output of this study can be used by operators and policy makers to improve the bus services in other urban areas. Since Johor Bahru is big city in Malaysia, authorities and the government can use the outcome from this study as their inputs to measure the effectiveness of public transportation in urban area. Future studies on this topic by other researchers are highly recommended.

5. References

[1] Abdullah A A 2013 Rapidkl Bus Service in City Center , Kuala Lumpur , Malaysia : An Epitome of Good Service ? *Inter. J. of Academic Res. in Business and Social Sci.* **3(4)** 333–349

[2] Adebola O, Samuel O, Feyisola A, and Eno O 2014 An Assessment of Public Transport Security and Safety: An Examination of Lagos Bus Rapid Transit (BRT), Nigeria, *J. Civil and Environment Research,* **6(4)** 105–117

[3] Azadi M, Shabani A, Khodakarami M and Farzipoor S R 2015 Reprint of Planning in feasible region by two-stage target-setting DEA methods: An application in green supply chain management of public transportation service providers *Transportation Research Part E: Logistics and Transportation Review*, Elsevier **74** 22-36

[4] Aziz A and Mohamad J 2013 Urban Public Transport in Penang: Some Public Considerations *Conf. of Eastern Asia Society for Transportation Studies* **9** 1-9

[5] Bekhet H A and Ivy Y L L 2014 Highlighting energy policies and strategies for the residential sector in Malaysia *Inter. J. Energy Econ. Policy* **4(3)** 448–456

[6] BusInterchange.net. 2021 *Bus Routes in Johor Bahru* Retrieved on June 10, 2021 from https://businterchange.net/privacy-policy.html

[7] Borhan M N, Ibrahim A N H, Syamsunur D and Rahmat R A 2019 Why Public Bus is a Less Attractive Mode of Transport: A Case Study of Putrajaya, Malaysia *Periodica Polytechnica Transportation Engineering* **47**(1) 82-90

[8] Calvo F, Eboli L, Forciniti C and Mazzurall G 2018 Factors influencing trip generation on metro system in Madrid Spain *Transportation Research Part D Transport and Environment* **67** 156-172

[9] Chow M F and Bakar M F A 2019 Environmental benefits of green roof to the sustainable urban development: A review, *Lecturer Notes in Civil Engineering* **9** 1525-1541

[10] Cui J, Gao Y, Cheng J and Shi L 2020 Study on the selection Model of Staying Adjustment Bus Lines along Rail Transit *J. of Adv. Transportation* **2020** 1-12

[11] Department of Statistics Malaysia, Official Portal 2010 *Population distribution and basic demographic characteristic* (Malaysia)

[12] Eboli L, Forciniti C and Mazzulla G 2018 Spatial variations of the perceived transit service quality at rail stations *Transportation Research Part a Policy and Practice* **114** 67-83

[13] Eboli L and Mazzulla G 2012 Performance indicators for an objective measure of public transport service quality, *J. of Public Transport* **(51)** 1–21

[14] Ensor J D 2004 Malaysia Transport Pricing Strategies, Measures, And Policies Inception Report *Malaysia Transport Research Group Massachusetts Institute of Technology*

[15] Esmailpour J, Aghabayk K, Vajari M A and Gruyter C D 2020 Importance - Performance Analysis (IPA) of bus services attributes: A case study in a developing country, *Transportation Res. Part A* **142(2020)** 129-150

[16] Margaret J D L 2018 Spatial dynamics of tour bus transport within urban destinations *Tourism Management* **64** 129-141

[17] Mazzulla G and Eboli L 2006 A Service Quality experimental measure for public transport Gabriella *European Transport* **34** 42–53
[18] Minhas A, Chatterjee A and Popli S 2020 Public perceptions: an important determinant of transport users' travel behaviour, *J. of Studies and Res. in Human Geo.* **14(2)** 177-196
[19] Ministry of Transport, Malaysia 2019 *National transport policy 2019-2030* Retrieved on March 29, 2020 from https://rb.gy/bo7avg
[20] Ngadiman N I, Roslan N F, Rahman A F, Ahmad F, Daud M M S and Abdullah R 2020 Survey on Quality of Service (QoS) at Larkin Central Terminal (LCT), Johor Bahru, *Inter. J. of Advanced Trends in Comp. Sci. and Eng.* **9(1)** 229-232
[21] Norhisham S, Bakar M F A, Tajri A A, Mohamed D, Yatuka S, Masjuki S A, Shukri N S M and Zaini N 2019 Statistical overview on quality bus service in Klang Valley, *Inter. J. of Advanced Sci. and Tech.* **28(10)** 370-380
[22] Rohani M M, Wijeyesekera D C and Karim A T A 2013 Bus operation, quality service and the role of bus provider and driver *Procedia Engineering* **53** 167–178
[23] Ponrahono Z, Bachok S, Osman M M, Ibrahim M and Abdullah M F 2017 Public Bus Level of Service Performance in Peninsular Malaysia: Correlation Analyses on Level of Service (LOS) and Passenger's Satisfaction Level, *J. of The Malaysian Institute of Planners* **15(1)** 229 – 234
[24] Purba A 2015 *A study on evaluating urban bus service performance in developing countries: case study of medium - sized cities in Indonesia* PhD dissertation (Yokohama National University)
[25] Saberi M, Ali Z K, Feng W and El-Geneidy A 2013 Definition and Properties of Alternative Bus Service Reliability Measures at the Stop Level *J. of Public Transport* **16(1)** 97–122
[26] Shukri M N S, Ponrahono Z, Sunoto N Y and Ghazali W W N W 2020 Evaluation of Level of Service (LOS) on Routes of Smart Bus in Kajang, Selangor *J. of the Malaysia Institute of Planner* **18(3)** 241-250
[27] Soh K L, Chong C L, Wong W P and Hiew Y H 2014 Proclivity of University Students to Use Public Bus Transport Service *Comprehensive Research J. of Education and General Studies* **2(2)** 24–34
[28] Suh W, Park S, and Lee E 2011 Fault Tolerant intelligent transportation systems with an agent *3rd Inter. Conf. Advanced Comp. Sci. and Info. Tech.* (Korea)16-25
[29] Transit Capacity and Quality of Service Manual (TCQSM) 2013 *Transit Cooperative Research Program, The Federal Transit Administration* TCRP Report 165 3rd edition (Washington: Transportation Research Board)
[30] Wahjono I S, Bachok K Y M, Marina A and Mochklas M 2017 The importance of MPIS on RK for futher ITS implementation in Malaysia *Inter. J. of Advanced and App. Sci.* **4(9)** 53-6
[31] Wang J and Cao Y 2017 Operating Time Division for a Bus Route Based on the Recovery of GPS Data, *J. of Sensor* **2017** 1-8
[32] Wei M, Liu T, Sun B and Jing B 2020 Optimal Integrated Model for Feeder Transit Route Design and Frequency-Setting Problem with Stop Selection *J. of Advanced Trans.* **2020** 1-12
[33] Yaakub N and Napiah M 2011a Public Transport: Punctuality Index for Bus Operation World Academy of Science *Engineering and Technology* **60** 857–862
[34] Zhang S, Yuan Z and Cao Z 2017 A Frequency-Based Assignment Model under Day-to-Day Information Evolution of Oversaturated Condition on a Feeder Bus information **8(19)** 1-19

Acknowledgement
Authors would like to thank Yayasan Canselor Universiti Tenaga Nasional for providing research grant for this study (202101015YCU) and Innovation Research Management Centre (IRMC), Universiti Tenaga Nasional.

Review of Runway Pavement Monitoring

I Sholichin[1*], E Ahyudanari[2] and M S Muntini[3]

[1] Doctoral Program Students, Department of Civil Engineering, Institut Teknologi Sepuluh Nopember, Indonesia
[2] Department of Civil Engineering, Institut Teknologi Sepuluh Nopember, Indonesia
[3] Department of Physics, Institut Teknologi Sepuluh Nopember, Indonesia
*Corresponding author: ibnu1609@mail.com

Abstract. Research on runway pavement is very local, meaning that it is strongly influenced by the quality of the material, the quality of implementation in the field and the climate in which the runway is built. In Indonesia, there are not many runway pavement studies. On the other hand, most airports in Indonesia only have one runway. If there is any damage, the airport should be closed. Based on these conditions, it is necessary to research the behaviour of runway pavement. On the other hand, research on runway pavement has been widely carried out in several developed countries. This study examines various existing studies to be used as a methodology in similar research in Indonesia. A review of the runway behaviour research is carried out by identifying the methodology related to using sensors to describe the pavement reaction in accepting the load. These methods were evaluated to assess their possible application in Indonesia. With the obtained results related to the pavement behaviour monitoring method, it is hoped to provide an initial step to identify runway pavement behaviour in Indonesia.

Keywords: Runway Pavement, Climate, Sensor, Pavement Behaviour

1. Introduction

A runway pavement structure consists of one or several layers of pavement [1][2]. Its function is to support the weight of aircraft traffic loads without causing significant damage to the construction itself [3]. The runway pavement structure consists of several layers with different hardness, and bearing capacity, each layer of pavement must be guaranteed its strength and thickness so that it will not experience distress, which is a change in shape because it is unable to withstand loads and does not quickly become critical or failure [4][5].

The strength of the runway pavement structure will decrease with increasing pavement age and aircraft traffic loads [6]. Aircraft traffic is getting denser and growing in line with economic development in an area [7][8]. The pavement age is generally determined based on the cumulative number of standard track loads expected to pass through the pavement, calculated from the start of the pavement being made and used until the pavement is categorized as damaged (finished service value). Several research has shown that the leading causes of runway pavement damage are construction quality[9][10], bad drainage [11], high pavement temperatures [12] and aircraft wheel friction with the runway [13]. One of the factors that cause early damage to the pavement is the drainage factor that is not functioning correctly. The runway transverse slope is a minimum of 0.5% and a maximum of 1.5%, but in reality, it is less than required [14]. As a result, rainwater pooled on the runway. Puddles of rainwater will result in reduced bonding power between asphalt and asphalt and asphalt with aggregate. Runway pavement becomes easy to peel off. Other impacts are damage to the runway before the technical age of the plan is reached, reduced safety levels (airplanes can skid), disruption of aircraft traffic services, and decreased environmental quality [15][16]. Runways pavement damage that occurs is a combination of several interrelated factors. From several studies, the quality of runway pavement depends on:

- Runway Pavement Design
- Load on Runway Pavement
- Material Selection
- Construction Process

1.1. Runway Pavement Design

Flexible pavement design at airports, one of them is the Design & Maintenance Guide 27 (DMG 27). DMG 27 is a guide made for designing and evaluating airport pavement thickness. This guide was produced by the UK Ministry of Defense's Construction Support Team to serve as a reference guide in designing and evaluating airport pavements in the UK. This guide also follows the provisions of the International Civil Aviation Organization (ICAO). In carrying out flexible pavement planning using DMG 27, several things need to be known, namely [17]:

- Basic Soil Carrying Capacity
- ACN Value
- Load
- Coverages and Load to Coverages Ratio
- Mixed Airplane Traffic
- Flight Frequency
- Runway Pavement Thickness

Runway pavement planning is a fairly complex problem that takes into account several variables. This variable depends on the location where the airport will be built. Currently, there are several airport planning methods, including methods and software developed by the FAA (Federal Aviation Administration). Some of the software used for planning airport pavement structures are COMFAA, FAARFIELD, and LEDFAA [18][19].

The FAA (Federal Aviation Administration) is a United States government agency whose job is to regulate all kinds of matters related to aviation and navigation in America. FAA issues rules for calculating runway designs air, namely Advisory Circular (AC) No. 150/5320-6D by manual [20] and Advisory Circular (AC) No. 150/5320-6E using FAARFIELD (Federal Aviation Administration Rigid and Flexible Iterative Elastic Layered Design) software [21].

In accordance with the function of the pavement to spread the load to the subgrade, the deeper the pavement layer, the smaller the stress received. For this reason, it is necessary to conduct research to determine the bearing capacity or strength of the subgrade and the properties of the subgrade. The characteristics of the subgrade greatly determine the total pavement so that the better the strength of the subgrade, the smaller the thickness of the pavement. The result of this subgrade investigation is the strength characteristics of the subgrade to be used in the FAA (Federal Aviation Administration) method, which consists of the CBR (California Bearing Ratio) value. Pavement planning procedures have been implemented in the FAA program. FAARFIELD applies elastic layer and finite element procedures to plan new pavements for runway pavements. The basic principle of calculating FAARFIELD pavement thickness based on FAA150/5320-6E:

- Enter the type of runway user aircraft into the design aircraft.

- The distance of the main landing gear of each aircraft from the centre line of the runway affects the cumulative CDF (Cumulative Damage Factor). CDF is numbers that show structural fatigue due to service period operational.

Another method in runway pavement design is ACN/PCN. The International Civil Aviation Organization (ICAO) developed a method to describe the strength of a pavement by comparing the Aircraft Classification Number – Pavement Classification Number (ACN/PCN). This method makes it possible to quickly describe the effect of an aircraft on a pavement with a numbering that varies according to the weight of the aircraft and its configuration, the type of pavement, and the strength of the subgrade. In this case, it is called ACN. On the other hand, the bearing capacity of pavement can be expressed by various numberings without specifying a specific type of aircraft or detailed information about the pavement structure. This definition is called PCN.

The ACN/PCN method has a structured system so that it can support a pavement with a certain PCN value, without weight restrictions on aircraft that have an ACN value equal to or less than the PCN value on a pavement. The PCN and ACN values are very important to determine the pavement performance of operating aircraft. ICAO has recommended this method for evaluating runway strength against operating aircraft (Aerodrome Manual Design Part I, ICAO). In an airport pavement, both flexible and rigid pavement, the ACN value should not exceed the current PCN value (ACN ≤ PCN). The use of the ACN/PCN method only applies to aircraft with an All-up Mass of more than 12,500 lbs or 5,700 kg. As for the pavement surface, which can only be used by aircraft weighing less than 12500 lbs/5700 kg, the form is only Maximum Allowable Aircraft Mass and Maximum Allowable Tire Pressure [22][23][24][25].

1.2. Load on Runway Pavement

Runway pavement planning has the same basic concept with pavement planning highways, where planning is based on workload and material strength used to support the load work. However, in the application, there are differences on runway pavement design and highways are [26]:

- Highways are designed for vehicles which weighs about 9000 lbs (4082.33 kg), while the runway designed to carry loads very heavy plane more significant, which is about 100,000 lbs (45,359,237 kg). This is the basic difference between runways and highways.
- The highway is planned to be able to serve the load repetition (repetition) of 1,000-2,000 trucks per day. While the runway is planned to serve load repetitions of 20,000 to 40,000 times during the design life.
- Tire pressure on the vehicle is around 80-90 psi. Whereas on the runway, tire pressure ranges up to 400 psi.

From the first, second, and third differences, you could say the runway is ten times stronger than the highway. Runway pavement is a pavement that generally uses a mixture of asphalt as the surface layer and grains material as the layer underneath (Figure 1).

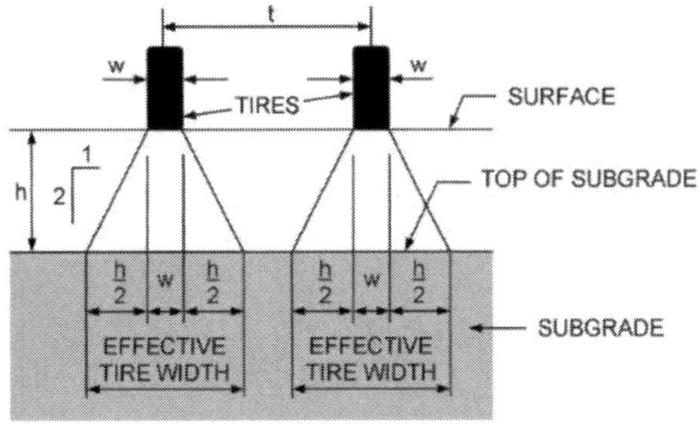

Figure 1. Load Transmission on Runway Pavement [27].

The aircraft load is transferred to the pavement through wheel contact with an evenly distributed load. This load is received by the surface layer and distributed to the subgrade. Runway pavement construction consists of layers placed on top of the compacted subgrade. This layer serves to receive the aircraft's load and pass it on to the layer below it. Overloading will increase the destructive force and shorten the service life of the runway pavement.

From Figure 2 below, the runway pavement can last up to the design life, but due to repeated loads, overloads, and environmental factors, the design life is generally not achieved. To achieve the planned life, periodic maintenance is required. If routine maintenance is not carried out, the costs needed for flexible pavement rehabilitation will be even greater.

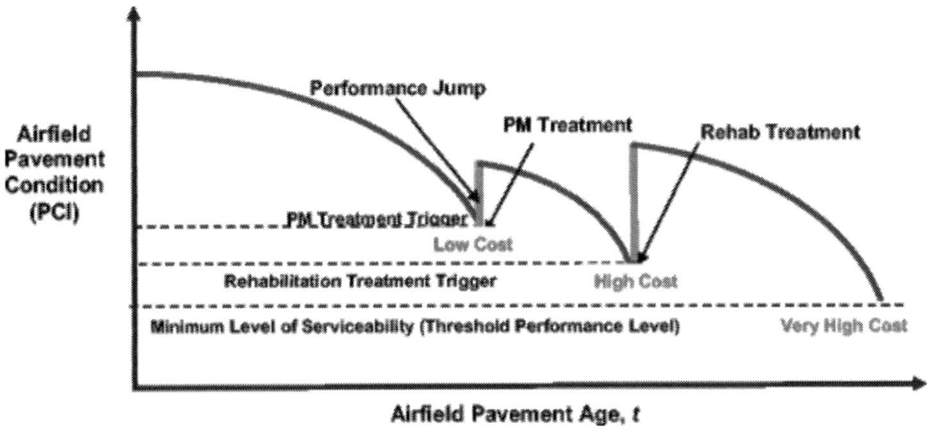

Figure 2. Runway Pavement Performance History [28].

1.3. Material Selection

Aggregate material is the main component of flexible pavement. Aggregates cover 90-95% by weight percentage and 70-75% by volume percentage, so the selection of aggregate material is very important. Aggregate strength and durability are influenced by gradation, aggregate size, clay content, dryness and resistance, grain shape, and surface texture. In addition to aggregate material, asphalt material also plays an important role. Good asphalt must have properties such as resistance to weathering during service life, good adhesion and cohesion properties, sensitivity to temperature, and asphalt hardness. There are many studies on asphalt modifications for highways, while asphalt modifications for runways are still limited. Modification of asphalt for runways has a specificity, namely asphalt has resistance to high temperatures. This high temperature is caused by the additional high temperature of the aircraft jet engine during take-off and landing. In addition, because the airport runway is a layer of asphalt concrete with a wide surface, so it is very easy to absorb heat, and the temperature is added due to friction between the aircraft wheels and the runway [29].

IOP Conf. Series: Earth and Environmental Science **971** (2022) 012017 doi:10.1088/1755-1315/971/1/012017

Conventional asphalt that has been used so far, the selected characteristics are based on the value of viscosity and penetration. The performance of conventional asphalt so far has produced a good performance. However, stripping, deflection, groove closure and premature ageing due to high-temperature changes have caused many airports to use modified asphalt. Asphalt modification is generally with the addition of polymers or other additives.

1.4. Construction Process

The construction process in runway pavement work affects the quality of the pavement. Good construction work starts with good planning. Good planning and construction must be followed by excellent and continuous pavement maintenance [30]. Some of the runway pavement damage caused by poor planning, construction, and maintenance are shown in Table 1.

Table 1. Runway Pavement Damage [26].

Type of Damage	Impacts on	Affected by	Level of Risk
Rutting	Skid resistance Surface smoothness	Deformation resistance	M
Shoving	Skid resistance Surface smoothness	Deformation resistance	H
Top down cracking	Loose material generation Water tight surface	Fracture resistance	M
Fatigue cracking	Loose material generation Water tight surface	Fracture resistance	L
Groove closure	Skid resistance	Deformation resistance	H
Bleeding	Skid resistance Deformation resistance	Volumetric composition air void content	L
Stripping	Deformation resistance	Moisture susceptibility Aggregate grading	M
Premature ravelling	Loose material generation	Aggregate grading Bitumen ageing	M
Aggregate breakdown	Loose material generation	Aggregate properties	L

2. Monitoring System in Runway Pavement

The aviation sector is a sector that demands excellent service quality and safety so that airport operations become more efficient. The complex system at the airport, especially on the airside, requires monitoring in accordance with national and international regulations.

A good runway pavement construction must ensure safe aircraft operations both during take-off and landing, for this reason, continuous monitoring is needed to ensure that everything is safe. This monitoring is also to reduce the cost of maintenance and rehabilitation, which is more expensive. There are several approaches to airport monitoring systems, including:

- Airport Pavement Management System (APMS)
  APMS is a system for monitoring runway pavement conditions, determining priority for handling damage, planning and managing human resources according to predetermined rules. This APMS has been widely implemented at international airports to facilitate airport operations and maintenance. The runway pavement monitoring system for aircraft traffic requires consideration and physical analysis of the runway pavement as well as determining state indices over time. The advantages of this APMS reduce the number of vehicles and personnel in monitoring. According to the ICAO Airport Services Manual, the parameters needed for runway pavement monitoring are:

   - Value of Aircraft Classification Number (ACN) and Pavement Classification Number (PCN). This ACN-PCN value is obtained from the size of the Falling Weight Deflectometer (FWD)
   - International Roughness Index (IRI) and deviation of 3 m Rolling Straight Edge (RSE) value
   - RUT value
   - Skid Resistance
   - Pavement Condition Index (PCI)

- Micro-Electro-Mechanical Sensor (MEMS)/Nano-Electro-Mechanical Sensor (NEMS)
  Aircraft loads, variations in temperature and humidity are some of the factors that cause runway pavement damage. While the causes of aircraft accidents include skid resistance, FOD, puddles of water on the runway [31]. For this reason, it is necessary to carry out effective and efficient runway pavement health monitoring to prevent aircraft accidents due to poor runway performance.

Rapid technological developments, especially sensor and wireless technology, have also penetrated the field of air transportation. The runway pavement structure can be monitored with sensor technology by continuously observing the pavement response and environmental conditions. Sensors will provide early warning if problems occur on the runway so that repairs can be carried out more quickly and reduce costs. In the long term, the sensor can also estimate the repair schedule at regular intervals [32], as shown in Figure 3.

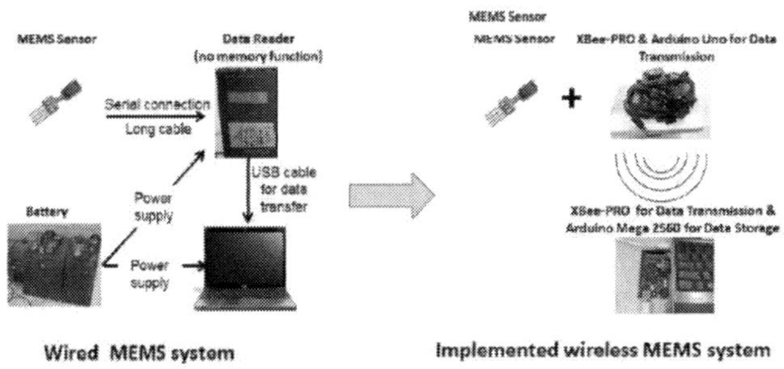

Figure 3. Wireless MEMS System [32].

2.1. The Remaining Service Life of The Runway Pavement

Runway pavement condition assessment stated by RSL is defined as the estimated number of years measured based on the condition survey from initial to projected requiring runway pavement rehabilitation activities, as shown in Figure 4.

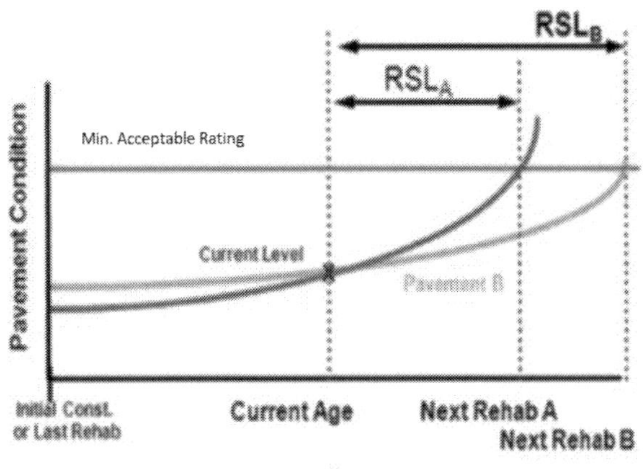

Figure 4. Remaining Service Life [33].

RSL can be used as a pavement condition measurement method for runway pavement maintenance. Pavement with a high RSL value indicates that the runway pavement is capable of serving aircraft traffic in accordance with the design age. Periodic RSL measurements will reduce the runway damage is greater, and the runway design life will be achieved. The more important goal is to ensure the safety of the aircraft in taking-off and landing [34].

RSL calculations on runway pavements are usually used for periodic maintenance or reconstruction. There are three methods to determine the RSL on a runway pavement, namely:

- Layered Elastic Design Federal Aviation Administration (LEDFAA)
- Regression Model China Airport Pavement Design Method (APDMC)
- Pavement Condition Index (PCI)

3. Conclusions

Based on the review of runway pavement above, several conclusions can be drawn, including the following:

- The quality of runway pavement depends on the pavement design factors, the load, the selection of aggregate and asphalt materials, and the construction process.
- Runway pavement design has developed rapidly, a lot of software has been used to calculate the thickness of the runway. In addition to calculating the thickness of the runway, there are several software used for runway pavement monitoring so that monitoring work becomes more efficient and inexpensive.
- The choice of aggregate and asphalt material determines the quality of the asphalt concrete mixture. The aggregates' selection must consider the factors of gradation, clay content, aggregate size, surface texture, grain shape, durability, and aggregate hardness. Meanwhile, the selection of asphalt needs to pay attention to durability, adhesion and cohesion properties, sensitivity to temperature, and asphalt hardness.
- In the implementation of runway pavement construction, the proper implementation stages must be considered. Indiscipline in carrying out the stages of work will make the quality of the flexible pavement low.
- Research on pavement service life and monitoring system is still wide open. Research using sensors and software still needs to be developed.

4. References

[1] Santoso S E, Sulistiono D and Mawardi A F 2017 Comparison study on flexible pavement design using FAA (Federal Aviation Administration) and LCN (Load Classification Number) code in Ahmad Yani international airport's runway *IOP Conf. Series: Mater. Sci. and Eng.* **267(1)** 012028

[2] Zhang D, Cai L and Zhou S 2019 An airfield soil pavement design method based on rut depth and cumulative fatigue *J. Adv. Transp.* **2019**

[3] Guyot X 2021 High Performance Asphalt for Airfield Pavement *IOP Conf. Series: Mater. Sci. and Eng.* **1075(1)** 012001

[4] Airport Pavement Essentials (AAA) 2017 *Airport Practice Note 12* Retrieved from https://airports.asn.au/airport-practice-notes/

[5] Federal Aviation Administration (FAA) 2016 *Airport Pavement Design and Evaluation, Advisory Circular 150/5320-6F* (US Department of Transportation, Federal Aviation Administration, Office of Airport Safety and Standards: Washington, DC)

[6] Wesolowski M and Iwanowski P 2021 Evaluation of Natural Airfield Pavements Condition Based on the Airfield Pavement Condition Index (APCI) *App. Sci.* **11(13)** 6139

[7] Tolcha T D, Bråthen S and Holmgren J 2020 Air transport demand and economic development in sub-Saharan Africa: Direction of causality *J. of Trans. Geo.* **86** 102771

[8] Parry C 1996 Ready for take-off **237(1)**

[9] White G 2016 Potential causes of top-down cracking of Australian runway surfaces In *27th ARRB Conf.*

[10] Su J, Xiao F, Liu Z, Kong L, Wang P and Li B 2021 Research and Analysis of Civil Airport Asphalt Concrete Pavement Damage Cause Evaluation and Countermeasure *IOP Conf. Ser. Earth Environ. Sci.* **787(1)**

[11] Yanti, Sunarjono S, Riyanto A, Hidayati N and Magfirona A 2019 Visual assessment deterioration analysis of runways at Sultan Aji Muhammad Sulaiman Sepinggan Airport Balikpapan *In AIP Conf. Proc.* **2114(1)** 050001

[12] Cook K, Garg N, Singh A and Flynn M 2016 Detection of delamination in the HMA layer of runway pavement structure using asphalt strain gauges *J. Transp. Eng.* **142(1)**

[13] D'Apuzzo M, Festa B, Giuliana G, Mancini L and Nicolosi V 2012 The evaluation of runway surface properties: a new approach *Procedia-Social and Behavioral Sci.* **53** 1192-1201

[14] Ong G P and Fwa T F 2009 Runway geometric design incorporating hydroplaning consideration *Transp. research record* **2106(1)** 118-128

[15] Federal Aviation Administration (FAA) 2013 *Advisory Circular. Airport Drainage Design* (US Department of Transportation, Federal Aviation Administration, Office of Airport Safety and Standards: Washington, DC) pp 1-4

[16] Pagar S R and Talikoti R S 2016 Analysis of Box Culvert for Storm Water Drainage System for Runway under Aircraft Loading at Airport 2350–2354

[17] Woodman G R 1992 Commentary on A guide to airfield pavement design and evaluation *Proc. Inst. Civ. Eng. Transp.* 95(3) 161–172

[18] Attia M I E 2019 Evaluation of Airport Flexible Pavement HMA Fatigue Using FAARFIELD

[19] Airports Council International 2013 Guide for maintaining runways in accordance to Annex 14 of ICAO

[20] Federal Aviation Administration (FAA) 2016 *Airport Pavement Design and Evaluation Advisory Circular* (US Department of Transportation, Federal Aviation Administration, Office of Airport Safety and Standards: Washington, DC) pp 1-4

[21] Yavari M R and Balali A 2015 Evaluation of Runway Pavement Design Software and Application of Modified Asphalt Overlay on Airfields *Evaluation* **5(6)** 257–264

[22] Loizos A, Abacoumkin C and Charonitis G 1998 Investigation of the ACN–PCN method for the structural evaluation of flexible airfield pavements *Technika Chronica, Scientific J. of Technical Chamber of Greece. I* **18(2)** 75-85

[23] Qassim G J 2012 Pavement Strength Evaluation of Selected Iraqi Airports Depends on ICAO

(ACN/PCN) Method *J. Univ. Babylon* **4** 1166-1179

[24] White G 2017 Limitations and potential improvement of the aircraft pavement strength rating system to protect airport asphalt surfaces *Int. J. Pavement Eng.* 18(12) 1111–1121

[25] Mohamady Abdallah Wahba A 2017 ICAO Overloading Practice versus Airport Pavement Design Life Using FAARFIELD 1.3 and COMFAA 2.0, 3.0 *Am. J. Civ. Eng. Archit.* 5(2) 57–65

[26] White G State of the art: Asphalt for airport pavement surfacing *Int. J. Pavement Res. Tech.* 11(1) 77–98

[27] Federal Aviation Administration (FAA) 2009 Pavement Design for Airplanes Weighing More Than 30000 Pounds AC *150/5320-6E* (US Department of Transportation, Federal Aviation Administration, Office of Airport Safety and Standards: Washington, DC)

[28] Irfan M, Khurshid M B, Iqbal S and A Khan 2015 Framework for airfield pavements management—an approach based on cost-effectiveness analysis *Eur. Transp. Res. Rev.* **7(2)**

[29] White G and Embleton K 2015 Next generation binder for airport asphalt *16th AAPA Int. Flex. Pavement Conf. Gold Coast, Queensland, Australia* 13–16

[30] Di Mascio P and Moretti L 2019 Implementation of a pavement management system for maintenance and rehabilitation of airport surfaces *Case Stud. Constr. Mater.* **11** e00251

[31] Ang L 2013 Research and design of an airfield runway FOD detection system based on WSN *Int. J. Distrib. Sens. Networks* **2013**

[32] Yang S, Ceylan H, Gopalakrishnan K and Kim S 2014 Smart airport pavement instrumentation and health monitoring *Civil, Constr. Environ. Eng. Conf. Present. Proc.*

[33] Mack J W and Sullivan R L 2013 Using remaining service life as the national performance measure of pavement assets 2 In Proc. of the Annual Meeting of the Transportation Research Board (Washington, DC, USA) 13-17

[34] Zhao H, Ma L and Zhang J 2018 Estimating the remaining life of airfield flexible pavements considering environmental factors *Trans. Hong Kong Inst. Eng.* 25(3) 208–216

Acknowledgments

Thank you to the Department of Civil Engineering, Institut Teknologi Sepuluh Nopember Surabaya and the head of the Civil Engineering Department of the Universitas Pembangunan Nasional "Veteran" Jawa Timur for the support so far.

Performance of Bus Services in Urban Area: A Case Study in Penang, Malaysia

S Norhisham[1, 2], M F Abu Bakar[1], S F Zamree[1*], N N I M Azlan[1], A M Mohamad[5] and H Y Katman[1,2]

[1]College of Engineering, Universiti Tenaga Nasional, 43000 Kajang, Selangor, Malaysia
[2]Institute of Energy Infrastructures (IEI), Universiti Tenaga Nasional, 43000 Kajang, Selangor, Malaysia
[3]School of Law and Centre for Testing, Measurement and Appraisal (CeTMA), Universiti Utara Malaysia, 06010 Bukit Kayu Hitam, Kedah
*Corresponding author: shuhairy@uniten.edu.my

Abstract. Since the escalated industrialisation and urbanisation process began traffic congestion has been a serious problem in many Malaysian cities. One of the solutions for the traffic congestion is by promoting public transport. Public transport is needed to manage the transport in cities. In 2007, a corporate body owned by the Malaysian Government named Prasarana Malaysia, introduced the Rapid bus service in Penang. Currently, Rapid bus is the main public transport operator in Penang. However, the bus service had some setbacks and passengers had publicly voiced their dissatisfaction. Therefore, to encourage more people in using the public transport, a better public transport system should be provided and improved occasionally. Previous comprehensive studies were inadequate, and thus a detailed analysis of the bus service performance in Penang is needed. The *Transit Capacity and Quality of Service Manual* (TCQSM) was chosen for guidelines. Based on reviews from journals and papers three (3) objectives which focused on service frequency, on time performance and passenger load factor were chosen for the study. The study was conducted to evaluate the bus service performance in Penang. The study results obtained the quality of service (QOS) of D for the Rapid bus service performance in Penang, which was regarded as satisfactory. This showed that buses in Penang needed to be greatly improved to change the passenger's perceptions.

1. Introduction

To encourage road users to change from using private vehicles to public transport, the public transport system should be developed and improved. Moreover, it should be updated occasionally to ensure that the condition of the public transport is always at its best. In the Malaysian context, particularly in Penang, bus is the main public transport [2]. The buses do not operate on a specifically designed lane, but the route is shared amongst other traffic buses [6][15]. The term 'rapid' refers to fast public transport in and around the city centre, such as the metro system. This research is focused on the public transport buses, particularly on facility issues of public transport in urban areas.

The public transport has many modes; hence, several things should be taken care of as they can affect the quality of service (QOS), which can be defined by several aspects, such as service coverage, frequency of services, service hours and service reliability [7][16]. These are determining factors that affect the ridership volume. In 2013, the Transportation Research Board reported [13] that there are two

IOP Conf. Series: Earth and Environmental Science **971** (2022) 012018 doi:10.1088/1755-1315/971/1/012018

aspects which affect the public bus ridership, namely external factor and internal factor [5]. The external factors have direct influence on transit demand and supply, for example, increase in population growth and service demand. On the other hand, internal factors can be controlled by the bus service operator, which is quality of service [4][13].

Time spent on waiting for a bus and the boarding time is a measurement of efficiency for QOS. Bus facility improvements, such as air-conditioning, hygiene, comfort, and Wi-Fi service provision on certain buses can also improve the QOS [8]. Besides, increase in ridership for bus can also increase traffic congestion time. Traffic congestion starts as early as 7am in morning because people are moving to their offices. In the evening, traffic congestion starts 4pm as people moving back after office hour.

2. Methodology

Based on previous studies, this research was conducted based on the TCQSM method. Four attributes were conducted from the guidelines. To cover all routes from Georgetown to all areas in Balik Pulau and via versa 48 routes were selected for the study [3]. Since most routes were slightly redundant, only 17 routes were selected.

Weld Quay Bus Terminal is the main bus station in Georgetown, which is the capital city. The longest and shortest routes were listed for research convenience [6]. Then, all trip schedules of selected routes were obtained from the local authorities. Data for service frequency range for each particular route were gathered from the schedule during the on-time performance evaluation. Then, several tables were produced to record the data of respective objectives.

2.1. On-time performance

Weld Quay Bus Terminal was chosen as the research location. The buses departure time were recorded and compared with the scheduled departure time of service operator. All tabulated data comprised on time departure, early departure, late departure, no departure and total departure. The departure time of buses in all routes were recorded [8][9]. The data were compared with the schedule departure. Then, the on-time performance percentage was calculated [11][12]. According to TCQCM [11][13], the on time refers to less than 5 min departure from the schedule. Table 1 shows percentage that represents the QOS grades.

Table 1. QOS for on time performance percentage [13].

Service Quality	A	B	C	D	E
On Time Performance Percentage (%)	95 – 100	90 – 94	80 – 89	70 – 79	< 70

2.2. Services Frequency

The service frequency of bus service was determined in accordance to the method stated in TCQSM [13]. The QOR can be analysed by the average headway between two buses. Firstly, buses which applied the same route were identified and a particular bus stop was chosen. Weld Quay Bus Terminal was the fixed location. Departure time between the first bus and second bus to obtain the headway was recorded. Then, data on departure time and headways were tabulated. The person who recorded the data needed to sit in a fixed location whereby he could see all the departed buses. This is to ensure that data from the earliest departure to the last departure of all routes were not missed during research. In addition, the bus driver must not know about the chosen location and observer's existence. Another easier method to obtain the headways was by using the on-time performance data based on the departure time recorded between buses. The average headway was calculated. Table 2 represents number of average intervals to represent the QOS grade.

Table 2. QOS for service frequency [13].

Quality of Service	A	B	C	D	E	F
Average Departure Interval (min)	<10	10 – 14	15 – 20	21 – 30	31 – 60	>60

2.3. Load Factor

The method used required records of bus departure from the starting stop and the number of passengers who got off and boarded the bus at each stop until it reached the end of route and turned back to the starting stop [17][18]. Weld Quay Bus Terminal was chosen as the starting stop. Due to limited time only a few selected routes were assessed. Recorded data were tabulated according information on bus stop, passengers hop on, passengers hop off and temporary load. Table 3 shows the passenger load factor which refer the QOS grade,

Table 3. QOS for load factor [13].

Quality of Service	A	B	C	D	E	F
Load Passenger (Passenger/seat)	0.00 – 0.50	0.51 – 0.75	0.76 – 1.00	1.01 – 1.25	1.26 – 1.50	>1.50

2.4. Hour of Services

As for analysing the hour of services, the operation hours were identified from the authorities and bus provider office. The total hours in one day were identified. The number of hours ere compared in Table 4 to identify the QOS grade in terms of service hours [14].

Table 4: QOS for hour of services [13].

Quality of Service	A	B	C	D	E	F
Hours of Service (Hr)	19 –24	17 – 18	14 – 16	15 – 12	11 – 4	0 – 3

3. Results

3.1. On time performance

There were 476 bus daily trips in the whole of Penang. The recorded data were from 17 routes. There were 110 early departure trips and 230 on time departure trips. According to the guideline early and on-time trips were grouped as on-time departure, which were 330 trips. Meanwhile, 136 trips were late because of traffic congestion. Some buses had skipped trips, which may affect the study. The on-time performance obtained for Route 101 and Route 10 was 95.45% and 100%, respectively, which is a QOS of A. Meanwhile, only route CAT obtained a QOS of B, and route 200 and 502 obtained a QOS of C. Three routes, which were 201, 202 and 11, obtained QOS of D. The remaining routes obtained a QOS of E. These numbers give an average for Rapid Penang an on-time percentage of 66.10%, which was a QOS of E. This means that passengers made a return trip, on every working day and without experiencing the not on- time bus service.

Overall, route with higher number of on-time departure led to higher score of QOS obtained. For example, starting with the highest number of on-time departure, 43 on-time departures in Route 101 obtained QOS of A, followed by 41 on-time departures in Route CAT obtained QOS of B, followed by 32 on-time departure in Route 200 which obtained QOS of C. The lowest number of on-time performance caused low score of QOS, for example, in Route 206 which had only three on-time departures that obtained QOS of E.

Based on this study, late departure and early departure were mostly caused by bus drivers who departed more than 5 min late and skipped trip which did not depart at the scheduled time. This issue was also caused by congested road during peak hours, making the buses to arrive at the bus stop late, and thus made them skipped the next trip due to the driver's rest time or lunch time. The relation between no departure and early departure was when there were buses which skipped the trip, making the bus departed earlier than the scheduled time.

Table 5. On-time performance for buses in Penang.

Route	Number of Late Departure	Number of On Time Departure	Number of Early Departure	Total Departure	On Time Percentage (%)	QOS
101	3	43	20	66	95.45	A
200	7	32	6	45	84.44	C
201	3	8	3	14	78.57	D
202	3	7	4	14	78.57	D
203	9	7	0	16	43.75	E
204	14	10	3	27	48.15	E
206	7	3	4	14	50.00	E
CAT	7	41	33	81	91.36	B
10	0	5	1	6	100.00	A
11	5	8	6	19	73.68	D
12	7	8	2	17	58.82	E
301	9	13	2	24	62.50	E
302	18	6	7	31	41.94	E
303	13	7	3	23	43.48	E
401E	16	8	11	35	54.29	E
401	10	5	1	16	37.50	E
502	5	19	4	28	82.14	C
		Average			66.10	E

3.2. Services frequency

Service frequency is a study to determine the headways between each bus departure . A lower value of headways between departures is considered as an efficient bus for that particular route as the bus have the minimum waiting time for passengers to board the bus. A summary of average headway values are generated in the Table 6. A comparison was made between the 17 routes. From the result, the lowest average headway was in Route CAT with QOS of C, followed by Route 101 and Route 200 with QOS of D. Next, Route 204, Route 11, Route 301, Route 302, Route 303, Route 401E and Route 502 had QOS of E, followed by Route 201, Route 202, Route 203, Route 206, Route 10, Route 12, and Route 401 had the QOS of G.

According to the manual, QOS of C gives relatively frequent service to passengers, and make them check on the scheduled departure times to minimise their waiting time at the station or bus stop, and thus leads to maximum desirable waiting time for the next service if they missed the bus. For QOS of D, passengers have to adapt their travel with the transit schedule, and often results in a less-than-optimal departure or departure times for them. QOS of E provides a minimum service level to meet the basic travel to passengers. The passenger must adapt their travel with the transit schedule, which usually results in less-than-optimal departure and departure times for them. Meanwhile, the QOS of G gives the undesirable for urban transit service due to typical long waits for return trips and when a bus is missed by passengers.

Based on this study, the causes to the higher time of headways was the departure of buses were not following the scheduled departure. Traffic congestion also contributed to the higher time of headway as on average every 100m–200m, there was a traffic light in a particular route. However, there were limitations and weaknesses to the data obtained. For example, the data outcome could be influenced by the traffic condition of that particular day only.

Table 6. Service frequency field table for all routes.

ROUTE	AVERAGE HEADWAY (MINS)	QUALITY OF SERVICES FOR BUS
101	16	D
200	20	D
201	64	G
202	77	G
203	65	G
204	40	E
206	83	G
CAT	13	C
10	180	G
11	60	E
12	67	G
301	46	E
302	34	E
303	49	E
401E	32	E
401	70	G
502	34	E
Total Average Headway = 56		E

3.3. Load Factor

Passenger load is the factor value for the number of seats being seated during the bus trip. The least number of seats being seated by passengers, the higher was the QOS score. There were six routes analysed for passenger load. Only Route 11 managed to get a QOS of A with the least number of passenger load. Next, Route CAT obtained QOS of C with the highest number of passenger load, while the other route was got QOS of B. Table 7 shows average passenger load for the selected routes.

According to the manual, the passenger will experience no sitting next to another for QOS of A and all passengers can sit perceived travel time up to 1.1 times of actual travel time for QOS of C. Meanwhile, for QOS of B, the passenger can have some freedom in choosing their sit and perceived travel time that is equal to the actual time as same as QOS of A. However, if there are less passengers during each trip it will cause unproductive service if the condition occurs at the maximum load point in the peak direction.

Based on this study, there were few causes for the higher passenger load in the particular trip or bus. Firstly, buses that skipped their trip led to the increase in number of waiting passengers at certain bus stops. Once the bus arrived at the bus stop, the seats were occupied by the large number of passengers. The peak hour of weekdays also led to higher passenger load during particular trips.

Table 7. Average passenger load for selected routes in Penang.

Route	Load	Load Factor (%)	Quality Of Services For Bus
CAT	24	96	C
11	7	28	A
101	13	52	B
204	14	56	B
302	15	60	B
401E	16	64	B
AVERAGE	15	60	B

3.4. Hours of Service

This section explained the study outcome to evaluate the QOS of bus performance. Four specific attributes were collected. Table 8 shows a summary of service hours for all routes which departed from the Weld Quay terminal.

The result showed that the average of QOS for service hours at Weld Quay Terminal was classified as QOS of C. There are 12 routes which achieved QOS of B, which were routes that operated for more than 16 hours of service. The first trip departed was recorded as early as 5.30 am and the last trip departed was recorded at 12.00 am.

Table 8. Hours of service for buses in Penang.

Route	**Route Number**	**No**	**Starting hours**	**Ended hours**	**Service hours**	**LOS**
Teluk Bahang	101	1	5.30 am	23.20 pm	17.50	B
Masjid Negeri	200	2	6.30 am	20.45 pm	14.15	C
Pekan Air Itam	201	3	9.45 am	23.30 pm	13.45	C
Paya Terubong - Farlim	202	4	6.50 am	23.30 pm	16.40	C
Pekan Air Hitam – Farlim	203	5	7.00 am	23.30 pm	16.30	D
Pekan Air Hitam	204	6	5.50 am	23.30 pm	17.40	B
Tesco	206	7	6.00 am	23.30 pm	17.30	B
CAT	CAT	8	6.10 am	24.00 am	17.50	B
Botanical Garden	10	9	6.30 am	21.30 am	15.00	C
Jln Tan Sri The Ewe Lim	11	10	5.30 am	23.00 pm	17.30	B
Jelutong Promenade	12	11	6.00 am	22.40 pm	16.40	B
Relau	301	12	5.40 am	23.30 pm	17.50	B
Batu Maung Jln Bkt Gambir	302	13	5.40 am	23.30 pm	17.50	B
Bukit Gedung	303	14	5.30 am	23.30 pm	18.00	B
Balik Pulau - Queensbay	401E	15	5.30 am	23.30 pm	18.00	B
Balik Pulau - Bayan Baru	401	16	5.40 am	23.30 pm	17.50	B
Terminal Balik Pulau	502	17	5.30 am	20.30 PM	15.00	B
					Average	**B**

3.5. Overall Summary

Four attributes were conducted to determine the QOS for bus performance in Penang. These studies were done by using service ratings based on TCPRM. Four different attributes were selected in these studies, which were hours of service, passenger load, service frequency and on-time performance. A total of 17 routes based from Weld Quay bus terminal determined the performance of the QOS for bus services. Passenger load factor for bus services in Penang obtained B. It was clear that there were sufficient passenger seats to cater the ridership. But for on-time performance several improvements were need to be done. As the results that were E, which means that only more than 70% of bus departures were more than 5 min late. Services hours were considered moderate to have 12– 14 hours of operation in a day. Since people did not use the bus at night before the office hour, services frequency for Penang was around 15 min waiting time, which was considered adequate, as indicated by results of C.

Table 9. QOS score vs. QOS attributes.

Attributes	QOS	QOS Score	Mean QOS	Overall QOS
Hours of Service	C	3	3.75	D
Service Frequency	E	5		
Passenger Load	B	2		
On Time Performance	E	5		
Total		15		

4. Conclusion

In conclusion, the overall QOS for all routes in Penang is D, which is acceptable, but certain areas need improvement. Outcome of these studies will benefit the authorities and bus provider in identifying significant counter measures to improve the QOS in bus performance. To improve the result obtained in the on-time performance, the actual timetable of bus needs to be supplied by the bus operator since they not publish it publicly. In this study, the estimated time of departure was based on service frequency. It was used according to the departure at particular bus stops. As for service frequency, it is recommended to use the correct departure schedule of on-time performance so that the data taken are reliable. At the same time, to improve the data obtained for passenger load, it was favourable to obtain the data on different days, as there could be difference in traffic volume on those routes. The outcomes could be influenced by the number passenger who used the public buses on that particular day. The above recommended improvements are for future researchers who are interested to continue this study. All suggestions are based on the results and analysis of this research. With the implementation of these recommended steps, it is aimed that the results obtained will be more precise and accurate. Therefore, it will heighten the perception, interest of passengers and improve the QOS of bus system in Penang or other parts of Malaysia.

5. References

[1] Department of Statistics Malaysia 2018 *Current Population Estimates Malaysia* Retrieved on July 31, 2018 from https://www.dosm.gov.my

[2] Opalyn M 2016 *Massive projects in place to alleviate urbanisation in Penang 29* The Malay Mail (October 29)

[3] Rasagam G 1999 A "step" towards sustainable transport: a case study of Penang, Malaysia *Transport and Communication Bulletin for Asia and the Pacific* (New York: United Nations) 68 33-47

[4] Sia M L 2003 *Factors determining public support for the Penang Outer Ring Road (PORR): An application of logit analysis* Research project for Master of Economic Management (Penang: Universiti Sains Malaysia)

[5] Eboli L and Mazzulla G 2008 A Stated Preference Experiment for Measuring Service Quality in Public Transport *Transportation Planning & Tech.* **31**(5) 509–52

[6] Murray A T 2002 A Coverage Model for Improving Public Transit System Accessibility and Expanding Access *Annals of Operations Research* **123**(1–4) 1 43–156

[7] Al Mamun M and Lownes N 2011 A Composite Index of Public Transit Accessibility, *J. of Public Trans.* **14**(2) 69–87

[8] Kim Y, Park J, and Kim E 2005 A development of punctuality index for bus operation, *J. of the Eastern Asia Society for Trans. Studies* **6** 492–504

[9] Arhin S A, Noel E C, and Dairo O 2014 Bus Stop On-Time Departure Performance and Criteria in a Dense Urban Area *Inter. J. of Traffic and Transportation Engineering* **3**(6) 233–238

[10] Suwardo, Napiah M B and Ibrahim B K 2009 On-Time Performance and Service Regularity of Stage Buses in Mixed Traffic *World Academy of Sci., Eng. and Tech.* **3**(7) 1137–1144

[11] Suwardo W, Napiah M and Kamaruddin I 2008 Punctuality and Expected Waiting Time of Stage Buses in Mixed Traffic, *J. Transportasi* **8**(3) 213–226

[12] Yaakub N and Napiah M 2011 Public Transport: Punctuality Index for Bus Operation *World Academy of Sci. Eng. and Tech.* 60(12) 12–23

[13] Transportation Research Board (TRB) 1999 *A Handbook for Measuring Customer Satisfaction and Service Quality* TCRP 3rd Edition Report 47 (Washington: National Academy Press)

[14] Eboli L, Forciniti C and Mazzulla G 2014 Service Coverage Factors Affecting Bus Transit System Availability *Procedia - Social and Behavioral Sciences* 111 984–993

[15] Piriyawat C, Candidate D and Author P (n.d.). Alternative Methods for Investigating Bus Service Quality and User Dissatisfaction in Bangkok and Its Vicinity Alternative Methods for Investigating Bus Service Quality and User Dissatisfaction in Bangkok and Its Vicinity

[16] Redman L, Friman M, Gärling T and Hartig T 2013 Quality aspects of public transport that attract car users: A research review *Transport Policy* **25** 19–127

[17] Leong L V, Ibrahim W H W and Sadullah A F M 2006 Passenger car equivalents and saturation flow rates for through buses at signalized intersections in Malaysia. *ARRB Conference – Research into Practice 2006 Canberra* (Australia)

[18] Nathan E and Alex P 2006 Reality mining: sensing complex social systems, *J. Personal and Ubiquitous Computing* **10**(4) 255–268

Acknowledgments

Authors would like to thank Yayasan Chancellor Universiti Tenaga Nasional for providing research grant for this study (202101015YCU) and Innovation Research Management Centre (IRMC), Universiti Tenaga Nasional.

Rainfall-Induced Hydraulic Properties for Unsaturated Soil in Klang Valley

F A Kechik[1], I S Izham[1], A Ibrahim[1*], Z A Hassan[1], M Mukhkisin[2], S J Matlan[3], and N Kasim[4]

[1]Department of Civil Engineering, Faculty of Engineering, Universiti Pertahanan Nasional Malaysia (UPNM), Sungai Besi Camp, 57000, Kuala Lumpur, Malaysia.
[2] Department of Civil Engineering, Politeknik Negeri Semarang, Semarang, Indonesia.
[3] Kulliyah of Engineering, International Islamic University Malaysia, Gombak, Malaysia.
[4] Faculty of Engineering, Universiti Malaysia Sabah, Malaysia.
*Corresponding author: aniza@upnm.edu.my

Abstract. Klang Valley area is one of the most landslide-prone areas in Malaysia, especially at Hulu Kelang, Kuala Lumpur. The area has been frequently hit by landslide since 1990s. Soil instability is agreed by researchers occurred due to high precipitation and long duration of rainfall which cause property damage and leading to injury and fatality. Slope failure is also triggered by the antecedent rainfall leads to infiltration of rainwater into soil. Therefore, study of rainwater infiltration is vital to relates soil – water interaction and soil behaviour for varies of rainfall intensities and duration for unsaturated soil. The objective of this paper is to determine and compare soil water characteristic curve (SWCC) which is one of the soil hydraulic parameters for Klang Valley area. Samples were collected to determine the soil hydraulic properties at Hulu Kelang area, Universiti Kebangsaan Malaysia (UKM) and Universiti Pertahanan Nasional Malaysia (UPNM) campuses. SWCC was obtained by pressure plate extractor apparatus experiment and the analysis was performed using Van Genuchten equation. Result of parameters obtained shows significant differences of soil at Hulu Kelang area compared to soils at UKM and UPNM campuses. This research is relevant to supports national slope master plan 2009-2023.

1. Introduction

Landslides occur due to decreasing of soil shear strength that leads to instability of the slope. The slope instability is dependent on the cohesion movement of the soil particles and with internal friction of the soil, geometry of the soil, slope angle and hydraulic behaviour of the soil, triggered by the rainfall infiltration. Previous research had shown perched water table development, increment level of groundwater table, existence of surface erosion and the denser in soil unit weight are depending on the soil moisture content from the rainfall process [1]. The factor of safety is the influential of soil slope stability analysis especially in geotechnical structure as well as in slope engineering design. The factor of safety is defined as the ratio of the shear strength to the shear stress. The unstable slope gives the safety factor output of less than 1. The factor of safety also represents the ratio of resisting forces to driving forces. The slope failure would be quick sliding when the resistance safety factor is below the allowable safety factor [2]. The volumetric water content of the slope increasing extremely followed with the decreasing of the safety factor during the raining season.

It is observed that the soil structure changed due to the precipitation of rainfall infiltration. The rainwater infiltrates into the soil structure through the voids of the particles. The major geotechnical problems to the structure are influenced by the presence of water. The history of the landslides zone is occurred from the rainwater, leaking pipe from the drainage and the presence of water through the void of soil particles. In consequence, water can cause a disaster to the slope instability, hence it induced the hydraulic properties of the soil matric suction. As the result, the slope stability formed a negative impact since water can pass through porous space and fracture in soil.

The rainfall intensity reduces the factor of safety marginally [3]. It was introduced that water in the soil categorized into water of gravity, water of capillary and water of hygroscopic [4]. The freely movement of water due to the force of gravity through the soil is known as water gravity. Water of capillary exists when the water moves due to the potential of capillary. Meanwhile, the adhesion force is required to hold very tightly the films formed by water of hygroscopic around the soil particles surface. The rainfall influenced the variation of the soil behaviour accommodate with the higher the rainfall intensity, the lower of effective cohesion and internal degree of friction, the lower or even disappearance of matrix suction, and the appearance and higher of pore water pressure [5].

The rainfall infiltration sustained by gravity and capillary action from the soil particles that able to collect the intensity of rainwater and creates groundwater table. The rainwater infiltrate into the ground while the capillary action promotes the capillary potential and capillary conductivity between the void of the soil particles. The relationship between the capillary potential with condition for water transmitted and stored when the soil in the unsaturated state was measured by E. Buckingham [6]. The increment of the pore air pressure together with the decrement of matric suction occurred due to the advancing wetting front during the process of rainwater infiltration to the soil slope of unsaturated zone. This phenomenon leads to decline of slope stability [7].

The presence of water in the soil particles is the major problems for geotechnical structure. Pores with variation geometry, sizes, and connectivity in the porous system of soil are reflected from the results of the soil hydraulic properties [8]. The weathering process of the soil triggered the movement of soil particles. Climate in Malaysia is equatorial throughout the year with hot, humid, and rainy. Soil slope becomes less stable when the matric suction decreases while the unit weight of soil increases due to the rainfall infiltration. From the previous observation, it is found that hydraulic conductivity of soil is the main role in rainfall-induced failures of the slopes. Moreover, it was stated that residual soil slopes have experienced a very long and severe weathering process consequently these slopes are potentially susceptible to instability [9].

1.1. History of Affected Area

In the area of Hulu Kelang, rainwater is the major factor influence the slope stability. The factor of safety decreases when the rainfall intensity increases. The annual rainfall can achieve as high as 4500 mm. The intense chemical weathering and formation of thick residual soil profiles influenced by the high temperatures combined with the annual rainfall. It was found that landslides are the deadliest catastrophic other than flood [10]. The landslides often happened during monsoon seasons each year.

Hulu Kelang region is one of the most prominent landslides tragedies in Malaysia. The historical of the avalanche zone demonstrates that various slope failures have happened to ensure the safety for the residents. The major landslides knocked in 1993 where the building of Highland Tower apartment was collapsed. There are 48 people killed during the tragedy. In 1999, many people were trapped in their homes when a landslide occurred near the Athenaeum Tower condominium and reoccurring until 2000. In another year, the instability slope is found in Taman ZooView, Ulu Klang which happened twice in October and November of 2001. The Taman Hill View is another location occurrence with landslides in November 2002. Later, in 2003, the Oakleaf Park Condominiums was having a major landslide on November and followed by the Jalan Bukit Mulia in five days after. In 2008, a landslide buried 14 bungalows in Taman Bukit Mewah. In 2017, the landslide in the Riverdale Park Apartments caused shock to the residents although there were no casualties or structural damages. In 2018, the backyard of Bukit Antarabangsa Mosque damaged due to the landslides. Therefore, mentioned tragedies provides evidence to investigate the location of Bukit Antarabangsa as one of location prominently with major landslides.

This paper is to investigate the correlation of the soil basic properties and its hydraulic properties which is SWCC in which indicates the soil behaviour towards soil stability of each location. The results from this study are intended to provide future basic risk management plan.

2. Methodology

2.1. Location of study

This This research study was conducted by using three previous papers about unsaturated soil. Previous researchers have studied the hydraulic conductivity impacts to the soil where the slope failure affected by the rainfall based on current condition for every year of rainfall event. The slope stability may change every year from the rainwater infiltration. The pore water pressure distribution in unsaturated zone is influenced strongly by the saturated or unsaturated permeability. Thus, this study is focused on the comparison of previous papers of the same method used which is soil water characteristic curve (SWCC) but at different locations which are the area with landslides tragedy (Hulu Kelang), the area of rainforest reserved (UKM) and the area of the development (UPNM).

It was discovered that 24 locations from Hulu Kelang soil slope region [11]. The major landslides of Highland Tower have been observed that it caused by the leakage of pipe under the ground [12]. However, the further question is whether the main problems of pipe leakage are the main cause of all the landslides in Hulu Kelang region. This question has previously never been addressed because the investigation from the causes of the landslide's disaster is still in observation. From previous researchers, landslides in Malaysia are triggered by earthquakes, rainfall, and volcanic activities, changes of groundwater and slope profile or combinations of all these factors. Therefore, further investigation is needed to find out the factors of the landslides in Hulu Kelang region.

The investigation of the effect rainfall infiltration into unsaturated soil using column located at Universiti Kebangsaan Malaysia (UKM) has been done by A. Ibrahim et al. [13]. The UKM main campus in Bangi is an area covered with the permanent reserved forest. The area is officially known as Bangi Forest Reserve (BFR) and act as a catchment area. The forest acts as a big sponge that soaking the rainwater while the infiltration process also occurred at the roots of the trees water seep into the ground. The slope of this rainforest reserved is considered stable.

R.M Ammer [14] studied the hydraulic properties parameter at Sungai Besi Camp Area. This paper is chosen because the area of Sungai Besi Camp especially in Universiti Pertahanan Nasional Malaysia (UPNM) is in progress with the development. Thus, one of the locations in Sungai Besi Camp area had been investigated and the hydraulic properties of the soil have stated in the results of this paper.

2.2. Instrumentation

All data were obtained through undisturbed samples that have been saturated. Samples were placed in pressure plate extractor machine and varies of pressures were applied to the samples. Each sample was weighted before increment of subsequent pressure applied to the sample. Then, results are interpreted using graph of suction versus volumetric water content. To obtain the parameters, the results measured through laboratory works were fitted using Van Genuchten [1] equation. It is suggested by M. Mukhlisin et al. [15] that this equation is widely used model and the equation is expressed by:

$$Se=[\frac{1}{1+(\alpha h)^n}]^m \quad (m=1-1/n) \tag{1}$$

where, Se is the effective water content, h is the suction head, α and n are the fitting shape parameters of SWCC [16].

IOP Conf. Series: Earth and Environmental Science **971** (2022) 012019 doi:10.1088/1755-1315/971/1/012019

3. Results

3.1. Basic properties of soil

It is noted that the soil consists of particle of minerals, organic matter, water molecules and air void where the combination of all these can determine the basic properties. Those properties of soil which are used in the identification and classification of soil are known as Index Properties. The variety Index Properties of soil are water content, in-situ density, specific gravity, particle size and consistency. The texture of the soil and the structure of the soil together with the types of soil also determined the soil properties. The basic properties of soil were obtained from the laboratory by testing the soil for moisture content, porosity, bulk density, and soil classification.

Based on the previous studies by T.H Guan [11], A. Ibrahim et al. [13] and R.M Ammer [14], soil of Hulu Kelang area is classified as very clayey sand while reserved forest soil in UKM is classified as poorly graded gravel with sand. Soil of UPNM development area is classified as silt and clayey sand. These three distinctive groups of soil will give better representation on hydraulic properties of every soil respectively.

3.2. Hydraulic properties of soil

Hydraulic properties of soil are defined as the behaviour of the soil. It can be measured by using soil water characteristic curve (SWCC). Hydraulic conductivity defined the movement of water in the soil that passes through the interconnected void spaces and to show the type of the soil which is silt, sand, or loam. According to A. R. Dexter [8] the pores with variation geometry, sizes and connectivity in the porous system of soil are reflected from the results of the soil hydraulic properties. The soil hydraulic properties reflect the movement of water for saturated and unsaturated soil condition through the void of soil particles. The pore size distribution reveals the hydraulic properties of soil due to the present of watercourse at the porosity of soil which relates with the definition of hydraulic conductivity, *K* [17]. Figure 1 show graphical presentation of SWCC plot derived from data of soil of every location. Table 1 shows the parameter obtained from the graphical data of SWCC of every soil.

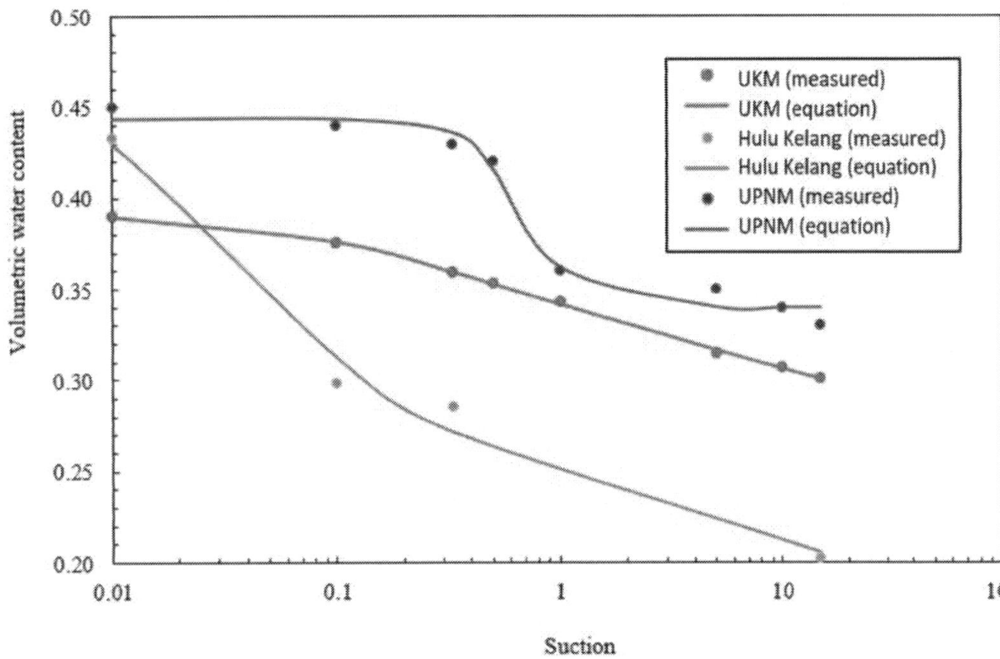

Figure 1. Graph of SWCC derived from data of UKM [13], Hulu Kelang [11] and UPNM [14].

The type of soil can affect the values of volumetric residual water content, θr and pore-size distribution, *n* obtained from SWCC function which are shown by higher values of θr and *n* of clayey Hulu Kelang and UPNM soil compared to gravelly sand of UKM soil. Results also prove that Hulu Kelang has the highest value of saturated water content, θs compared to the soil at UKM and UPNM. It shows that the highest water interaction, the highest soil porosity, the higher the density of soil to become saturated promoting the soil slope failure in Hulu Kelang.

Table 1. Hydraulic soil properties for all locations.

Parameter	Hulu Kelang	UKM	UPNM
Volumetric saturated water content, θ_s	0.5227	0.3919	0.4434
Volumetric residual water content, θ_r	0.1727	0.14636	0.3395
Inverse to air entry suction, α	229.88	12.471	1.6712
Pore-size distribution, n	1.2918	1.0886	3.7839
R^2	0.9932	0.9994	0.9824

4. Conclusion

The soil water characteristic curve (SWCC) is the best method to investigate the hydraulic properties for unsaturated soil because it is closely related with the matric suction and the volumetric water content of soil. The results from the hydraulic parameter of SWCC is calculated using Van Genuchten [1] equation formula and the graphs are constructed with the output of best fit curve representing the volumetric water content verse matric suction. It shows that the unsaturated soil is partially saturated because the reading of SWCC for unsaturated soil shows there is existence of water content in the soil particles. In contrast, the soil moisture contents can be decreasing with high temperature and formed an unsaturated soil. Therefore, it can be concluded that the movement of water infiltration in the soil changing from dry soil to wet soil and returned to dry soil may influence the movement of soil particles before leading to landslides phenomenon.

Further studies are needed in determining the behaviour of the landslides slope especially for the unsaturated soil because the unsaturated soil usually can turn into saturated soil during rainfall and the weathering process for the soil to return unsaturated is existed. Thus, the authorities must check overall area to ensure the safety for the residents. Besides, the landslides can be overcome by preparing the chemical agents so the slope can be reinforced. On the other hand, further studies can be continued by modifying the geometry of the slope. The installation of retaining walls, piles, grouting rock joints and fissures together with the rerouting surface and underwater drainage also can reduce the disaster of landslides.

5. References

[1] Lei W, Dong H, Chen P, Lv H, Fan L, and Mei G 2020 Study on runoff and infiltration for expansive soil slopes in simulated rainfall *Water (Switzerland)* **12**.
[2] Shinoda M *et al.* 2015 Dynamic behavior of slope models with various slope inclinations *Soils Found* **55** pp 127–142.
[3] Kristo C, Rahardjo H, and Satyanaga A 2017 Effect of variations in rainfall intensity on slope stability in Singapore *Int. Soil Water Conserv. Res.* **5** pp 258–264.
[4] Zhai Q, Rahardjo H, Satyanaga A, and Dai G 2019 Estimation of unsaturated shear strength from soil–water characteristic curvE *Acta Geotech.* **14** pp 1977–1990.
[5] Rahardjo H; Lee T T; Leong E C; Rezaur R 2005 Response of a residual soil slope to rainfall *Can. Geotech. J.* **42** pp 340–351.
[6] Buckingham E 1907 Studies on the movement of soil moisture *U.S. Dept. Agric., Bur. Soils*

Bull. **38**.
[7] Jeong S, Lee K, Kim J, and Kim Y 2017 Analysis of rainfall-induced landslide on unsaturated soil slopes *Sustain.* **9** pp 1–20.
[8] Dexter A R 1988 Advances in characterization of soil structure *Soil Tillage Res.* **11** pp 199–238.
[9] Suradi M 2016 Rainfall-induced failures of residual soil slopes with spatial variability of hydraulic conductivity *INTEK J. Penelit.* **3** pp 1–7.
[10] Abdul Rahman H and Mapjabil J 2017 Landslide disaster in Malaysia: an overviewN*Heal. Environ. J.* **8** pp 58–71.
[11] Guan T H 2010 *Kajian ciri-ciri asas tanah dan kesan porositi terhadap kejadian tanah runtuh di Ulu Klang* BSc Thesis (Malaysia: Universiti Teknologi Malaysia).
[12] Kazmi D, Qasim S, Harahap I S H, and Vu T H 2008 Analytical study of the causes of the major landslide of Bukit Antarabangsa in 2008 using fault tree analysis *Innov. Infrastruct. Solut.* **2** pp 1–11.
[13] Ibrahim A, Mukhlisin M, and Jaafar O 2018 Effect of rainfall infiltration into unsaturated soil using soil column *AIP Conf. Proc.* **1930**.
[14] Muhammad Ammer R 2019 *Comparison of soil hydraulic properties of unsaturated soil slopes at Sungai Besi Camp area* BSc Thesis (Malaysia: Universiti Pertahanan Nasional Malaysia).
[15] Mukhlisin M and Taha M R 2011 Effect of soil porosity and slope gradient on the stability of weathered granitic hillslope *J. Kejuruter.* **23** pp 57–68.
[16] Van Genuchten M T 1980 A Closed-form Equation for Predicting the Hydraulic Conductivity of Unsaturated Soils *Soil Sci. Soc. Am. J.* **44** pp 892–898.
[17] Nimmo J R 2013 *Porosity and pore size distribution.* (Elsevier Inc.).

Acknowledgement
The authors would like to express their appreciation to the financial supports provided by National Defense University of Malaysia through UPNM/2019/GPJP/2/TK/8.

A Preliminary Study: Influence of Sub-Surface Ground Condition on the Tunnel Boring Machine Performance

K Ramanathan[1], R A Abdullah[1*], A Rahim[1] and S N Jusoh[1], C M Khoo [2], X Fua[2], Y Xin Ban[2] and Q Xie[2]
[1] Department of Geotechnics and Transportation, School of Civil Engineering, Faculty of Engineering, Universiti Teknologi Malaysia, UTM Skudai, 81310 Johor, Malaysia
[2] Mass Rapid Transit Corporation Sdn. Bhd., Malaysia
*Corresponding author: asnida@utm.my

Abstract. Tunnelling through various ground condition affects the performance of Tunnel Boring Machine (TBM). Previous researches have conducted studies to explain on this relationship. However, a more localised influence of Malaysian Kenny Hill Formation subsurface condition towards TBM drive are yet to be carried out. This study will assess this relationship at areas nearby Museum Station during the Klang Valley Mass Rapid Transit (KVMRT) Line 1 with a length span of 540m. After the data collection from TBM and borehole report, data analysis was carried out namely using cutter head, thrust cylinder and subsurface profiling. These data were plotted based on the chainage and tunnel ring number. It is observed that subsurface profile at Ring 9 and Ring 11 required a higher thrust force due to its cohesive soil condition and inclusion of sandstone rock. However, Ring 10 is vice versa as there were more non-recovered soil samples at the specified area. The reduction of thrust force eventually increases the speed of TBM penetration. Thus, a clear relationship showing that subsurface condition highly influences the TBM drive.

1. Introduction

Tunnelling activities have been major alternatives to reduce the traffic congestion and limited land space in major cities in Malaysia which forbids future development. With the increasing population and demand, Malaysia has opted to this alternative according to Sustainable Development Goals that specify life on land and sustainable cities [1].

However, during the tunnelling process, the problem arises when the subsurface conditions are not so favourable [2]. This will affect the Tunnel Boring Machine's (TBM) performance. Therefore, tunnel linings are installed once the excavation process is completed. Nevertheless, geological conditions play an important role in influencing the excavation process. Various researchers conducted studies on the influence of ground conditions on the TBM performance. Yet, a localised study in Malaysian ground conditions is relatively limited in terms of preliminary studies. Hence, this study will show the results achieved via the objectives set alongside the relevant framework.

IOP Conf. Series: Earth and Environmental Science **971** (2022) 012020 doi:10.1088/1755-1315/971/1/012020

2. Objective
The main aim of this study is to assess the performance of the Tunnel Boring Machine (TBM) drive of the tunnelling project site. In order to solve this aim, these objectives must be achieved as follows:

- To identify and classify important parameters from the TBM report
- To analyse the nearest Soil Investigation report
- To assess the factors influencing the TBM drive

3. Methodology
The methodology used for this study is shown in Figure 1. A simple operational framework was conducted to solve the objectives by stages. Based on Figure 1, there are 4 stages to be carried out before completing this study. Therefore, this aids in completing the main aim of this paper.

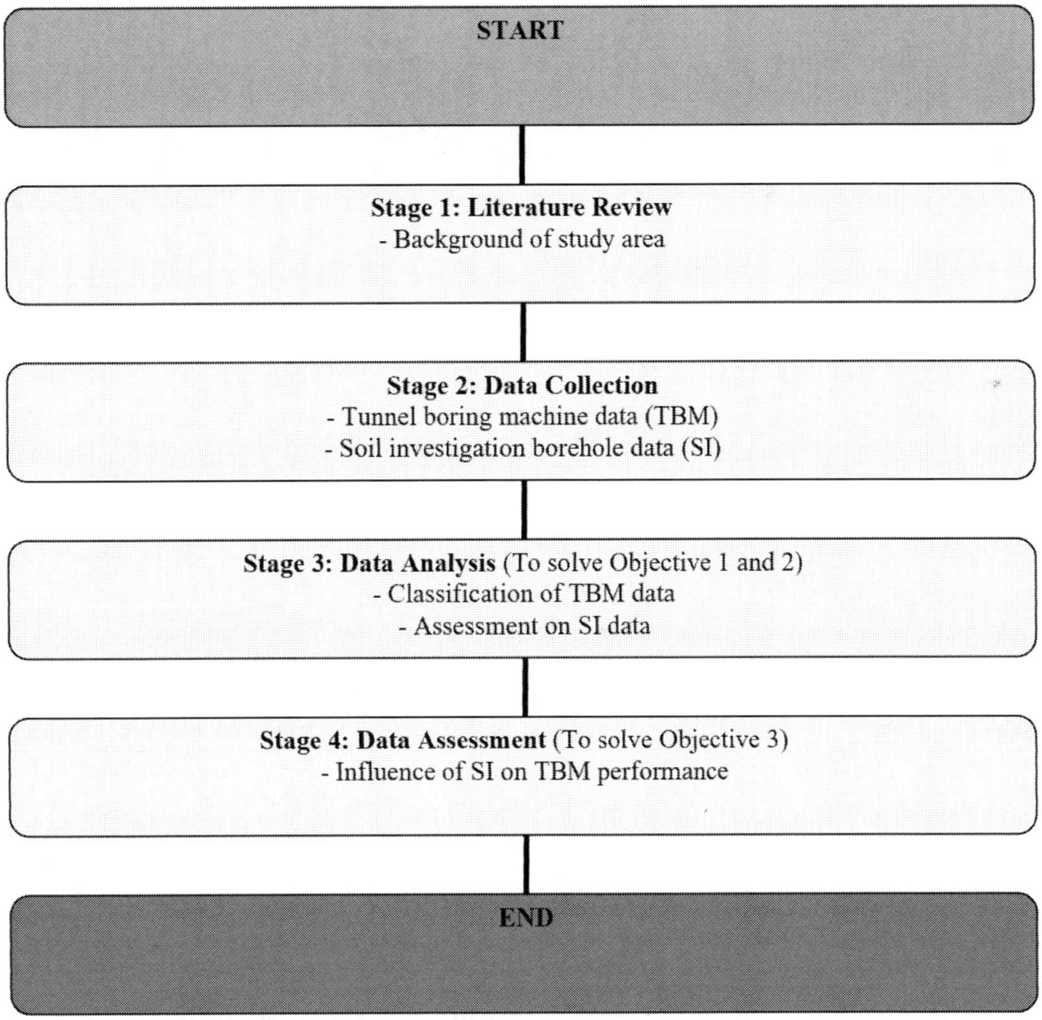

Figure 1. Operational framework for the purpose of this research.

4. Background of Study Area

This study is solely based on the secondary data received from the Mass Rapid Transit Line 1 (MRT Line 1) Technical team. This project is also known as the Semantan North Portal – Pasar Seni Klang Valley Mass Rapid Transit Line 1, which bounds in the North (NB) and South (SB) directions. The total length of the bored tunnel using the Tunnel Boring Machine (TBM Serial Number: crec050) is approximately 520m starting from Gerakan Belia 4B Malaysia till Orang Asli Craft Museum (refer to Figure 2). The tunnel diameter adapted in this project is uniform along the boring length is 6.35m. While the method of tunnel boring used is shield tunnel type associated with TBM.

Figure 2. Location of tunnel bored studies.

The TBM used involves using automated tunnel ring installation once the subsurface profile is bored out using the screw conveyor. The tunnel rings are numbered starting from R916 to R1121 for NB while R902 to R1109 for SB. These tunnel rings are located in between NB 2+384 to NB 2+830 and SB 2+360 to SB 2+800. Furthermore, detailed TBM numerical data was provided for the range of the tunnel rings stated previously. In addition to the TBM numerical data presented, the project location is underlain by Kenny Hill Formation. The subsurface profile here comprises of silt with Standard Penetration Test (SPT) N-value of 50 and weathered sandstone. The relationship between TBM data and these geological profiles will be discussed thoroughly in the next chapter.

5. Results & Discussion

Data entry and collection were carried based on the utmost important parameters, as shown in Table 1.

Table 1. List of data and parameters captured from the given TBM report of each rings.

Item	Data / Parameters		In Terms Of
1	Chainage		NB or SB
2	Ring Number		R
		Rotation	r/min
3	Cutter Head	Torque	kN.m
		Pressure	bar
4	Thrust Cylinder	Total force	kN
		Speed	mm/min

By referring to Table 1, the selection of data shows the overall operation of TBM during the subsurface profile boring process till the installation of the tunnel ring [3]. This dataset collected is then plotted graphically to observe TBM's trend with Item 1 and 2.

Besides that, Item 3 and Item 4, especially cutter head torque and total force of thrust cylinder are the most critical parameter that needs to be focused in this study. According to [4], these two parameters can evaluate the interaction of TBM during subsurface excavation works and the machine's overall performance. Theoretically, a weaker subsurface profile will result in higher torque

IOP Conf. Series: Earth and Environmental Science **971** (2022) 012020 doi:10.1088/1755-1315/971/1/012020

development when provided a constant thrust force [5]. A simple graphical correlation was plotted for each ring type at NB and SB, as shown in Figure 3 and Figure 4, respectively.

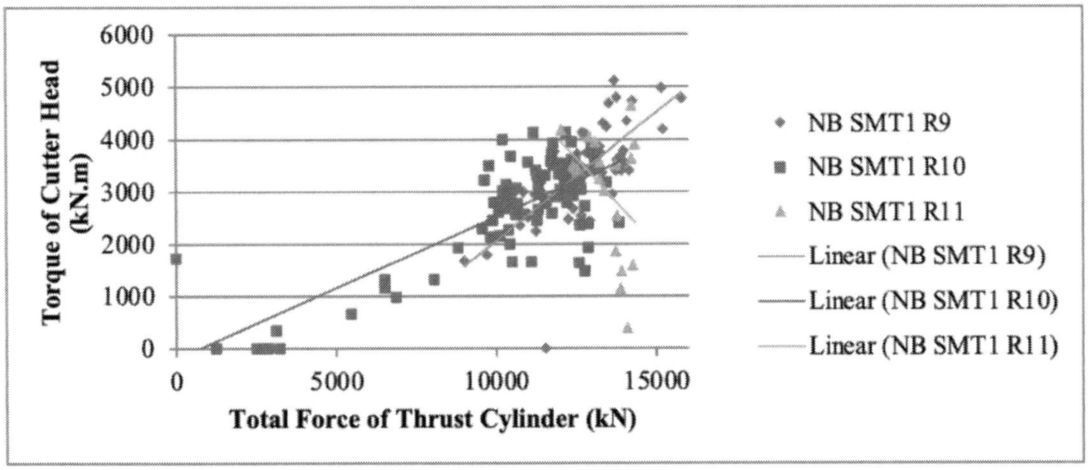

Figure 3. Relationship between the total force of thrust cylinder and torque of cutter head for NB.

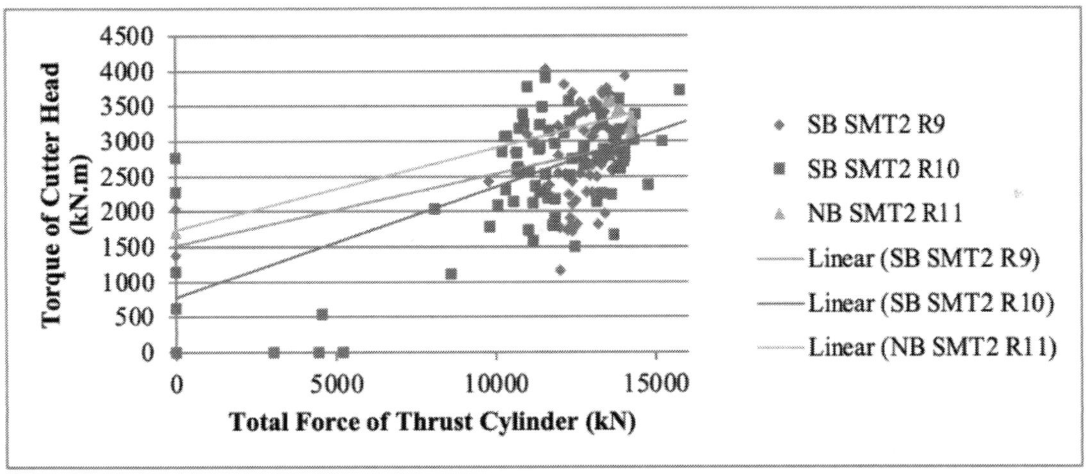

Figure 4. Relationship between the total force of thrust cylinder and torque of cutter head for SB.

From the graphs shown in Figure 3 and Figure 4, an increasing trend between the parameters was generally observed. A higher torque supplied by the more significant thrust force shows that the boring machine encountered a hard subsurface condition during the excavation works. This type of subsurface condition possesses a lower weakness plane which requires a greater force to be exerted on it to ease the penetration of TBM [6]. A similar trend line was observed during the correlation between thrust ranging from 4000 kN to 28000 kN and torque ranging from 0 to 10000 kNm of TBM where the coefficient of determination, R^2, was ranging from 0.5140 to 0.8682 [7].

However, a decreasing linear trend line was observed for NB SMT1 R11 as shown in the green coloured line in Figure 3. Numerically, Ring 11 (R11) at SB SMT2 shows a similar trend line if the first point of anomaly (0, 1698.15) is removed. This could be due to the weaker subsurface condition or infill cavities that required a lower thrust and torque for cutter head penetration. The lack of a complete dataset for R11; such as the nearest borehole data and TBM parametric report may also cause the inconsistent trend line. A clearer picture on the dataset obtained can be observed when comparing these parameters based on the reference number of ring, as shown below in Figure 5 and Figure 6.

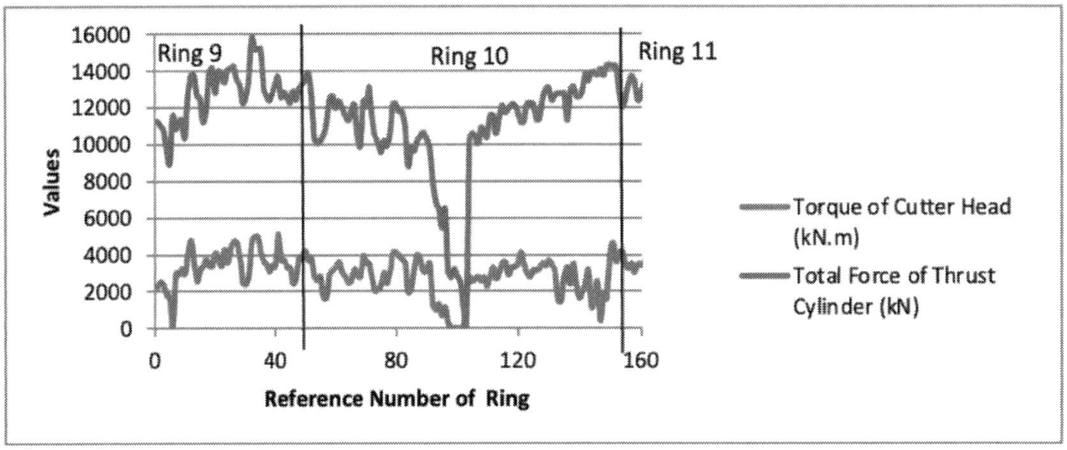

Figure 5. Excavation torque of cutter head and total force of thrust cylinder with reference to the ring number at NB SMT1.

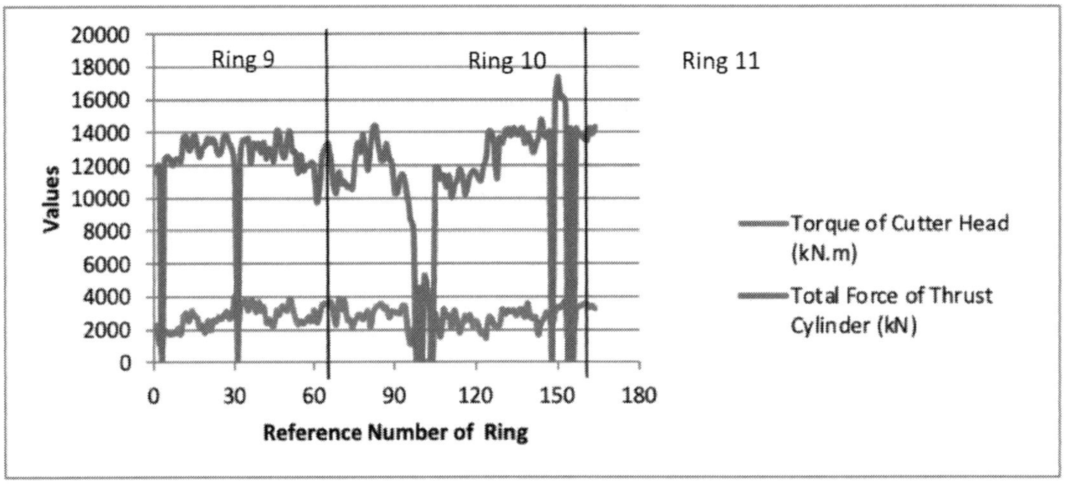

Figure 6. Excavation torque of cutter head and total force of thrust cylinder with reference to the ring number at SB SMT2.

Based on Figure 5 and Figure 6, Ring 9 and Ring 11 shows high cutter head torque and thrust cylinder force compared to Ring 10, where a significant drop in these values was observed. This can be related and explained based on its geological condition at the specified ring numbers. Since R11 has insufficient borehole data, the larger values observed at R9 are based on the Borehole 11 Report 40 (Table 2). While, lower values of R10 are due to the Borehole 12 Report 40 and Borehole 14 Report 40 as shown in Table 3 and Table 4 below, respectively.

Table 2. Borehole report for Ring 9 at NB SMT1 and SB SMT2.

Data	NB SMT1 and SB SMT2 Ring 9 (BH11 Report 40)
Ground level	32.414 mRL
SPT-N 50 Silt	23.414 mRL
Hit rock level	10.214 mRL
End of Borehole	3.214 mRL

Table 3: Borehole report for Ring 10 NB SMT1.

Data	NB SMT1 Ring 10 (BH12 Report 40)
Ground level	32.817 mRL
SPT-N 50 Silt	22.317 mRL
No recovery	16.317 mRL
No recovery + SPT-N 50 Silt	13.317 mRL
End of Borehole	4.067 mRL

Table 4: Borehole report for Ring 10 at SB SMT2.

Data	SB SMT2 Ring 10 (BH14 Report 40)
Ground level	37.693 mRL
SPT-N 50 Silt	31.693 mRL
No recovery	24.193 mRL
1st Hit rock level	22.693 mRL
No recovery + SPT-N 50 Silt	18.193 mRL
2nd Hit rock level	12.193 mRL
End of Borehole	2.893 mRL

Table 2 to Table 4, the yellow highlighted box denotes the location of tunnel bored using the TBM. The tunnel centerline is along the 15 mRL mark, where the tunnel diameter of 6.35 m shows the tunnel crown and toe is at 18.175 mRL and 11.825 mRL, respectively, as shown in Figure 7. While the overlay of the tunnel alignment on the subsurface profile are shown on the bar chart in Figure 8.

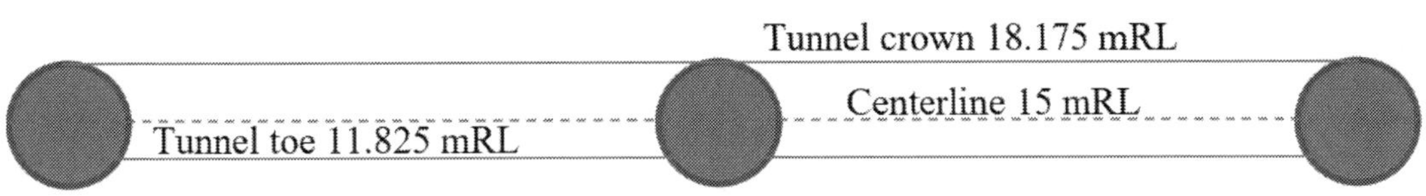

Figure 7. Tunnel dimension and levels for the studied project site.

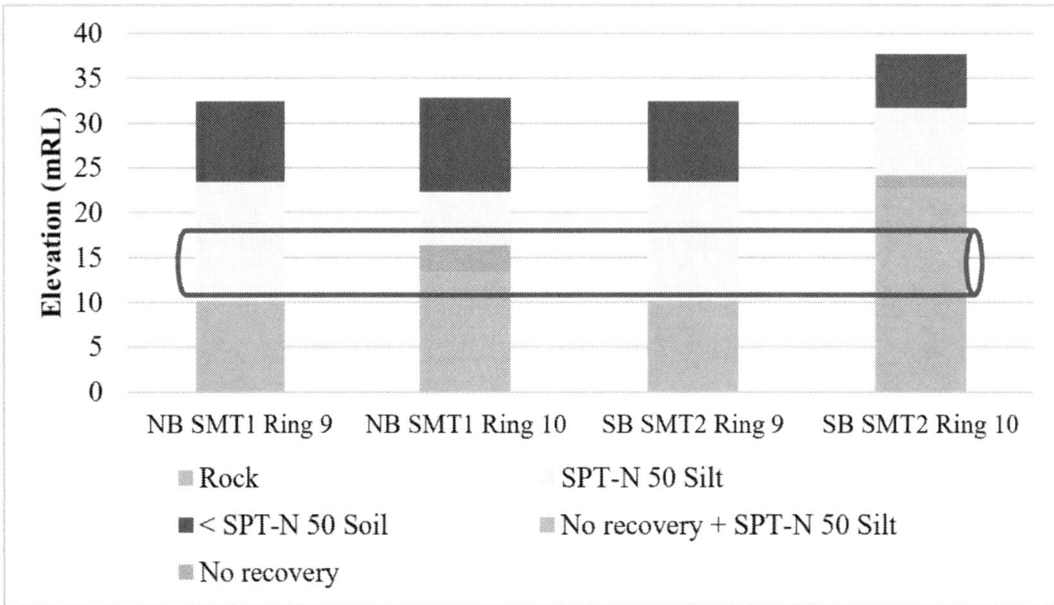

Figure 8. Overlay of tunnel alignment onto the subsurface profile obtained from the borehole data.

Based on the borehole reports, there are some differences in terms of geological conditions from Gerakan Belia 4B Malaysia till Orang Asli Craft Museum albeit the short span of length. According to Wallis (2015) [8], the main geological formation that underlain these study areas are the Kenny Hill Formation (Figure 9). This formation comprises interbedded clastic sedimentary rocks such as sandstone, siltstone, and shale.

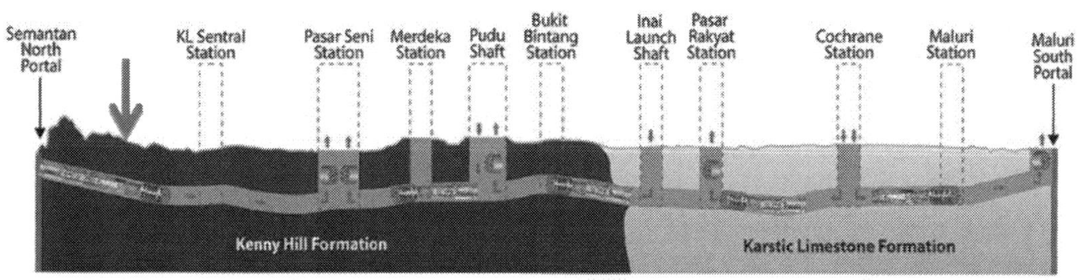

Figure 9. Tunnel drive and an indication of Museum Station via the blue arrow [8].

The differences in the subsurface profile mentioned previously are observed in Figure 8. The NB TBM goes through various profiles starting with silt with Standard Penetration Test (SPT-N) of 50 blows and a mixture of profiles with no sample recoveries. A similar TBM drive was observed at SB TBM, but it ends at profiles consisting of sandstone rock and no sample recoveries of soil.

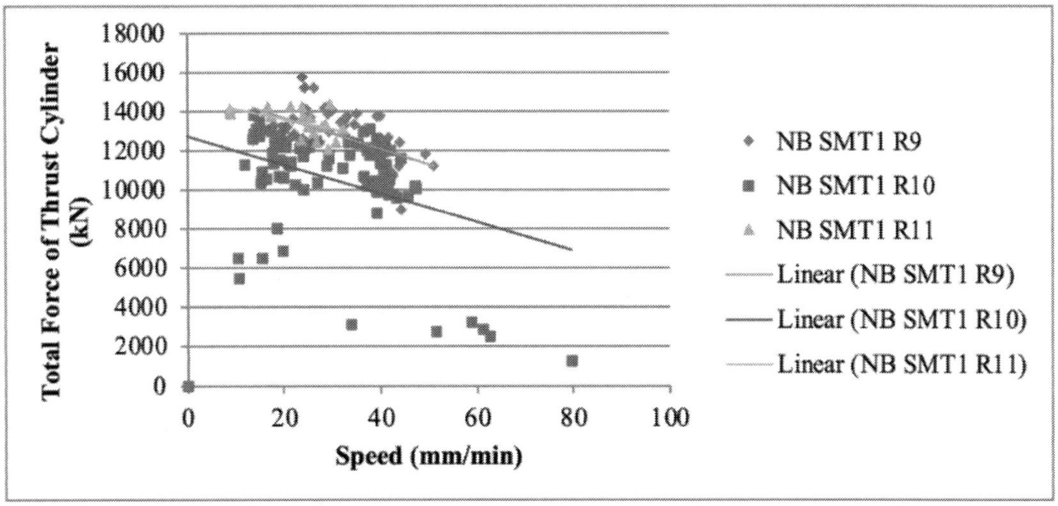

Figure 10. The relationship between total force and speed of thrust cylinder during TBM drive.

The inclusion of subsurface profiles consisting of the non-recovery samples shows that the soil type is non-cohesive and easily lost during the wash boring process. A similar process is done during the TBM movement. This eventually reduces the force needed by the TBM to penetrate through these profile types. Thus, the drop of TBM drive at Ring 10 for both NB and SB are proven in this matter. Theoretically, the decline of TBM thrust force will increase the penetration speed, which can be shown in Figure 10. This is due to the lesser work done required to penetrate weaker subsurface profiles. The validation of this statement are shown via the decreasing trend of about 18% was observed between the thrust force ranging from 26000 kN to 34000 kN and advance rate ranging from 16 mm/min to 66 mm/min during the Beheshthabad water conveyance tunnel excavation [9].

6. Conclusion

This study was based on secondary data along the alignment of completed tunneling sites surrounding the Museum Station of the KVMRT Line 1. This study has answered all the 3 objectives outlined in this paper. Firstly, the TBM data that were classified important in this study were cutter head and thrust cylinder parameter that focuses on the TBM performances. Secondly, the subsurface profile was plotted and analyzed using the SI borehole data. In general, the SPT-N 50 Silt, no recoveries (N/R) samples, and highly weathered sandstone were observed in the plotted profile. Thirdly, this TBM performance was successfully related to the influence of the geological profile. A higher torque and thrust force was needed whenever TBM penetrates through harder subsurface conditions like sandstone and cohesive SPT-N 50 silt. This situation was observed at Ring 9 and Ring 11 of both NB and SB TBM. Besides that, a sudden drop was observed in Ring 10 in relation to these values. This is due to the major non-recovered weak samples in the subsurface profile during TBM excavation. Furthermore, this drop in TBM drive causes a decline in total force but an increase of penetration speed. In a nutshell, the overall results obtained from the relationship between SI and TBM data are complementing and in line with each another. Thus, a subsurface condition highly influences the TBM performance in a localized area based on this study.

7. References

[1] Economic Planning Unit 2021 *Sustainable Development Goals* Retreived on January 31, 2021 from https://epu.gov.my/en/sustainable-development-goals.

[2] Park H, Oh J, Kim D and Chang S 2018 Monitoring and Analysis of Ground Settlement Induced by Tunnelling with Slurry Pressure-Balanced Tunnel Boring Machine", *Advances in CivilEngineering* ArticleID: 5879402, 10 pages

[3] Stypulkowski J B, Bernardeau F G and Jakubowski J 2018 Descriptive Statistical Analysis of TBM performance at Abu Hamour Tunnel Phase I, *Arab J Geosci.* **11** 191

[4] Faramarzi L, Kheradmandian A, and Azhari A, 2020 Evaluation and Optimization of the Effective Parameters on the Shield TBM Performance: Torque and Thrust—Using Discrete Element Method (DEM), *Geotech Geol. Eng.* **38** 2745–2759

[5] Kim K, Kim J, Ryu H, Rehman H, Jafri T H, Yoo H and Ha S Estimation Method for TBM Cutterhead Drive Design Based on Full-Scale Tunneling Tests for Application in Utility Tunnels. *Appl. Sci.* **10** 5187

[6] Zhang K, Yu H, Liu Z and Lai X Dynamics Characteristic Analysis of TBM Tunneling in Mixed-Face Conditions, *Simulation Modelling Practice and Theory.* **18** 1019–1031

[7] Cardu M, Catanzaro E, Farinetti A, Martinelli D and Todaro C 2021 Performance Analysis of Tunnel Boring Machines for Rock Excavation. *Appl. Sci.* **11** 2794

[8] Wallis S 2015 *Final breakthrough ends KVMRT TBM tunnelling.* Retrieved from https://tunneltalk.com/KualaLumpur-MRT-16Apr2015-Final-TBM-breakthrough.php.

[9] Mohammadzamani D, Mahdevari S and Baherpour R 2019 Evaluation of required thrust force based on advance rates in shielded TBMs under squeezing conditions, *Journal of Geophysics and Engineering*, Volume 16, Issue 5, October 2019 pp 842-861

Acknowledgments

This work was supported and funded by the Ministry of Higher Education under Fundamental Research Grant Scheme (Ref: FRGS/1/2020/TK0/UTM/03/5) Behaviour of Surface Settlement Induced By Tunnelling Under Non-Greenfield Condition and Collaborative Research Grant National (PY/2019/03147). The author acknowledges numerous lecturers in the civil, geology and geotechnical fraternity from Universiti Teknologi Malaysia (UTM) for their constructive comments on this study. Besides that, special thanks to the Mass Rapid Transit (MRT) team for providing the tunnelling data and permitting the publication of this article.

IOP Conf. Series: Earth and Environmental Science **971** (2022) 012021 doi:10.1088/1755-1315/971/1/012021

Effect of voids in kaolin stabilised by ground granulated blast furnaces slag mixtures

A M Awad Mohammed[1*], N Z Mohd Yunus[1], M A Hezmi[1], A S A Rashid[1], S N Jusoh[1] and J Arliansyah[2].

[1]School of civil Engineering, Faculty of Engineering, Universiti Teknologi Malaysia, 81310 Skudai, Johor Bahru, Johor, Malaysia.
[2]Department of Civil Engineering, Faculty of Engineering, University of Sriwijaya, Indonesia
*Corresponding author: khatib.kaka@yahoo.com

Abstract. Kaolin is known as problematic soil which has a low strength that needs to improve before construction. Chemical stabilization with cement is used widely to stabilize different clay types; however, using it comes with disadvantages associated with carbon dioxide emission and sustainability issues. As an alternative to cement, Ground Granulated Furnaces Blast Slag (GGBS) has been applied to stabilise kaolin. The increment in the strength is due to the formation of new products that fill the soil voids. The formation of those new products is verified using Field Emission Scanning Electron Microscopy (FESEM). However, the changes in pore space cannot be quantified using FESEM. Thus, this study uses Image J to calculate voids area detected by (FESEM) for different kaolin-GGBS mixtures with different content of GGBS cured for different curing period. The book-like structure for the raw kaolin have changed slightly when mixed with GGBS. This is due to the formation of hydrate gels as a result of the pozzolanic reaction. The results from image J indicates that the void decreases as the content of GGBS increases as well as the curing period.

1. Introduction

Kaolin is widely distributed around the world, and it is considered the most common clay minerals. It is a problematic soil with low strength and relatively high compressibility. Such a soil needed to be stabilized before construction [1, 2, 3]. Cement is known to be the common binder used for soil stabilisation. However, using cement comes with disadvantages associated with large amount of carbon dioxide (CO_2) emission during the manufacturing, producing 1 ton of cement leads to the emission of 0.95 ton of CO_2 [4,5]. Generally, global cement production is responsible for 7% of the total amount of artificial CO_2 emission [2]. Globally annual emission of carbon dioxide increased from 17.78 billion tons in 1980 to reach 32.1 billion tons in 2015, this increment is about 100% in 35 years and it is expected to reach double this value in 2035. In Malaysia, CO_2 emission increased by 69% from 2000 to 2020 reached 285.7 million tons [6].

Finding alternative to traditional stabilisers (cement and lime) have been challenging to researchers. Nowadays a lot of by products materials have been used to replace cement and lime either fully or partially. By product materials such as silica fume [7], rice husk ash [1], calcium carbide residue [8, 9] and demolish concrete [3], fly ash and bottom ash [10] have been used to stabilise different soil type. Silica fume is used to replaced lime partially, it has improved soil characteristics by decreasing permeability and increasing strength. Replacing cement by up to 10% rice husk ash (RHA) have shown an increment of CBR value. However, the addition of more than 10% RHA decreases the CBR value [1]. Other additives such as calcium carbide residue and demolish concrete are used alone to replace cement and lime in soil stabilisation.

Nowadays, recycling waste materials becomes essential for reducing the landfilling areas. It helps to reduce the landfilling areas, to achieve a more sustainable environment and to reduce the use of other non-sustainable material such as cement and lime. Current study demonstrates the use of ground granulated blast furnaces slag (GGBS) as a sustainable material to stabilise kaolin. GGBS is a by product material from iron manufacturing [11, 12], and it is widely available in Malaysia. GGBS can be used with OPC or other pozzolanic materials to produce durable concrete [13]. GGBS has an advantage over ordinary Portland cement, the production of 1 ton of GGBS leads to the emission of 0.07 ton CO_2 [14, 15]. It is also known that GGBS can capture CO_2 while enhancing soil strength [16].

This paper concerns on the effect of the GGBS content and the curing period in enhancing the soil strength by filling the voids within kaolin-GGBS mixture. The voids were determined by applying IMAGE J software on the images that have been produced from Field Emission Scanning Electron Microscopy (FESEM).

2. Methodology:

Slightly acidic brown kaolin was used in current study, it is a very fine powder with very little coarse particles; it has a brown color. Kaolin was bought from KAOLIN (MALAYSIA) SDN BHD factory, located in jalan Tapah – Bidor, 35000 Tapab, Parak, Malaysia. GGBS is a very fine off-white material with no course particle larger than 2mm. GGBS used in this study is a fine off-white material with no large particles (more than 2mm), it was collected from a local factory located in Johor Bahru, Johor, Malaysia. The physical properties for kaolin and GGBS are presented in Table 1 below.

Table 1. Properties of kaolin and GGBS.

Material	Kaolin	GGBS
Liquid Limit (%)	40.5	36.6
Plastic Limit (%)	22.5	-
Plasticity Limit (%)	18	-
Clay (%)	0.65	1.876
Silt (%)	92.6	98.124
Sand (%)	6.75	0
Specific Gravity	2.52	2.83
Optimum Moisture Content (%)	18	-
Maximum Dry Density (kg/m^3)	1640	-
UCS (kPa)	115	-
pH	4.33	10.6

Kaolin was mixed with 20% and 25% GGBS from its dry weight, then 18% distilled water was added to the mixture and mixed properly to ensure homogeneous sample. The GGBS contents chose were based on the suitable amount for kaolin stabilisation at earlier curing (7 days) which were 20% and 25% [17]. The mixture was placed in a cylindrical mould with 76mm and 38 mm diameter. Then, it was extracted and sealed properly to ensure no lost in moisture occurs. The samples were left to be cured for 0 and 7 days in a humidity chamber.

Once the curing period is finished, the sample was placed on the oven for 24 hours, then crushed to small pieces and sieved. The particles with diameter less than 2 mm was collected and used for the FESEM analysis, to take microstructure image. Those images were used for image J analysis.

Image J is an image processing program that uses Java, and it was developed at the National Institutes of Health (NIH). The current study used this software to process the FESEM images and measure the voids in raw kaolin and other selected treated kaolin samples under ambient conditions. Image J involves two methods to measure the voids, the one used in this study is threshold method. Figure 1 shows the example of voids detected by image J using the threshold method. The voids are shown in this figure in black color. Afterward, the software calculates the voids area and presents it in Excel sheet.

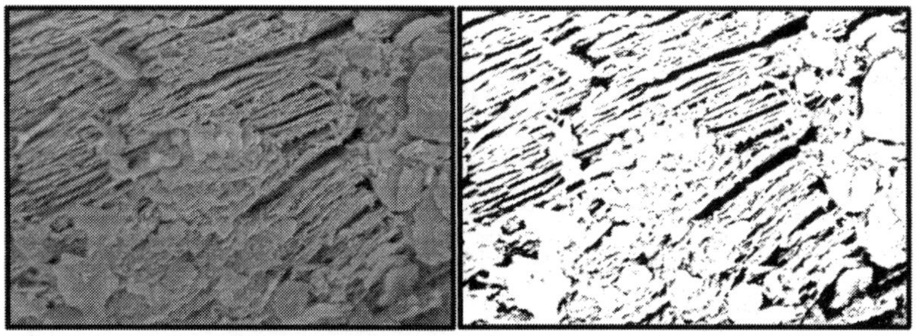

(a) (b)

Figure 1. Voids in kaolin clay detected by image J.

3. Result and discussion:

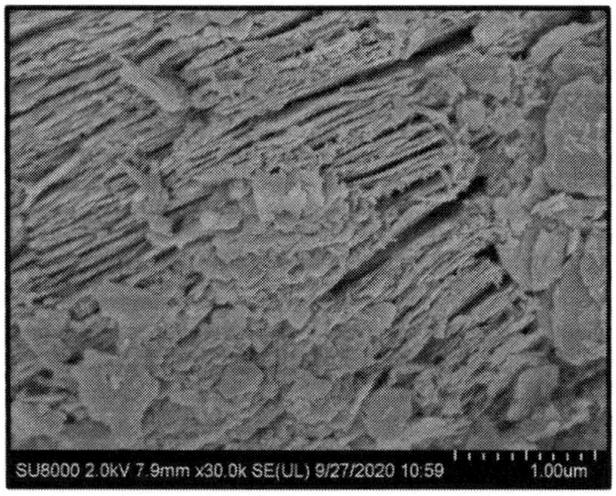

Figure 2. FESEM image for untreated kaolin.

Figure 2 shows the microstructure image for untreated kaolin taken at 30000 magnification. The shape of kaolin particles can be described as flaky and platy formation with smooth surface. The book like structure of kaolinite minerals is observed and it is considered the dominant feature of brown kaolin. Kaolin is known to have edge to face and edge to edge aggregate flocculated structure [18, 19, 20]. Based on the image, it can be seen that the structure of untreated kaolin contains large voids. The occurrence of voids leads to low strength.

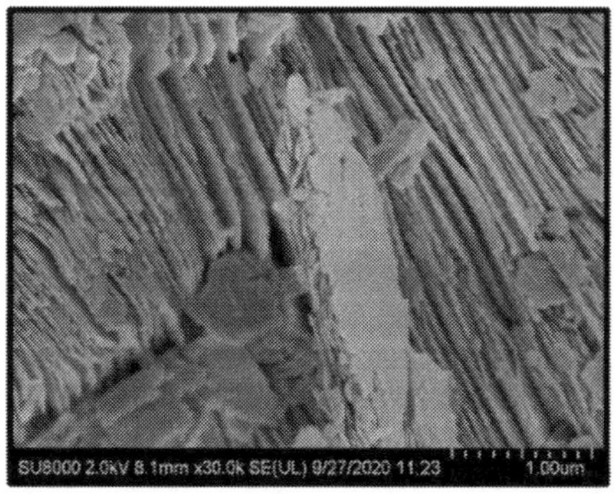

Figure 3. FESEM image of treated kaolin with 20% GGBS cured for 0 days.

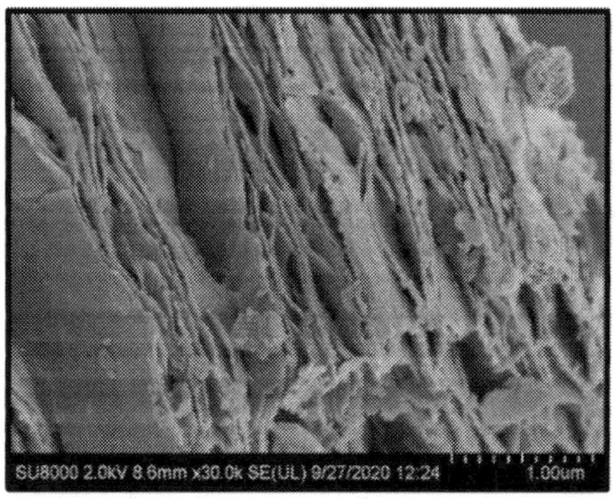

Figure 4. FESEM image of treated kaolin with 25% GGBS cured for 0 days.

Figure 3 and figure 4 show the image of treated kaolin cured for 0 days with 20% and 25% of GGBS. Similarly, images were taken at 30000 magnifications. The shape of micrographs does not differ compare to the untreated kaolin in which the dominant book like structure remains the same. Voids in the structure are still observed. In figure 4, for the treated kaolin with 25% GGBS calcium hydroxide ($Ca(OH)_2$) are observed, as a result the particles are slightly ponded and the voids became less .

Figure 5. FESEM image of treated kaolin with 20% GGBS cured for 7 days.

Figure 6. FESEM image of treated kaolin with 25% GGBS cured for 7 days..

Figure 5 and figure 6 shows the microstructure image at 30000 magnification for the treated kaolin with 20% and 25% GGBS cured for seven days. Changes start to occurs on the kaolin structure, the cementitious gels from pozzolanic reaction pond and coated the particles together, this makes the voids less. The kaolin structure changes from flaky structure to flocculated structure. Moreover, the voids in figure 6 are hardly detected.

In addition, Image J was used to determine the area of the available voids using the images detected from FESEM analysis.

Table 2. Voids measured in different kaolin-GGBS mixtures using Image J

Mixture	Total Area (um)	Voids Area (um)	Percentages of Voids (%)
Kaolin	11.84	2.11	17.8
20%GGBS-kaolin cured for 0 days	11.84	1.86	15.68
20%GGBS-kaolin cured for 7 days	11.84	1.69	14.27
25%GGBS-kaolin cured for 0 days	11.84	1.8	15.2
25%GGBS-kaolin cured for 7 days	11.84	1.37	11.57

Table 2 shows the calculate voids in different kaolin-GGBS mixtures using image J. The results illustrate that the voids in the raw kaolin represent about 17.8% from the total area. The percentage of voids decreases with the addition of GGBS. Adding 20% GGBS to kaolin reduces the voids by 2% while the addition of 25% GGBS to kaolin reduces its void by 2.5% at 0 days curing. Likewise, the voids have shown further reduction after 7 days of stabilization process. Significant drop on the percentage of voids from 0 to 7 days can be seen on 25% GGBS-kaolin mixture with 11.5% voids from the total area of the mixture. 25% GGBS-kaolin mixture shows less voids compare to 20% GGBS-kaolin and the raw kaolin may due to the availability of more calcium in 25% GGBS-kaolin mixture compare to other mixtures. Calcium reacts with silica and aluminum in kaolin to produce calcium silicate hydrate gels which bond soil particles hence reduce the voids. Higher curing period (7 days) results in a lower void compare to a lower curing (0 days) hence increase the strength more. 7 days curing allow GGBS to hydrate and form more gels that bond the particles together.

4. Conclusion:

Image J was used to determine the changes in voids within kaolin-GGBS. It was found that the addition of GGBS content decreases the voids available, is due to the pozzolanic reaction and the formation of the gels such as calcium silicate hydrate gels that bond the soil particles hence increases the soil strength. It is assumed that lesser voids can be observed with longer curing period, as it allows the calcium within the GGBS to hydrate and produce more gels to bond the particles.

5. References

[1] Gupta D, Priyadarshee A, Sharma S K and Singh H 2018 Behaviour of cement treated kaolin clay mixed with fiber and rice husk ash, *International Research Journal of Engineering and Technology.* **5** 250–255

[2] H Abdullah H, A.Shahin M and Sarker P 2018 Use of Fly-Ash Geopolymer Incorporating Ground Granulated Slag for Stabilisation of Kaolin Clay Cured at Ambient Temperature. Geotechnical and Geological Engineering, 2

[3] Mohd Yunus N Z, Ayub A, Wahid M A, Satar M K I M, Abudllah R A, Yaacob H, Hezmi M A 2019 Strength behaviour of kaolin treated by demolished concrete materials. *In IOP Conf. Series: Earth and Environmental Science* 220

[4] Du Y J, Wu J, Bo Y L and Jiang N J 2020 Effects of acid rain on physical, mechanical and chemical properties of GGBS–MgO-solidified/stabilized Pb-contaminated clayey soil, Acta Geotechnica. **15** 923–932

[5] Hung K and Johnson M U 2015 Utilization of ground granulated blast furnace slag as partial cement replacement in lightweight oil palm shell concrete *Materials and Structures.* **48** 2545–2556

[6] Jorat M E, A Aziz M, Marto A, ZainiN, Jusoh S N and Manning D A C 2018 Sequestration Atmosperic CO_2 Inorganically: A Solution for Malaysia's CO_2 Emission. Geosciences Journal, 8

[7] Hasan M, Alrubaye A J, Seng L K, Ideris M S and Marto A 2016 Undrained Shear Strength of Soft Clay Mixed with Different Percentages of Lime and Silica Fume. Jurnal Teknologi, 23–30

[8] Latifi N and Meehan C L 2017 Strengthening of Montmorillonitic and Kaolinitic Clays with Calcium Carbide Residue: A Sustainable Additive for Soil Stabilization. Geotechnical Frontiers 2017 154–163

[9] Latifi N, Vahedifard F, Ghazanfari E and A Rashid A S 2018 Sustainable Usage of Calcium Carbide Residue for Stabilization of Clays. *Journal of Materials in Civil Engineering,* **30** 04018099

[10] Jamaludin N, Mohd Yunus N Z, Mohammed A M A, Haron Z, Marto A, Mashros N, Abdul Kadir M A 2018 Comparison between cement and concrete waste on the strength behaviour of marine clay treated with coal ash, *MATEC Web of Conferences.* **250** 1–8.

[11] Mohammed A M A, Almthailee Y H Y Mohd Yunus N Z and Hezmi M A 2021c Effect of Ground Granulated Furnaces Blast Slag on Kaolin Properties Malaysian Construction Research Journal. Special Issue **13** 2

[12] Mohammed A M A, Mohd Yunus N Z, Hezmi M A, A Rashid A S and Horpibulsuk S. 2021d Carbonated ground granulated blast furnace slag stabilising brown kaolin. Environmental Science and Pollution Research

[13] Suresh D and Nagaraju K 2015 Ground Granulated Blast Slag (GGBS) In Concrete – A Review, *IOSR Journal of Mechanical and Civil Engineering.* **12** 76–82

[14] Islam A, Alengaram U J, Jumaat M Z and Bashar I I 2014 The development of compressive strength of ground granulated blast furnace slag-palm oil fuel ash-fly ash based geopolymer mortar, *Materials and Design.* **56** 833–841

[15] Mohammed A M A, Mohd Yunus N Z, Hezmi M A, Abang Hasbollah D Z and A Rashid A S 2021a Ground improvement and its role in carbon dioxide reduction: a review. *Environmental Science and Pollution Research* **28** 8969-8988

[16] Priyanga G, Divya Krishnan K, and P T Ravichandran 2018 Characteristics of Rubberized Soil with Ground Granulated Blast Furnace Slag as Binder Material. *In Materials Today: Proceedings* 8655–8661

[17] Mohammed A M A, Mohd Yunus N Z, Hezmi M A and Ahmad Rashid A S 2021b Sequestration of Carbon Dioxide Using Ground Granulated Blast Furnaces Slag and kaolin mixtures. *Global NEST Journal* **23** 105-111

[18] Saeed K A H, Kassim K A, Nur H and Yunus N Z M 2013 Characterization of Hydrated Lime-Stabilized Brown Kaolin Clay. *International Journal of Engineering Research & Technology.* **2** 3722–3727

[19] Saeed K A, Kassim K A and Nur H 2014 Physicochemical characterization of cement treated kaolin clay, *Gradjevinar.* **66** 513–521

[20] Saeed K A H, Kassim K A, Yunus N Z M and Nur H 2015 Physico-Chemical Characterization of Lime Stabilized Tropical Kaolin, *Jurnal Teknologi.* **72** 83–90.

Acknowledgment

The authors are grateful for the support of the Fundamental Research Grant Scheme (FRGS) No. FRGS/1/2019/TK10/UTM/02/21 & R.J130000.7851.5F197, Collaborative Research Grant National (CRG24.4) No. R.J130000.7351.4B496 and Matching Grant between Universiti Teknologi Malaysia and Universitas Sriwijaya, Indonesia No. Q.J130000.3051.02M

Carbon Dioxide Storage Potential in Malaysian Sandstone Aquifer: An Overview

I Z Abd Rahman [1], D Z Abang Hasbollah [2*], N Z Mohd Yunus[1], E H Kasiman[1] and A N Mazlan[1]

[1] School of Civil Engineering, Universiti Teknologi Malaysia, Johor Bahru, Malaysia
[2] Centre of Tropical Geoengineering, Universiti Teknologi Malaysia, Johor Bahru, Malaysia
*Corresponding author: dzulaika@utm.my

Abstract. Carbon dioxide (CO_2) emission rate from energy industries in Malaysia has increased exponentially in these few years. Based on this alarming increment of CO_2 emission rate from previous data, carbon dioxide sequestration in deep saline aquifer has been identified as one of the way to reduce and overcome the increment of CO_2 emission rate in Malaysian atmosphere. Malaysian geology consists of enormous amount of deep saline aquifer which most of the aquifer material is sandstone. Malay Basin has been identified as the most potential basin for CO_2 sequestration in Malaysia by previous researchers. There are four sites in Malay Basin which has potential to be CO_2 sequestration site such as Jerneh, Dulang, Tangga and Semangkok because they contain the formation of fine-grained sandstone, mudstone and coal properties. This paper aims to discuss CO_2 storage potential in Malaysian sandstone aquifer particularly Malay Basin in terms of potential injection location, properties of Malaysian sandstone aquifer, available technologies of injection, injection parameter as well as issues and challenges of carbon sequestration implementation in Malaysia.

1. Introduction

Power plant and industrial point are the main sources of CO_2 for carbon capture, storage and utilization (CCSU). According to [1], CCUS technologies has two important functions which to store the CO_2 in deep saline aquifers and to create a new important product from the CO_2 process. Moreover, the CCUS technologies are also using two resources from fuel combustion and industrial processes where all the CO_2 will transfer by using pipeline services or with water transportation which is by ship. Eighty-one percent of energy comes from the used of fossil fuels with 61% of total anthropogenic contributed to the greenhouse gas (GHG) emissions [2]. The second contribution is from industrial processes with 5% of global GHG emissions (as illustrated in Figure 3).

According to the projection of energy use worldwide, global CO_2 emissions are expected to increase by 55% between 2004 to 2030 (see Figure 1, World Data, 2019) CO_2 emission is projected to increase by one-third between 2012 and 2040 from 32.3 bil metric tons to 43.2 bil metric tons. Since Malaysia is one of the main oil-producing countries in the world, [3] the CO_2 that was generated from petroleum production or fossil fuel combustion has been identified as one of the main contributors to the emissions of CO_2 in Malaysia other than natural gas refining and cement manufacturing. Up to June 2019, 7.83 tonnes per capita of CO_2 has been emitted to Malaysia atmosphere and this scenario is anticipated to continue well in the future if there are no mitigation taken to manage CO_2 emission in Malaysia. Figure 2 also shows the increasing trend of emission rate has reduced slightly from 2016.

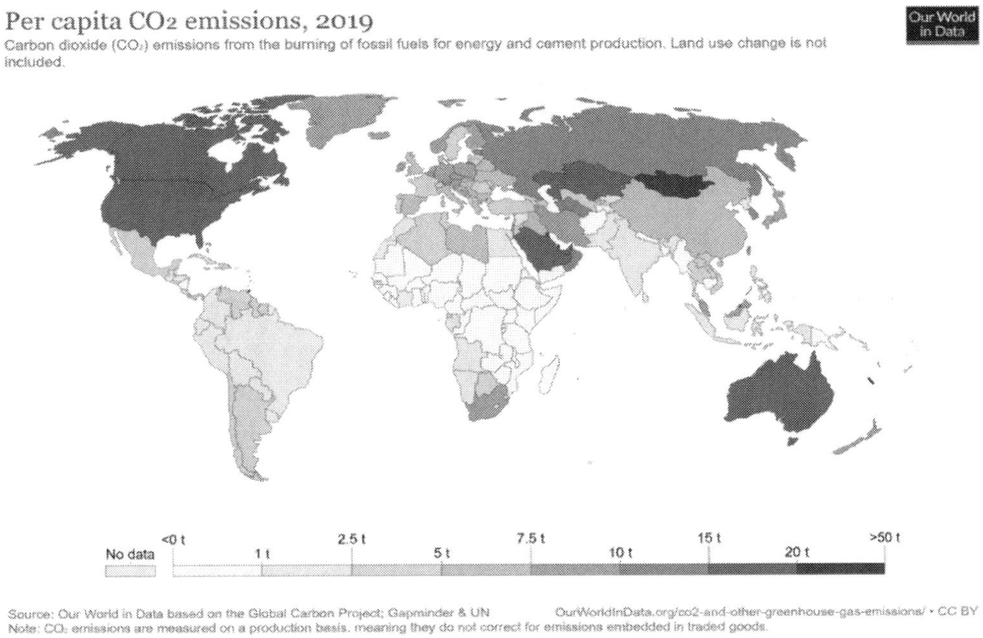

Figure 1. Global CO_2 Emission per Capita in 2019 (World Data).

According to [4], Malaysia, Thailand and Indonesia categories as a moderate emission around 10^2 Mt/year and storage prospectively which refer to the current CO_2 emissions.

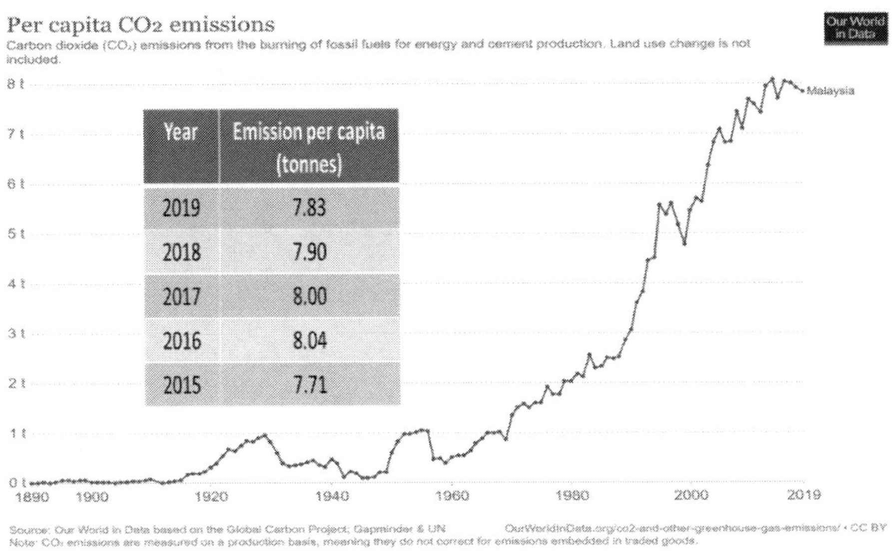

Figure 2. Trend of CO_2 Emission in Malaysia (World Data).

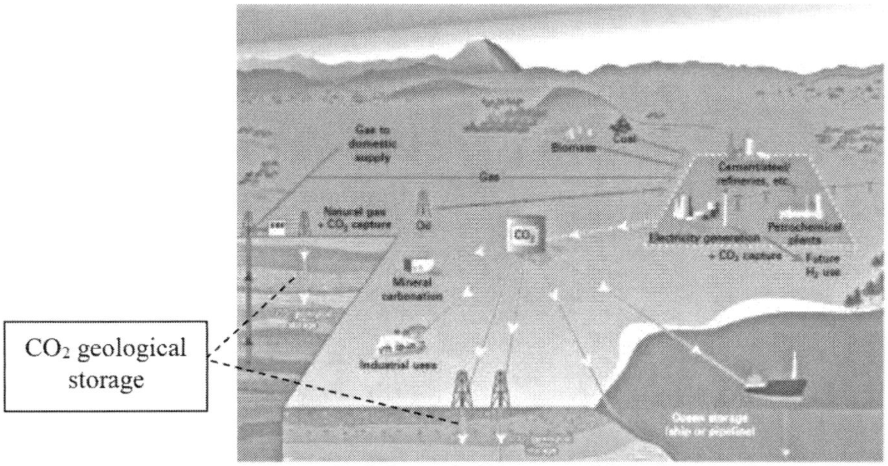

Figure 3. Schematic Diagram of CO_2 Storage System [5].

2. Potential Sedimentary Basins of Malaysia for CO_2 Sequestration

The previous study shows the high potential CO_2 storages which were located in three area of basin such as Malay Basin offshore in Terengganu, Central Luconia Province in Sarawak basin offshore and West Baram Delta also in Sarawak basin offshore was classified as the storage site among twenty-seven percent of geological storage in Malaysia [6]. These potential basins for CO_2 sequestration are not equally suitable in terms of its suitability. There are various factors that influence the evaluation of sedimentary basins mostly due to lack of crucial data for poorly explored basins. It is critical to determine if the sedimentary basins of Malaysia can provide a safe storage for CO_2 before commencing the sequestration as potential leakage and catastrophe escape may cause remnant of disputes in terms of environmental issues and might have some problems with public perceptions.

Table 1 and Figure 4 showed a ranking list of sedimentary basins in Malaysia which has been assessed in terms of their suitability for CO_2 storage capacity based on certain criteria and their location respectively. From the list, it is shown that Malay Basin scored the most and ranked as the most potential basin for carbon sequestration in Malaysia.

Table 1. Ranking for Sedimentary Basins in Malaysia [7].

Rank	*Basin*	*Score*	*Rank*	*Basin*	*Score*
1	Malay Basin	0.8113	*8*	Penyu Basin	0.5554
2	Central Luconia Province	0.7356	*9*	Tatau Province	0.4938
3	West Baram Delta	0.7041	*10*	West Luconia Province	0.4553
4	Balingian Province	0.6938	*11*	Tinjar Province	0.4200
5	Sabah Basin	0.6864	*12*	Northeast Sabah Basin	0.3543
6	East Baram Delta	0.6260	*13*	Southeast Sabah Basin	0.3370
7	Straits of Melaka	0.6200	*14*	North Luconia Province	0.2659

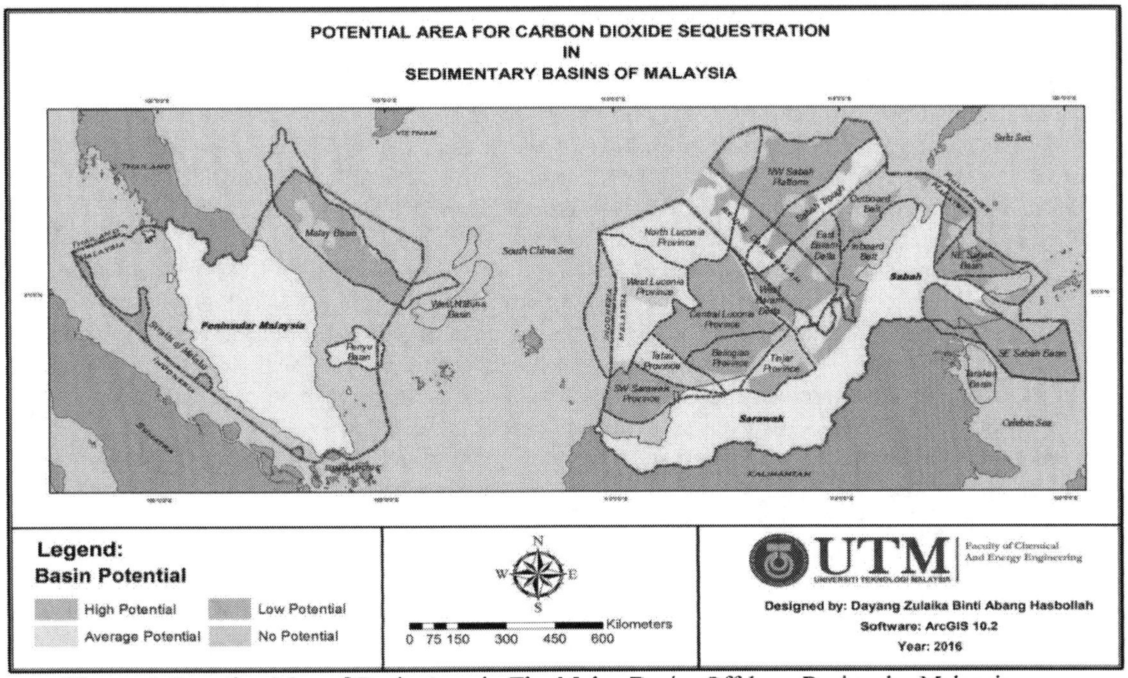

Figure 4. Location Map of Study Area in The Malay Basin, Offshore Peninsular Malaysia.

Malay Basin has the highest score because it fulfils most of the criteria of suitable basin for carbon sequestration. Detailed basin-scale assessment of Malay Basin was discussed profoundly by [8]. However, in terms of potential injection sites, the structure of Jerneh, Tangga, Dulang and Semangkok field are highly recommended to be explored for CO_2 sequestration area in the Malay Basin (Figure 5). The discussion on the suitability of geological structure for carbon sequestration in these four potential sites is focusing on the basin stratigraphy, permeability, porosity and capacity of storage system.

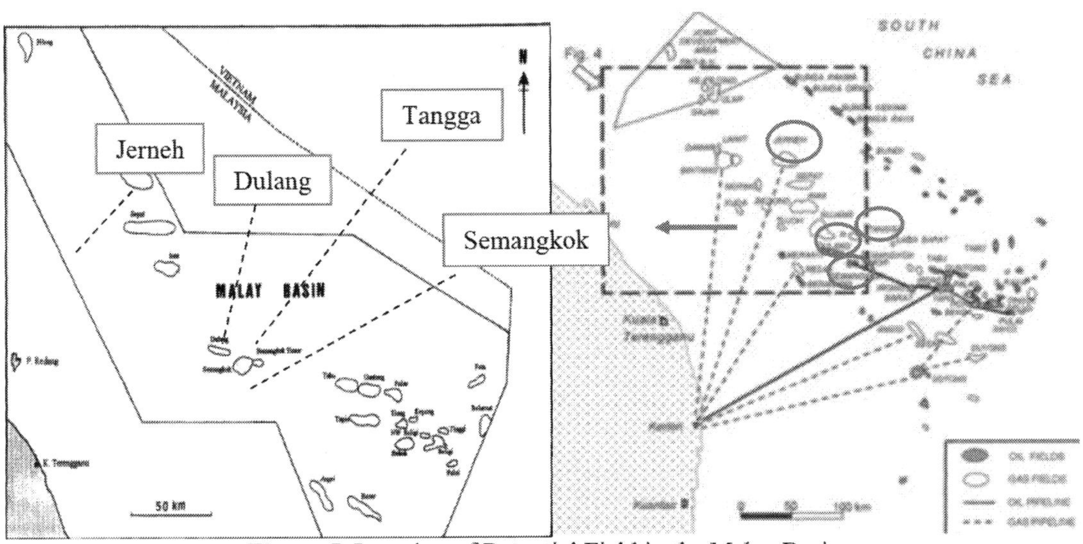

Figure 5. Location of Potential Field in the Malay Basin.

IOP Conf. Series: Earth and Environmental Science **971** (2022) 012022 doi:10.1088/1755-1315/971/1/012022

2.1 Basin Stratigraphy of Potential Injection Sites

Malay Basin comprises of numerous grabens and compressional anticlines that some of it has not been penetrated because of the great depths. Figure 6 showed the basin structure of Malay Basin which has been classified using alphabet A to M. Groups D down to K are found to be sandstone reservoirs where most of the hydrocarbons of this basin are found. The depositional environments of the sandstones vary with the stratigraphy. In older groups like K, L and M reservoirs are mainly consists of fluvial channels in a nonmarine – lacustrine setting. Meanwhile in younger groups such as J to I, the sandstones are predominantly fluvial – deltaic to estuarine channel complexes.

Sandstone reservoirs are well known to be a good medium for long-term CO_2 storage. Cleaner and coarser grained sandstones tend to have the best reservoir properties. Meanwhile, Groups D and E have been interpreted as tidal, deltaic to lower coastal plain deposits. The sandstones were interpreted as deltaic and include distributary mouthbar, shoreface and channel sediments formed during a Middle-Late Miocene regression. The classification of sedimentary rocks especially in group M show the types soil with fine particles until coarse particles, followed by shapes of subangular and subrounded form and comes from immature sandstone properties [9]. Thus, it is developed in major of feldspar formation (feldspathic arenite) and some micro-structure for examples silica, carbonate cements, authigenic kaolinite and subordinate illite can be defined by using SEM images.

Group E (Jerneh field) has the formation of fine-grained sandstone, mudstone and coal, while having a mineral for example litharenite and feldspathic litharenite cover the group E properties. According to [9] again, based on previous core data sandstones also are developed through modification of three types of rocks which are from metamorphic, igneous and combination of sedimentary rocks. Moreover, it is controlled by the syn-rift tectonics and developed by factors of certain climatic conditions and uncertainty of fluctuations in lake level. On the other hand, according to [10], there is a high potential sandstone formation in the Mesohellenic through with mineralogical shown calcite with 47% is the highest mineral recorded. Table 2 below show a property of mineral from previous study recorded by [10] which is determined using XRD analysis.

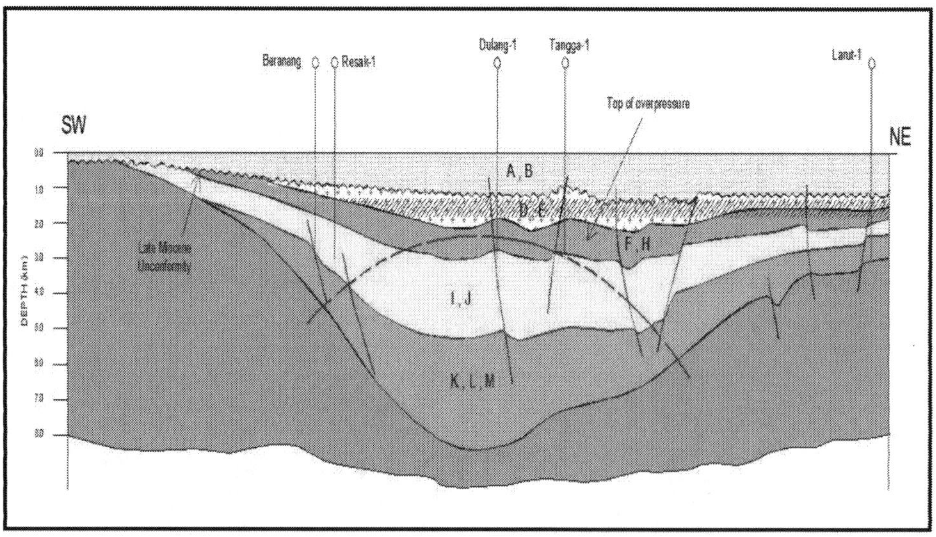

Figure 6. Basin Stratigraphy of Malay Basin.

IOP Conf. Series: Earth and Environmental Science **971** (2022) 012022 doi:10.1088/1755-1315/971/1/012022

Table 2. Mineral Properties of Group E in Malay Basin.

Mineral	Percentage
Calcite	47%
Quartz	16%
Feldspar	14% (12% plagioclace, 2% K-feldspar)
Clay minerals	8%
Mica	7%
Clorite	5%
Dolomite	3%

3. Injection Technology Development for Carbon Sequestration

The development of industry related to CO_2 sequestration has been updated by Global CCS Institute. According to [11], there are two types of facilities of CO_2 storage capacity. The large scale includes a facility related to the industrial sources and power generation consists a capacity of 400 and 800 ktpa. Besides, industrial and power generation sources which is not similar with large scale threshold are some of pilot and demonstration facilities criteria. Some examples of commercial CCS facilities for natural gas processing such as Terrell, Sleipner, Shute Creek, Core Energy, Snohvit, Century Plant, Lost Cabin, Petrobras Santos, Uthmaniyah, Cnpc Jilin, Gorgon, Qatar LNG CCS, Santos Cooper Basin and Abu Dhabi CCS 2.

According to Figure 7, the development of pipeline has shown a decreased especially in total capacity in year 2011 to 2017 due to global financial crisis change the scenario when they turn to focus in short term uses by two sectors of private and public. However, there are increasing in year 2018 onwards derived to achieve reduction on emissions especially related to greenhouse gas (GHG) in development countries/ cities. However, to overcome the financial issue because of high cost CCSU projects, Clean Energy Ministerial (CEM) at Abu Dhabi in April, 2011 conclude the cooperation within stakeholders to fund and develop CCSU operations by relationship with World Bank, Asian Development Bank (ADB) and World Resources Institute [2],[12]. CEM in London in 2012 also raised a fund worth US$100 million for development of CCSU in developing countries. Malaysia as a signatory of Kyoto Protocol and member of non-annexure I obligate to control carbon emissions by introduced National Policy on Climate Change in 2010. According to [13], here are some recommended from CCSU exploration and implementation:

- CCSU has the potential to reduce emissions in power, oil, gas and other industrial sectors
- The cost of electricity produced by fossil fuel plants with CCSU is equal to other low-emission power generating sources, example solar and wind
- CCSU encourage Malaysia to achieve low emission of 2005 by 2020 (approximately 40% reduction in 2020)

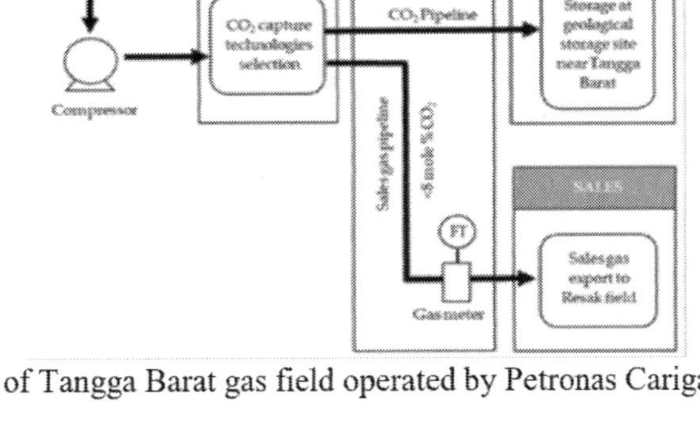

Figure 7. Concept of Tangga Barat gas field operated by Petronas Carigali Sdn Bhd (PCSB)

4. Parameter for CO_2 Injectivity Control

According to [14], if there any changes in injectivity system will affected the loss of reservoir/ caprock or may happen an overburden integrity. The effect from CO_2 injectivity to the saline aquifer are may fracturing the sandstone formations, it will definitely increase a local permeability and also enhance the injectivity. On the other hand, there are some factors affected the injectivity system to be study which is formation permeability, injectivity loss, wellbore design and injection well pressure. From the previous study [15], there was a problem at the beginning stage where it goes peak rate in the pressure from CO_2 injectivity. Then, the other way to resolve this issue is by adding of Monoethyleneglycol (MEG) which is good in high pressure and low temperature to reduce the hydration especially in the low temperature at the injection well [3].

4.1 Formation Permeability

Permeability of a rock/ sandstone formation in the Malay Basin show that how the rock will permit the passage of fluids. The formation of permeability affected on porosity and those relationship summarize that when it is become higher porosity simultaneously increase to the higher permeability (shown in Figure 8). Moreover, the connectivity between pore spaces is contribute by size and shape of grains, the grain size distribution and operation of capillary forces in the sandstone properties. According to [16],[17],[18], the main controlling parameter for CO_2 injection is from formation permeability. The higher permeability is suitable near to wellbore due to enhance an injectivity while to maximize the rate of residual trapping, dissolution and mineral trapping by having a lower permeability far from wellbore radius [3].

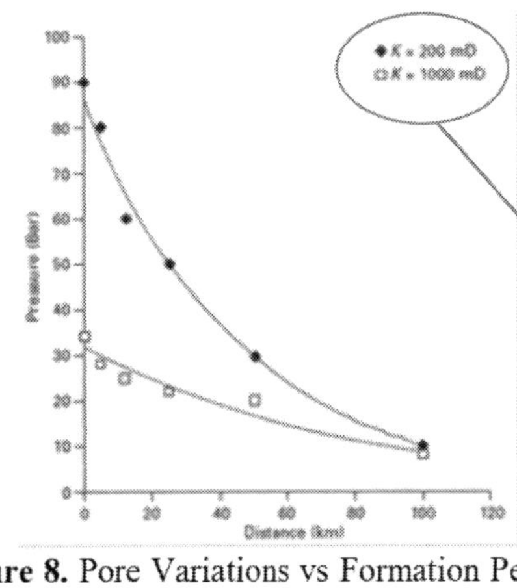

Permeability Value (mD)	*Classification*
<10	Fair
10 - 100	High
100 - 1000	Very High
>1000	Exceptional

Figure 8. Pore Variations vs Formation Permeability on CO_2 Injectivity Data from Generic Reservoir Simulations in Saline Aquifer [4].

4.2 Injectivity Loss

There are factors contribute to the loss of injectivity by process in the CO_2 reaction are comes from precipitation of halite reaction, the particles movement, incorporate and sweep along in its flow (fine particles) also reactions of chemical. According to [19], the evaporation effect happens when dry supercritical CO_2 has been injected and incorporate (similar time) at the accumulated of brine location. The effect will be continued to the reaction of dry formation waters to ensure the halite precipitation are happened. This is similar to [20], which is mentioned how the supercritical of CO2 also dissolves amount of water inside the storage of natural gas affected the injection rates. In addition, the particles will be spread and move (change in location) inside the pore throats with narrow condition due to the particle's movement [21].

Unfortunately, the mobilized will decrease a local permeability, other than that will completely plugging the rock formation. The loss injectivity by geochemical reaction products and effects happen when CO_2 injected react (dissolution) with certain reservoir matrix minerals. The effects from that, there were increase porosity and permeability in the initial stage after CO_2 saturated brine. Nevertheless, it will sediment downstream after there are not settle in a reaction product and potentially to have a plug pore throat. However, it is important to understand the minor effect of loss injectivity which comes from impurities in the injection stream. The impurities of gases inside the CO_2 injection will be affected the storage of CO_2 and give the consequences to the geochemical reactions especially for the purpose of plugging inside pore throats reservoir. The consequences of loss injectivity due to the impurities of gases depend on impurities category inside the injection rate and the condition of reservoir minerology (reactive nature).

However, the Malay Basin is showing the highest potential of CO_2 storage while having a good formation of caprock, will reduce any leaking (loss) of CO_2 injection. According to [8], Malaysia's sandstone has a criterion in terms of tectonic setting which found to be a foreland, limited faulting and does not have any seismic activity. In addition, with the properties of warm basin related to geothermal regime, average porosity of 17% and shale of caprock formation, the CO_2 injection will better in store for a long period (years) without shows any loss of injectivity.

4.3 Wellbore Design

Wellbore design are important to ensure it is suitable and in good condition while storing to make sure CO_2 injectivity in saline aquifers, preserve an environment surrounding also maintain the integrity of wellbore usually from operation time until it is completed an injectivity time [22]. Normally for CO_2 wellbore design are follow to the oil and gas industry especially in drilling and construction practices. For examples in Weyburn field in Canada especially for enhanced oil recovery (EOR) and Salah Project in Algeria where is inside the 20 m thickness with low permeability has been injected with CO_2 sequestration [23]. The wellbore design is separately into three types which in vertical alignment, horizontal and inclined well alignment. It is certified that the horizontal wells much more benefit compared to the vertical wells as mention below:

- It will reduce a quantity of horizontal wells as it is using a horizontal injector to spread the CO_2 into deep saline aquifers. Refer to Figure 9.
- It will impact to the connection through permeable sections of saline aquifers as long to enhance injectivity of CO_2.
- It will decrease the pressure needed for any given volume of fluid and as a benefit to reduce the load capacity on the CO_2 compressors then subsequent energy requirements.
- The horizontal wellbore design also would reduce potential flowing adverse effect by having a high permeability zone entering and create the injection profiles layout.

According to [24], there was collaboration between Japan Oil, Gas and Metal National Corporation (JOGMEC), JX Nippon Oil and Gas Exploration and PETRONAS to use core technologies from EOR instruments including design of wellbore, pipeline and others. From there, CCUS will develop with EOR technologies to ensure an initiative for long term purpose especially an economic value through recovery, storage of CO_2 and an environmentally friendly in resource development. Therefore, from this project of high CO_2 gas fields in Helang and Layang in Sarawak will enhance the perspective of CO_2 sequestration especially in the Malay Basin.

IOP Conf. Series: Earth and Environmental Science **971** (2022) 012022 doi:10.1088/1755-1315/971/1/012022

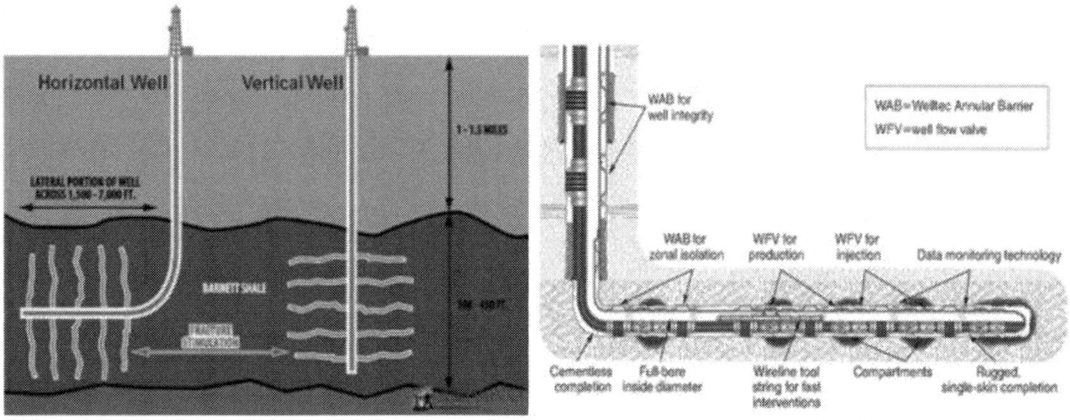

Figure 9. Illustration of Two Wellbore Design *(left)*, Horizontal Injector to Distribute CO_2 into Saline Aquifer (*right*).

4.4 Factors Affecting Well Integrity

4.4.1 Seismic Events

The certain quantity of stress in reservoir itself would affect the well integrity and will cause failure with commence of micro-earthquake or induced earthquakes. For examples, there are parameters which contributed to the induced earthquake such as hydraulic fracturing, gas storage, production (reservoir compaction), low-porosity reservoirs, high injection pressures, fluids consumer, stress regimes and natural fractures and faults. According to [25], the similar of gas which is for production and injection, there are seismicity happen when a main fault has been active in the gas field located at depth 2.5 km +- 0.5 km with magnitude of 1 to 3.5 richer scale. According to [26], the latest earthquake happens in 2004, Acheh in Indonesia with 8.9M earthquake which is triggered Tsunami and Ranau, Sabah recorded 5.9M earthquake in 2015. From the previous study, it is shows that the seismic activities are considered dormant event though there are increasing in seismic activities, high frequency and intensity in and around Malaysia. Then, there was shown the fault lines from seismic activities and did not disturbing the area of the Malay Basin.

5. Issues and Challenges in Malaysia

The implementation of CO_2 in Malaysia is facing issues and challenges for example regulatory framework, costing/ expenses (capital cost, generation cost, funding, impact on tariff), capacity building, technology issues, commercial framework across value chain, availability of sequestration sites and transportation of CO_2 to sequestration site [27]. The research and development in CO_2 storage has gained less interest in researcher, non-governmental organization (NGO) and companies in Malaysia due to the limited access of data for deep exploration in the CO_2 storage implementation and also quick tough in site accessibility for their data collection.

For example, in year 2020 a joint venture between Japan Oil, Gas and Metals National Corporation (JOGMEC) with JX Nippon Oil & Gas Exploration and Petronas to explore the potential gas field in Helang and Layang fields of offshore Sarawak shown a cooperation of experienced companies with expertise to access the location by their own technologies. The government also facing a challenge to raised funds to encourage the R&D of the CO_2 storage among the stakeholders. Nevertheless, the government need to strengthening their governance system especially the level of awareness on CO_2 storage benefit to the all stakeholders, NGO, University and community. The only way to overcome this challenge is by having a specific framework on CO_2 storage implementation in Malaysia.

The study from Petronas found that there are few challenges regarding to the CO_2 development in Malaysia with every stage recorded the barrier of implemented projects. For example, there are five

(5) stage indicated on how does Malaysia need to face the challenge or issue in future such as in field appraisal, development option, CO_2 separation, CO_2 transport and last but not least the CO_2 disposal and usage as presented in Table 3[28].

Table 3. Stage in CO_2 Challenges Implementation in Malaysia [28].

Field Appraisal	High CO2 content which is 85%
Development Option	a. Economic outlook for field development alone due to the return an investment and totally need some holistic approach b. Optimal development solution needs to balance between time to meet gas demand and most economic development option
CO_2 Separation	a. Technology not fully ready to deal with very high CO_2 content coupled with very high flow rate b. Large footprint and exotic materials lead to astronomical capital expenses
CO_2 Transport	a. High pipeline cost due to tolerance for CO_2 b. Users or disposal locations scattered and far from production area
CO_2 Disposal and Usage	a. Kyoto protocol becoming legally binding b. Massive CO_2 by products from field production c. Meeting EOR requirement

5.1 CO_2 Storage and Scenario

Joint venture among two participants from government which is Ministry of Energy, Green Technology and Water (KeTTHA) and Global CCS Institute has create a program such as Malaysian CCS Capacity Development. The aim is to guide stakeholders to prepare on the CO_2 sequestration activity in future. According to [2], it was summarised from participants in various types of expertise regarding the CCS implementation issues. The whole issues are as follows:

- Regulatory framework in Malaysia is not defined any CO_2 matters.
- The process of capture, transport and storage are not very clear in term of implementation by agency.
- CCS need some legislation to be applied for all of chain and sectors.
- There are no standards use as proper guideline especially how does CO_2 capture start until completed the injection for example the way to describe pipeline specification, proper site determining, how to select, monitor, measurement and verification.
- The safety part especially in every CCS chain need to be solve under Occupational health and safety department.
- For ensure all participants be responsible to their liability, need to create ownership system and can apply in other chain.
- The land acquisition and environment impact assessment (EIA) will be important criteria to access in the CO_2 sequestration and need coordination from public and awareness of technology.
- Relationship between cross boundaries country are serious aspects to be focus in term of regulation and management of CO_2 sequestration.

From all the major issues listed, there are some planning or further analysis to progress in next step without containing CCS policy mandate since there is no clear agency, legislation, policy and regulatory has been developed. The further planning that should be considered are describing and determining CO_2 whether it is under waste, pollutant, resource or GHG using existing regulatory framework. This is important to enhance the outcomes for the milestone in any CCS implementation. Besides, the stakeholders and agencies could play their role in the CCS chain and it is easier to handle for the capture, transport and storage activity. Besides, it would be additional value especially to the transboundary relationships while having a relevant expertise from geological field, experience with pipelines and engineering field. Other than that, a comparison in CCS legislation and regulation of framework (capture, transport and storage), without consider any factor of cost, suitability and implementation procedure should be done.

6. Conclusion

From the previous assessment, it was identified that the Malay Basin has the highest potential for CO_2 sequestration site since it fulfils the criteria from [29], which mentioned the characteristic of deep saline aquifers with formations greater than 800 m thick, caprock from shale properties layer as sealing function, good in porosity and permeability in high scale volumes and flow rates of pressure. In addition, it has been suggested that the EOR development and technologies could be adopted for CO_2 injection. Government of Malaysia and stakeholders should collaborate to develop the potential area such as the Malay Basin, establishing a master plan for offshore injection sites including pipeline systems and management of CO_2 sources.

7. References

[1] International Energy Agency *Global Energy Review 2021Official Report* Retrieved on July,20 2021 from https://www.iea.org/reports/global-energy-review-2021
[2] Global CCS Institute 2013 Malaysian CCS Legal and Regulatory Workshop *Workshop Report* Lumut, Malaysia, 27/2 – 01/3/2013
[3] Bachu S, Gunter W, D and Perkins E, H 1994 Aquifer Disposal of CO_2: Hydrodynamic and Mineral trapping *J. Ene. Con. & Man.* ***35*** 4 pp 269-279
[4] ICTPL 2005 CO_2 Storage Prospectivity of Selected Sedimentary Basins in the Region of China and South East Asia. *J. Introduction Assessment of Geologic storage potential of carbon dioxide in the APEC Region-Phase I.* pp 1-232.
[5] IPCC, 2005 – Bert Metz, Ogunlade Davidson, Heleen de Coninck, Manuela Loos and Leo Meyer (Eds.) Carbon Dioxide Capture and Storage, *Cambridge University Press*, UK. pp 431. The Edinburgh Building Shaftesbury Road, Cambridge CB2 2RU ENGLAND
[6] Hasbollah D and Junin R 2018 Mapping of Potential Sedimentary Basins for Carbon Dioxide Sequestration in Malaysia by Using ArcGIS. *Adv. Sci. Let.* 24 pp 4153-4157(5).
[7] Junin, Radzuan & Hasbollah, Dayang. (2015). A Preliminary Basin Scale Evaluation Framework of Potential Sedimentary Basins in Malaysia for Carbon Dioxide Sequestration. Chemical Engineering Transaction. 45. 1537-1542.
[8] Hasbollah D.Z.A., Junin R., Taib A.M., Mazlan A.N. (2020) Basin Evaluation of CO2 Geological Storage Potential in Malay Basin, Malaysia. In: Duc Long P., Dung N. (eds) Geotechnics for Sustainable Infrastructure Development. Lecture Notes in Civil Engineering, vol 62. Springer, Singapore. pp 1405-1410.
[9] Yusak S, A, M 2012 Sedimentological Characterization of Deeper Group M Reservoirs in Malay Basin. *Warta Geologi*. 38(2).
[10] Dimadis G, Bacasis I and Karampasis M 2017 Impact of CO_2 on Mechanical Properties of Sandstone during Sequestration for Climate Change Mitigation. *10TH International Conference on Energy and Climate Change.* 89-98 Athens-Greece
[11] Integrated Carbon Observation System, Global Carbon Project 2020. *Supplemental Data of Global Carbon Budget 2020* Retrieved on July,20 2021 from https://doi.org/10.18160/gcp-2020
[12] Ahmadi M A, Pouladi B and Berghi T S. 2016 Numerical Modeling of CO_2 Injection Scenarionin Petroleum Reservoirs: Application to CO2 Sequestration and EOR. *J. Nat. Gas Sci. and Eng.* **30**, 38-49
[13] Kementerian Tenaga, Teknologi Hijau dan Air (KeTTHA) 2011 The National Green Technology Policy Retrieved from: https://www.ppj.gov.my/storage/pdf/20/20.pdf

[14] Global CCS Institute 2020 Retrieved (1st April 2021) from: https://www.globalccsinstitute.com/about/what-is-ccs/

[15] Shi J, Q, Imrie C, Sinayuc C, Durucan S, Korre A and Eiken O 2013 Snohvit CO2 Storage Project: Assessment of CO_2 Injection Performance Through History Matching of the Injection Well Pressure Over a 32-Months Period *J. Ene. Pro. Vol 37* pp 3267-3274

[16] Van Der Meer L, G, H (1995). The CO_2 Storage Efficiency of Aquifers. *Energy Conversion and Management vol 36(6-9)* pp 1155-1160.

[17] Law D, H, S and Bachu S 1996 Hydrogeological and Numerical Analysis of CO2 Disposal in Deep Aquifers in the Alberta Sedimentary Basin *Energy Conversion and Management vol 37(6-8)* pp 1167-1174.

[18] Ennis K, J and Paterson L 2001 Reservoir Engineering Issues in the Geological Disposal of Carbon Dioxide *Proceedings of the Fifth International Conference on Greenhouse Gas Control Technologies: GHGT5*, Collingwood, VIC, CSIRO Publishing: 290-295

[19] Spycher N and Pruess K 2005 CO_2-H_2O Mixtures in the Geological Sequestration of CO_2. II. Partitioning in Chloride Brines at 12-100oC and up to 600 bar *Geochemica vol 69(13)* pp 3309-3320.

[20] Kleinitz W, Koehler M and Dietzsch G 2001 The Precipitation of Salt in Gas Producing Wells', SPE 68953, *SPE European Formation Damage Conference*, The Hague, The Netherlands. pp 1-7

[21] Solomon S and Flach D, N, V. 2010 Carbon Dioxide (CO2) Injection Processes and Technology.
Research and Innovation. Woodhead Publishing Limited.

[22] EPA 2008 *Federal Requirements Under the Underground Injection Control (UIc) Program for Carbon Dioxide (CO2)* Geologic Sequestration (GS) Wells; Proposed Rule', Federal Register, *vol 73(144)* pp 43492–43541

[23] Riddiford F, A, Tourqui A, Bishop C, D and Taylor B 2003 A Cleaner Development The in Salah Gas Project, *Algeria 6th International Conference* pp 595-600

[24] Ahmed A, S, Rahman M, R and Bakri M, K 2021 A Review on Low-and High-Stream Global Carbon Capture and Storage (CCS) Technology and Implementation Strategy *J. App. Sci. & Pro. Eng. Vol 8 (1)*

[25] Eck T, V, Goutbeek F, Haak H and Dost B. 2006 Seismic Hazard Due to Small-Magnitude, Shallow-Source, Induced Earthquakes in The Netherlands *J. Eng. Geo vol 87(1)* pp 105-121

[26] Abd Razak S, M S, Adnan A, Che Abas M, R, Lin W, S, Zainol N, Z, Yahya N, Rizalman A, N and Mohamad M, E. 2018. A Historical Review of Significant Earthquakes in Region Surrounding Malaysia. *Proceeding of international conference on durability of building and infrastructures*. Miri, Sarawak. pp 242-245

[27] IEA International CCS 2012 4th CCS Regulatory Network Meeting Workshop *Workshop Report* Paris, 09/5 – 10/5/2012.

[28] Darman, N and Harun, A.R. 2006. Technical Challenges and Solutions on Natural Gas Development in Malaysia. The Petroleum Policy and Management (PPM) Project 4th Workshop of the China – Sichuan Basin Case Study. Beijing. Retrieved from 1 April 2020. http://www.ccop.or.th/download/PPM_finalrpt.pdf

[29] Benson S and Cole D. 2008. CO_2 Sequestration in Deep Sedimentary Formations *J. Elements. vol 4* pp 325-331

Acknowledgment
The authors would like to acknowledge UTM Research Grant-Tier 2 (Vot No: Q. J130000.2651.16J84) for financial support.

GEOTECHNICAL PROPERTIES OF TROPICAL RESIDUAL SOILS FOR SUSTAINABLE LANDFILLING

Y B Yamusa[1*], H I Mohammed[2], K Ahmad[3] and Z Majid[4]

[1] Department of Civil Engineering, School of Engineering Technology, Nuhu Bamalli Polytechnic, Zaria, Nigeria
[2] Department of Geography, Faculty of Science, Kaduna State University, Nigeria
[3] Department of Geotechnics and Transportation, School of Civil Engineering, Universiti Teknologi Malaysia
[3] Department of Geoinformation, Faculty of Built Environment and Surveying, Universiti Teknologi Malaysia
*Corresponding author: yamusabello@gmail.com

Abstract. To select a site for sanitary landfilling, it is paramount that geotechnical analysis is conducted to acquire relevant data for the purpose of studying soil properties. One of the basic requirements for the landfill site selection process is geotechnical criteria. This ensures that the chosen sites are founded on an impermeable soil deposit having adequate geotechnical characteristics that is also socially acceptable, economically cheap, and environmentally friendly. Geotechnical properties investigated in the laboratory for three sites samples comprises volumetric shrinkage, shear strength, and permeability. Groundwater contamination comes because of leachate. As such soil meant for landfill sites must have exceptional geotechnical properties able of restraining these leachates. Consequently, landfill soils must possess excellent ability to retain leachate in terms of permeability, an acceptable shear strength to withstand the waste load, and appropriate volumetric shrinkage to refute wetting and drying seasonal variations. The physical properties which showed that site A has a fines content of its soil up to 51%, site B with 29% and site C with 14%. Engineering property showed that site A has the lowest permeability of 2.1 x 10^{-9} m/s, followed by site B with 2.08 x 10^{-8} m/s, and then site C with the highest permeability of 1.5 x 10^{-7} m/s at *OMC*. The results show that only site A achieved maximum regulatory range of permeability value of 10^{-9} m/s. This implies that site A is the most suitable and sustainable, then followed by site B and then site C.

1. Introduction

The main elements of solid waste management (SWM) are collection, transfer, resource recovery, recycling, and treatment. It aims to abolish waste from the surrounding area to prevent threats to the public [1]. Furthermore, one of the primary goals of SWM is to ensure the protection of public health, to facilitate environmental quality, to ensure the development of sustainable waste disposal technique, and to offer support to economic productivity [2]. To attain these goals, there is a need for local authorities to collaborate with the public and private sectors to embrace strong solid waste management and to be able to provide adequate maintenance of the system. Presently, the solid waste generated in urban regions like developing nations is fast approaching the advanced countries where there is adequate solid waste management [3].

In an engineered landfill, waste collected are dispersed in shallow strata of compacted layers and secured with soil cover. Also, in present sanitary landfills the base is generally protected with an impermeable soil and geosynthetic liner. The liner safeguards the ground water from being polluted due leachate infiltration. Leachate from bottom is propelled for treatment. At the point when landfill is full it is covered with sand, clay and gravels to protect contamination of ground water. Various number of wells are dogged close to the landfill site to check if any spillage is affecting the ground water. The methane produced by anaerobic disintegration is collected and is been burnt and converted to power for electricity generation. In some developing countries, large volume of their generated wastes are thrown in open dumps or uncontrolled landfills [4]. However, engineered or sanitary landfills prevents the groundwater being contaminated because of liners in a way that can avoid environmental pollution and general well-being of the public [5].

Landfill liners are materials (both natural and man-made product) applied at the base edge and underneath all floors of the landfill site. Mostly, these liners comprise of a series of clay compaction and geosynthetic materials intended to serve as a barrier to stop the occurrence or transportation of landfill gas and leachate. The facilities used for landfill control incorporate compacted clay liners, system for the accumulation and retrieval of leachate and gas. Liners give the last line of barrier against groundwater pollution. Liners for waste control might be characterized into three; geosynthetic clay liners, composite liners and soil (clay) liners [6]. In this research, the soil liner is explored.

Landfill development and construction depends on the initial geospatial site location. Though, the site soil should be investigated to check its suitability as a liner material. The soil liner must possess the quality of impeding the migration i.e. the downward movement of landfill leachate to protect the groundwater. This soil liner when compacted should as well be able to sustain the load imposed on it, i.e. to have adequate shear strength. Moreover, the soil liner should have acceptable volume change that might arise due to seasonal weather variations which can cause shrinkage and swelling [7].

The guidelines in most regulatory agencies as well as researchers stated that 1×10^{-9} m/s maximum permeability, minimum of 200 kN/m^2 shear strength, maximum of 4% volumetric shrinkage, and minimum of 30% soil fines content are specified for soil liners [8-12]. Although, for tropical residual soils, 50% fines content is required [7, 13, 14]. Moreover, according to Prakash and Poulose [15], landfill liners are constructed based on some standards obtained from extensive researches. The soil liner materials are classified as suitable, marginal, and unsuitable. Greater than or equal to 12% plasticity index are classified as suitable, plasticity index between 6% and 12% are classified as marginal, and the soils with plasticity index below 6% are classified as unsuitable [16]. On this basis, all landfill sites selection must adhere to these criteria to render them suitable and usable without the fear of failure.

2. Methodology

To avoid organic matter on sample collection, the soils used in this study were obtained beneath 0.5 m ground surface. Oversized gravels were removed in each soil sample by passing them through BS 4.75 mm aperture sieve. These samples are collected from three different locations in Johor Bahru, Malaysia as presented in Table 1.

Table 1. Location of sample sites

Site	Latitude	Longitude
A	1°35'45.62"N	103°38'12.00"E
B	1°32'14.19"N	103°38'34.56"E
C	1°33'7.37"N	103°41'34.87"E

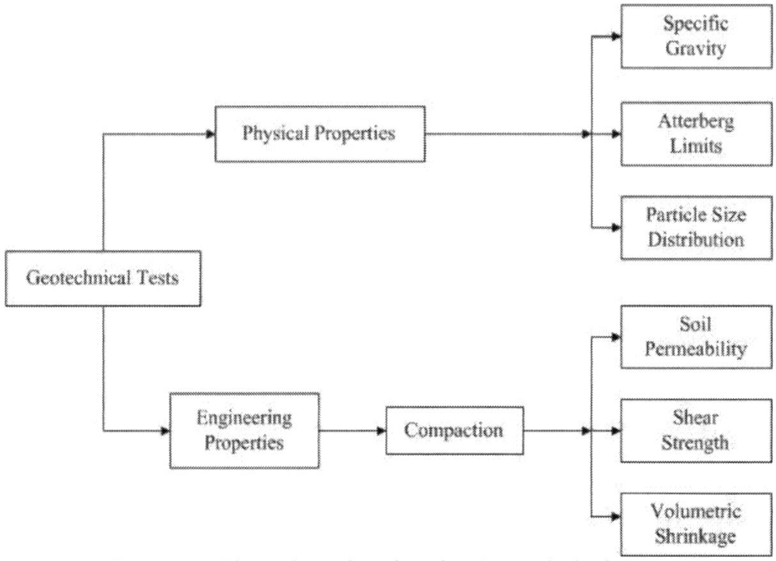

Figure 1. Flow chart showing the Geotechnical tests

The geotechnical property is divided into two; the physical properties tests and the engineering properties tests carried out using the BSI [17] as shown in Figure 1.

2.1 Physical Properties

Physical properties investigated in this research include specific gravity, Atterberg limit, and particle size distribution in order to classify the soils.

2.1.1 Specific Gravity

Determination of specific gravity of the soil samples was done using small pyknometer method as recommended in BS 1377: Part 2:1990. The specific gravity is the ratio of soil density to the density of water. About 5 g of air dried soil sample passing through 2 mm BS sieve size was used. Using a 50 ml density bottle the sample was placed into it and the masses measured including the soil, bottle, and stopper. The bottle was then opened, and distilled water added to the sample; its then shaken carefully to remove any trapped air in the bottle. Shaking could not be sufficient to eliminate all the trapped air, so a vacuum desiccator was employed without the stopper in which the bottle was placed inside for at an hour. The sample was removed from the desiccator carefully and vibrated. This procedure was repeated for the soil sample until no more air was evolving. Thereafter, weight was taken of the bottle, stopper plus soil and water. After discarding sample in the bottle, it was then filled with only water. The weight of the bottle, stopper and water were recorded. The average value for the specific gravity was obtained by repeating the whole procedures three times. Equation 1 was adopted in calculating the specific gravity of the soil.

$$G_S = \frac{W_2 - W_1}{(W_4 - W_1) - (W_3 - W_2)} \quad (1)$$

Where;
G_s = Specific gravity
W_1= mass of density bottle, stopper (g)
W_2= mass of density bottle, soil, and stopper (g)
W_3 = mass of density bottle, stopper, water, and soil sample (g)
W_4 = mass of density bottle, stopper, and water only (g)

IOP Conf. Series: Earth and Environmental Science **971** (2022) 012023 doi:10.1088/1755-1315/971/1/012023

2.1.2 Atterberg Limits

The Atterberg limit tests cover the determination of liquid limit (*LL*) and plastic limit (*PL*) of the soil. The difference between liquid and plastic limits equals to Plasticity Index (*PI*). Using these three values, sample is classified using Plasticity Chart of British Standard Classification System [18]. The detailed procedures followed in performing Atterberg limit tests were in accordance with BS 1377: Part 2: 1990.

2.1.2.1 Liquid Limit

The liquid limit (LL) is the water content at which a soil changes from plastic to liquid behaviour. To determine the *LL* of the soil, cone penetration test was conducted in accordance with Head [19]. The test method consists of dropping an instrumented cone, with the tip facing down, into the soil sample at a controlled rate of 2 cm/s.

About 300 g of soil sample passing through BS sieve size 425 μm was put on a glass plate, and distilled water added to achieve a homogeneous mix. By using the two spatulas start to mix the soil continually at least 10 minutes. The sample was placed into plastic container and left overnight to allow the water to permeate the soil particle. It was then removed from the plastic container and remixed it after adding a small amount of distilled water. The soil was put inside the cup carefully to ensure no air bubble was entrapped, and the final surface smoothened and flatten. The cup was then put under the cone and lowered the point of cone to touch the surface of sample and made the reading of gauge on zero. The cone was dropped to fall freely for 5 second by ensuring that the cone penetration is about 15 mm and recorded the reading of the gauge. About 10 g of the sample from the cup was taken and put it into a container of known mass to determine the moisture content. After that, weighed the sample with the container, and put it in oven for 24 hours to dry. Add water to the sample and repeat the procedures five times. A graph between the penetration depth and the water content was plotted and best line through the points drawn. The value of water content corresponding to cone penetration depth equal of 20 mm represent the liquid limit of the soil sample.

2.1.2.2 Plastic Limit

The plastic limit is the empirically established moisture content at which a soil becomes too dry to be plastic and the test procedures adopt the BS 1377: Part 2: 1990.

About 20 g of soil sample passing through 425 μm BS sieve was placed on a glass plate. A small amount of distilled water was added, and the sample mixed using two spatulas until it reached a plasticity enough to make a ball. Using fingers and palm the mixed sample was rolled and made into a ball and divided the ball into two portions each approximately 10 g. The divided portions were subdivided into four portions. By using fingers and thumb, each part was moulded into 6 mm diameter of homogenous thread. The threads were rolled between finger and surface of glass plate until they were started to crumble at about 3 mm diameter. Immediately, the crumbled threads were put into a container and weighed. Thereafter, they were put into an oven for 24 hours to dry and weighed the dry samples after cooling. The average result of the moisture content was then calculated as the plastic limit value of soil.

2.1.2.3 Plasticity Index

Using the expression in Equation 2, the plasticity index of the soil was calculated.

$$PI = LL - PL \tag{2}$$

2.1.3 Particle Size Distribution

Wet sieving and dry sieving tests were conducted to determine the particle size distributions of the soil specimens. BS 1377: Part 2:1990 described the wet sieving test that was adopted in this experiment. Dried soil sample weighing 200 g placed in a dish and soaked with water. The sample was then stirred, washed, and sieved through 63 μm sieve using water jet from distilled water bottle until water became clean. Particles retained in the sieve were collected into the crucible and oven dried for 24 hours. Afterward the oven dried soil was sieved, weighed, and recorded.

The size distribution of the soil particles was measured using dry sieve analysis. A stack of sieves with accurately dimensioned holes between a mesh of wires was used to separate the particles into sizes.

Dried soil weighing 200 g was put into the top of these stack of sieves arranged from coarse to fine (i.e. sieve with large aperture on top of sieve with small aperture and the receiving pan at the bottom). The stack of sieves was shaken for at least 10 minutes so that the particles are sorted into the respective sieves. The soils collected in each sieve was weighed and recorded.

2.2 Engineering Properties

Engineering properties include permeability, compressive strength, and volumetric shrinkage in order to select the soil that can be compacted on the site as soil liner for the sanitary landfill.

2.2.1 Compaction

In this research, the British Standard light (BSL) compaction tests were carried out as procedures outlined in BS 1377: Part 4: 1990. The connection concerning moisture content and dry density of soil is achieved through compaction tests. Dried soil sample of about 2 kg was mixed with 18 – 32% of water. The soil was divided into 3 portions. First portion was put inside mould and compacted given 27 blows using 2.5 kg rammer. Second and third portions followed the same procedure making 3 equal total layers. Thereafter, the collar of the mould was removed, and the compacted soil trimmed. The compacted soil in the mould was then weighted on a scale. A typical specimen was taken from the mould for determination of moisture content. The procedure was repeated four times making a total of five test at different water content. A graph of maximum dry density (*MDD*) versus optimum moisture content (*OMC*) was plotted. From the three suitable sites, five different moisture contents from dry to wet side of optimum were applied on a representative of each soil specimen. Similar compaction process was also applied for the permeability, shear strength, and volumetric shrinkage tests.

2.2.2 Permeability

Hydraulic conductivity or permeability is the average velocity of fluid through soil under its hydraulic gradient, with dimension as meter per second (m/s) [20]. Generally, fluid permeates through the interconnecting voids or pores of soil.

In this research, the falling head permeability test was used to measure the permeability as explained by Head and Epps [21]. Samples of the soil were compacted at moulding water contents dry (-2%), optimum (0%), and wet (+2%) of the *OMC*. Thereafter, the samples were drenched in a water tank for total saturation when its observed that no more air bubbles coming out from the sample. Then its connected to the permeameter containing water through a standpipe. The test parameters were then recorded, and the coefficient of permeability calculated. The procedure was repeated at different moisture contents. The coefficient of permeability was calculated from Equation 3.

$$k = 2.3\frac{aL}{At}\log_{10}\frac{h_1}{h_2} \quad (3)$$

Where;
k = Coefficient of permeability (m/s)
a = Area of the standpipe (m^2)
L = Length of sample (m)
A = Cross sectional area of sample (m^2)
h_1 = Initial height of water in standpipe (m)
h_2 = Final height of water in standpipe (m)
t = Time required to get head drop (s)

2.2.3 Shear Strength

Unconfined compressive strength (*UCS*) of a soil specimen is the force per unit area [22]. The BSL procedures was adopted for the compaction of soil samples. The size of each soil sample is 76 mm long and 38 mm in diameter. The BS 1377: Part 7: 1990 summarizes the procedures for the *UCS*. The sample compacted in the mould was then removed and positioned in-between platens on the pedestal of the

IOP Conf. Series: Earth and Environmental Science **971** (2022) 012023 doi:10.1088/1755-1315/971/1/012023

UCS machine. Next, specimen was compressed at an axial strain of 2% per minute. At regular intervals, force and axial deformation readings were recorded. The experiment was sustained until 20% axial strain of the soil specimen diameter was reached or the maximum value passed. Equation 4 was adopted in computing the unconfined compressive strength.

$$UCS = RC_r \frac{\left(1 - \left(\frac{L_1}{L_0}\right)\right)}{100\, A_0} \tag{4}$$

Where;
L_1 = final length (m)
L_0 -= initial length of specimen (m)
A_0 = initial cross-sectional area (m^2)
R = load ring reading (kN)
C_r= calibration of load ring

2.2.4 Volumetric Shrinkage

Volumetric shrinkage is expressed as a percentage decrease in soil volume to the initial soil volume. This occurs when the water content is reduced from a specified quantity to the shrinkage limit [20]. Outlined procedures in the BS 1377: Part 4: 1990 was applied to air dried compacted soils at different moisture contents with respect to the optimum moisture content. The compacted specimens in the cylindrical mould were then extruded out and kept on a lab deck at a temperature of 26 ± 2°C for a month to dry. Daily measurements of heights and diameters were recorded using a Vernier calliper for each sample of different moisture contents. Using the average diameters and heights, the volumetric shrinkage strain was computed by applying Equation 5 [23-25].

$$VSS = \frac{(V_0 - V_d)}{V_0} 100\% \tag{5}$$

Where;
VSS = Volumetric shrinkage strain (%)
V_0 = Volume of wet sample (m^3)
V_d = Volume of dry sample (m^3)

3. Result and Discussion

3.1 Physical Properties

Laboratory results for physical properties and sieve analysis curves are presented in Table 2 and Figure 2, respectively. According to the British Standard (BS) classification, the soil sample for site A is classified as sandy clay of high plasticity (MH). Site B soil sample is classified as very silty sand of intermediate plasticity (MH). Soil sample of site C is classified as very silty sand of very high plasticity (MV).

Table 2. Physical properties of the soils used in this study

Property	**Value**		
	Site A	**Site B**	**Site C**
Specific gravity	2.70	2.74	2.70
% Fines	51	29	14
% Sand	46	61	46
% Gravel	3	10	40
OMC, %	27.9	19.3	17.8
MDD, Mg/m^3	1.38	1.52	1.51
Liquid limit, %	64	50	76
Plastic limit, %	37	29	37
Plasticity Index, %	27	21	39
BS Classification	MH	MH	MV
Colour	Brown	Reddish Brown	Red

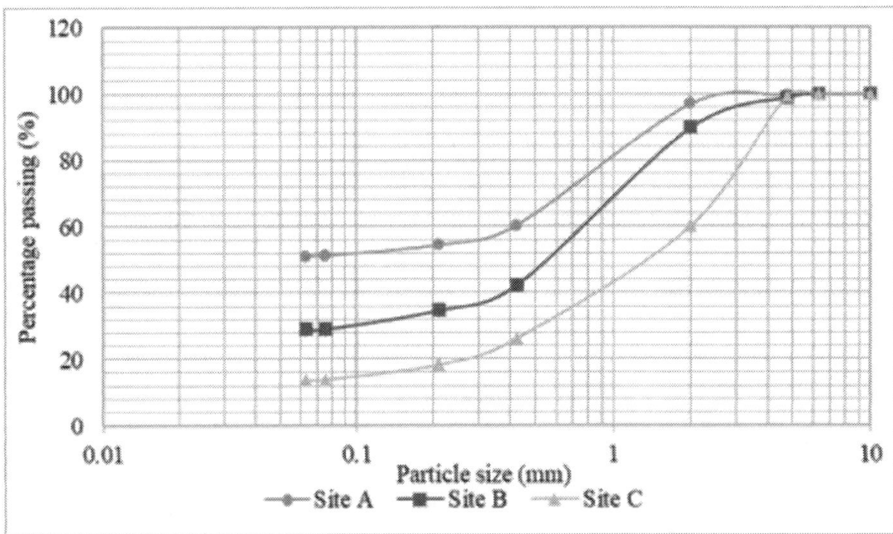

Figure 2. Particle Size Distribution Curves for Best Sites Soils

The following soil index properties as suggested in literature that: percentage of fines ≥ 30%, plasticity index ≥ 7 and liquid limit ≥ 20% can be specified for landfill liners [6, 26]. Looking at the liquid limit and plasticity index it could be assumed that all the site samples have met these two criteria. Whereas, comparing the index properties of the sites soil used in this study and that of the literature shows that the soil is suitable to be use as liner material for site A only. Sites B and C have their fines content less than 30% and they might not likely impede the flow of leachate through them. Moreover, Yamusa, Yunus [27] explained the morphology and effects of fines content on laterite residual soil as liner, that at least 50% fines content of soil should be used. From Table 2, it showed that site A soil sample has 51% fines. Although, the main criteria are discussed based on the engineering properties.

IOP Conf. Series: Earth and Environmental Science **971** (2022) 012023 doi:10.1088/1755-1315/971/1/012023

3.2 Engineering Properties

The engineering properties in this research is discussed with respect to the soil compaction. Because the compacted soil on which the sanitary landfill is to be constructed as liners must be able to hinder the permeability of the leachate generated, have adequate strength to be able to carry the loads imposed on it, and has low volume change due to seasonal variation.

3.2.1 Soil Compaction

Compaction of soils by means of mechanical process produce a homogeneous soil mass that is voids free, enhance strength and density, and decreases its permeability. The compaction test carried out purposely to determine the optimum moisture content (*OMC*) at which the maximum dry density (*MDD*) is achieved. By determining the *OMC,* variation of the moisture contents on the dry side (-2%) and the wet side (+2%) then take effect. The effect of compaction on *MDD* and *OMC* of the best soil samples are presented in Figure 3.

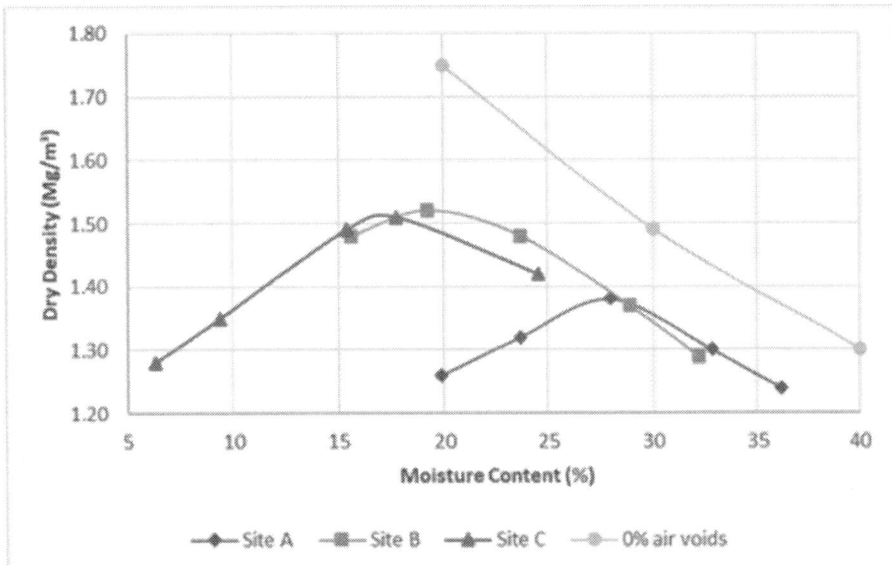

Figure 3. Compaction Curves for Best Sites Soils

3.2.2 Soil Permeability

The performance of liners or hydraulic barriers are majorly affected with respect to permeability also known to as hydraulic conductivity. When soils are susceptible to high permeability then, they are not fit for lining landfills. This is because the penetration of water through the soil is great thereby polluting underground water. To protect the groundwater, it is recommended that clayey soils should be use due to their very low permeability [28]. The permeability is discussed with relation to dry density and moulding water content in this research.

3.2.2.1 Effect of Moulding Water Content

Generally, hydraulic conductivity decreases with increase in moulding water content for all soil samples as shown in Figure 4. This trend is in line with Ahmad, Yamusa [29] and Bello [30]. Samples from site A has the lowest permeability of 2.1 x 10^{-9} m/s, followed by site B with 2.08 x 10^{-8} m/s, and then site C with the highest permeability of 1.5 x 10^{-7} m/s at *OMC*. The results show that only site A achieved maximum regulatory range of permeability value of 1 x 10^{-9} m/s. Attaining this range of regulation makes site A practically impermeable to leachate. Sites B and C did not meet the hydraulic conductivity criteria which are vulnerable to leachate permeation.

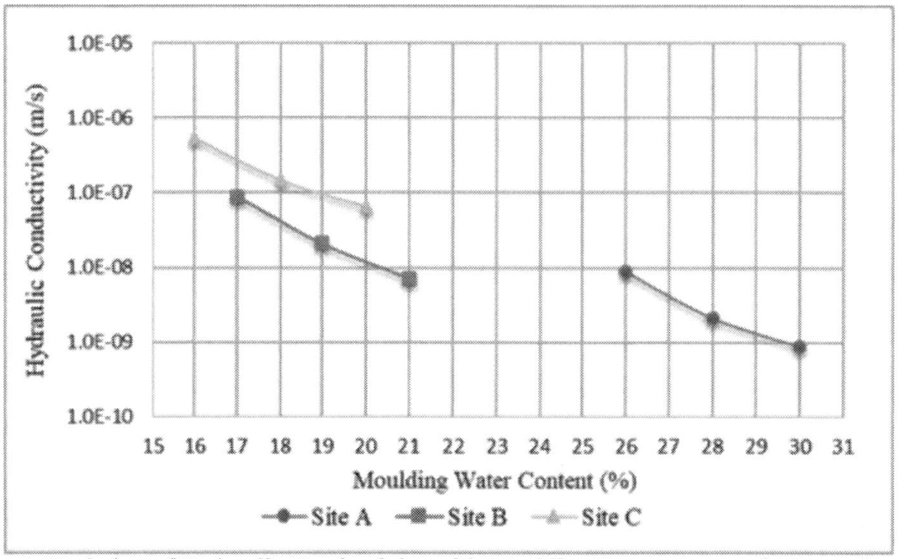

Figure 4. Variation of Hydraulic Conductivity with Moulding Water Content for Best Sites Soils

3.2.3 Shear Strength

In the design of compacted soil liners, shear strength is determined in form of unconfined compressive strength (*UCS*). *UCS* is crucial because the liner must possess acceptable strength to carry the solid waste weight imposed on it and its own self-weight [22]. The minimum *UCS* value of 200 kN/m^2 is adopted to sustain the liner integrity against stresses by construction vehicles as well as slopes and imposed loads [9, 31]. In this research, the unconfined compressive strength is discussed relation to dry density and moulding water content.

3.2.3.1 Effect of Moulding Water Content

The shear strength decreased with increased in moulding water content, agrees with the findings of other researchers such as Osinubi, Moses [32], Amadi, Eberemu [31], Osinubi, Bello [22] and Bello [33]. The soil fabrics were loosened due to increase in water content which resulted in loss of soil cohesion [34, 35]. Hence, the reduction in soil strength is experienced at higher water content. Figure 5 presents the variation of *UCS* values with moulding water content. On the dry side of optimum, all the soil samples have the recommended *UCS* of 200 kN/m^2. At the *OMC*, site A and B have their *UCS* of 200 kN/m^2, while site C does not meet the requirement. Therefore, site A has the highest *UCS*, followed by site B, and then site C which can be used as soil liners in sanitary landfill.

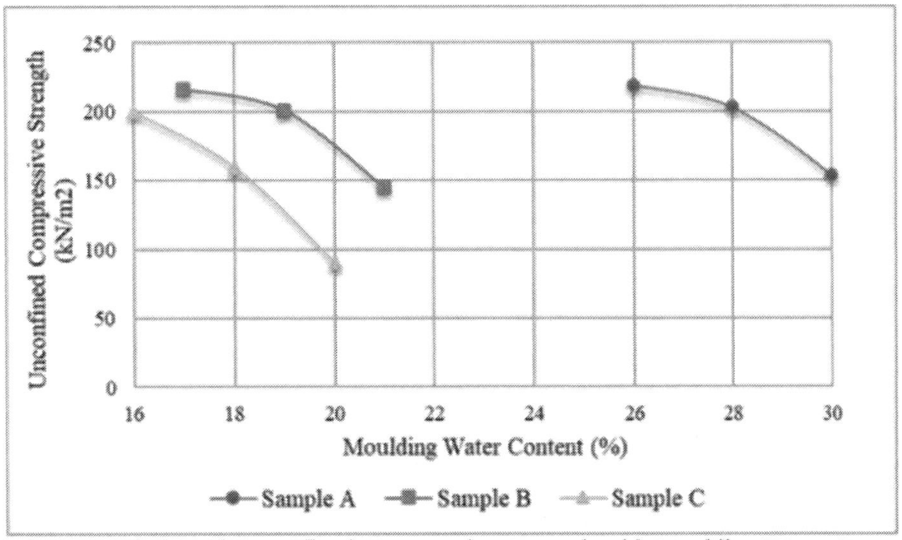

Figure 5. Variation of Unconfined Compressive Strength with Moulding Water Content for Best Sites Soils

3.2.4 Volumetric Shrinkage

Shrinkage in soils takes place when there is a decrease in water content. Depending upon the volume change, it can cause severe structural damage. In this research, the volumetric shrinkage is discussed with relation to dry density and moulding water content.

3.2.4.1 Effect of Moulding Water Content

The volumetric shrinkage strain (*VSS*) was found to increase as the moulding water content increase, and the results is consistent with the results of Yamusa, Kamarudin [36], Amadi [37], Osinubi and Eberemu [38] and Osinubi and Nwaiwu [25]. As a result of capillary menisci, pore water tension is developed in fine-grained soils. The compacted soils shrink more when dry if are moulded at higher moulding water content [23, 25, 38]. The pore spaces in compacted soils which literally is a storage for water. The volume of water leaving these pores is directly proportional volumetric shrinkage [39]. Therefore, soil samples compacted wet of optimum have higher shrinkage limit when dry. Sample A with higher fines content has the highest *VSS*, followed by sample B, then sample C. All the soil samples recorded lower *VSS* values that attain the minimum recommended of $\leq 4\%$ that are used as hydraulic barriers in sanitary landfill. So, seasonal variation may not affect the volumetric shrinkage of these soils, i.e. the shrinkage limit is quite minimal that can resist leachate penetration. The variation in volumetric shrinkage strain at different moulding water contents is presented in Figure 6.

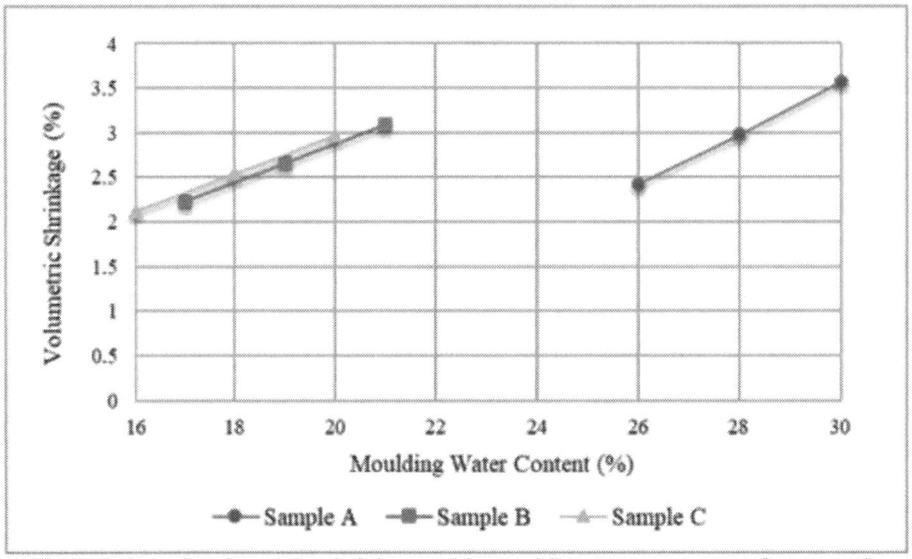

Figure 6. Variation of Volumetric Shrinkage with Moulding Water Content for Best Sites Soils

3.2.5 Overall Acceptable Zone

In modern geoenvironmental practice, these certain range of water content and dry density are used in compacting soil liners. The water content and dry density of soil are determined at an acceptable range and this played a critical step for compacted soil liner design. It is primarily crucial to achieve a minimum dry density for factors governing the performance of compacted soil liners particularly the permeability [40]. Three design parameters i.e. $k \leq 1 \times 10^{-9}$ m/s, *VSS* ≤ 4%, and *UCS* ≥ 200 kN/m^2 at their permissible limits in relation to dry density with moulding water content must intersect at an acceptable zone. The average values of density where *k*, *UCS* and *VSS* are attained on the compaction plane for the test specimens is the overall acceptable zone [41, 42]. These three parameters must be met collectively. Failure in one factor would result in making the liner unusable as it will not impede the infiltration of leachate through the soil.

Table 3 shows ranges of acceptable moulding water contents for the three sites soil samples. Samples for sites B and C did not meet the *k* requirement, rather met the *UCS* and *VSS* requirements. Hence, sites B and C are not suitable as liner materials. Only site A samples qualify the *k*, *VSS* and *UCS* moulding water contents within the ranges acceptable. Figure 7 show the overall acceptable zone for site A indicated by the hatched zone portion. In this experiment, the three parameters on compaction plane zone is illustrated in From Table 3. The overall acceptable range of moulding water contents ranged between 26 - 28% and dry densities ranged between 1.30 - 1.38 Mg/m^3 and this is the overall acceptable zone.

Table 3. Acceptable ranges of moulding water contents

Engineering Criteria		**Sample**	
	Site A	Site B	Site C
		Moulding Water Content Range, %	
***k*, m/s**	26-30	-	-
***UCS*, kN/m^2**	26-28	17-19	16
***VSS*, %**	26-30	17-21	16-20
OAR			26-28

OAR – Overall Acceptable Range

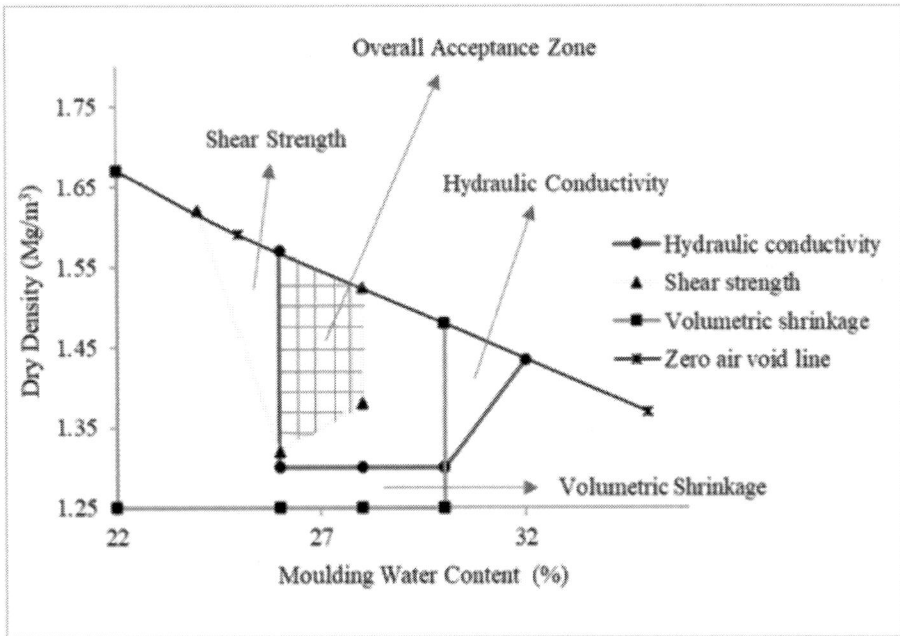

Figure 7. Overall Acceptable Zone for Best Site A Soil

4. Conclusion

In the geotechnical technique, two properties of the soils were investigated. First, the physical properties which showed that site A has a fines content of its soil up to 51%, site B with 29% and site C with 14%. Research shows that soils with 30% fines content should be used as soil liners in sanitary landfill. This implies that site A is the most suitable, then followed by site B and then site C, respectively.

Second, the engineering properties of these three sites soils. Relating to the engineering properties, permeability, shear strength, and volumetric shrinkage where investigated. The permeability of site A attained the minimum requirement, while site B slightly met the requirement at wet of optimum moisture content. Site C failed to meet the permeability requirement. Equally, the shear strength of the soils was investigated. Site A has the highest shear strength, followed by site B, with site C having the lowest. Similarly, volumetric shrinkage of the site soils revealed that all of them have lower shrinkage capacity that falls within the recommended minimum, and therefore, can be used as soil liner materials. Checking the overall acceptability of soil to be within the specified requirements of permeability, shear strength, and volumetric shrinkage, only site A met the requirements. Site B and site C can be used only when the soils are modified with additives such as bentonite, cement, lime, pozzolanas, etc, which will significantly increase the cost of construction. Also, the energies of compaction can be increased, and the depth of the soil liner can be increased to make these soils usable.

References

[1] Shekdar, A.V., *Sustainable solid waste management: an integrated approach for Asian countries.* Waste management, 2009. **29**(4): p. 1438-1448.

[2] Marshall, R.E. and K. Farahbakhsh, *Systems approaches to integrated solid waste management in developing countries.* Waste Management, 2013. **33**(4): p. 988-1003.

[3] Henry, R.K., Z. Yongsheng, and D. Jun, *Municipal solid waste management challenges in developing countries--Kenyan case study.* Waste Management, 2006. **26**(1): p. 92-100.

[4] Yamusa, Y.B., K. Ahmad, and N.A. Rahman, *Gradation Effect Review on Hydraulic Conductivity and Shear Strength Properties of Compacted Laterite Soil Liner.* Proceedings of The 11th International Civil Engineering Postgraduate Conference - The 1st International Symposium on Expertise of Engineering Design (SEPKA-ISEED'16), 2016. 26-27th

September, 2016. Faculty of Civil Engineering, Universiti Teknologi Malaysia, 81310 Johor Bahru, Johor, Malaysia.

[5] Yamusa, Y.B., K. Ahmad, and N.A. Rahman, *Hydraulic Conductivity and Volumetric Shrinkage Properties Review on Gradation Effect of Compacted Laterite Soil Liner.* Malaysian Journal of Civil Engineering 2017. **29**(1): p. 153-164.

[6] Daniel, D.E., *Geotechnical practice for waste disposal.* 2012: Springer Science & Business Media.

[7] Yamusa, Y.B., *Effects of Gradation on Engineering Properties of Compacted Laterite Soil Liner.* 2018, PhD. Thesis, School of Civil Engineering, Faculty of Engineering, Universiti Teknologi Malaysia.

[8] UKEA, *Earthworks in landfill engineering: LFE4.* Design, construction and quality assurance of earthworks in landfill engineering. UK Environment Agency., 2014.

[9] Amadi, A. and A. Eberemu, *Characterization of geotechnical properties of lateritic soil-bentonite mixtures relevant to their use as barrier in engineered waste landfills.* Nigerian J Technol, 2013. **32**(1): p. 93-100.

[10] USEPA, *Solid Waste Disposal Facility Criteria: Technical Manual. EPA 530-R-93-017. United States Environmental Protection Agency, 1-84,* 1993.

[11] USEPA, *Design and construction of RCRA/CERCLA final covers.* Soils Used in Cover Systems, EPA/625/4-91/025. United States Environmental Protection Agency, 1-25, 1991.

[12] USEPA, *Requirements for hazardous waste landfill design, construction, and closure.* EPA/625/4-89/022. United States Environmental Protection Agency, 1989: p. 53-74.

[13] Yamusa, Y.B., K. Ahmad, and G. Moses, *Acceptable Zone for Green Construction of Compacted Tropical Laterite Soil,* in *Fossil Free Fuels.* 2019, CRC Press. p. 85-101.

[14] Yamusa, Y.B., et al., *Engineering Characteristics of Compacted Laterite Soil as Hydraulic Barrier in Waste Containment Application.* Journal of Engineering Science and Technology, 2020. **15**(1): p. 508-523.

[15] Prakash, A. and E. Poulose, *Kuttanad Clay Amended Laterite as a Landfill Liner for Waste Disposal Facilities.* International Journal of Scientific Engineering and Research (IJSER), 2016. **4** (3): p. 75-78.

[16] Jones, R., et al., *Selection of clays for use as landfill liners.* Waste disposal by landfill-GREEN, 1995. **93**: p. 433-438.

[17] BSI, *Methods of Testing Soil for Civil Engineering Purposes (BS 1377:Part 1-9). British Standards Institute, London.* 1990.

[18] BSI, *Code of practice for site investigations (BS 5930). British Standards Institute, London.* 1999.

[19] Head, K.H., *Manual of soil laboratory testing, third edition. Whittles publishing, Dunbeath, Caithness KW6 6EY, Scotland, UK.* 2006.

[20] Punmia, B. and A.K. Jain, *Soil mechanics and foundations.* 2005: Firewall Media.

[21] Head, K.H. and R.J. Epps, *Manual of soil laboratory testing, volume 2. Whittles publishing, Dunbeath, Caithness KW6 6EY, Scotland, UK.* 2011.

[22] Osinubi, K., et al. *Characterization of shear strength of abandoned dumpsite soils, Orita-Aperin, Nigeria.* in *Proceedings of the 15th African Regional Conference on Soil Mechanics and Geotechnical Engineering. Resource and infrastructure geotechnics in Africa: putting the theory into practice, Maputo, Mozambique, 18-21 July 2011.* 2011. IOS Press.

[23] Bello, A.A., *The use of Standard Proctor for the determination of Shrinkage Properties of Reddish Brown Tropical soil.* Leonardo Journal of Sciences, 2011. **44**(19): p. 57-68.

[24] Eberemu, A.O., *Desiccation Induced Shrinkage of Compacted Tropical Clay Treated with Rice Husk Ash.* International Journal of Engineering Research in Africa, 2011. **6**: p. 45-64.

[25] Osinubi, K.J. and C.M.O. Nwaiwu, *Desiccation-induced Shrinkage in Compacted Lateritic Soils.* Geotechnical and Geological Engineering, 2008. **26**(5): p. 603-611.

[26] Benson, C.H., H. Zhai, and X. Wang, *Estimating hydraulic conductivity of compacted clay liners.* Journal of Geotechnical Engineering, 1994. **120**(2): p. 366-387.

[27] Yamusa, Y.B., et al., *Effects of fines content on hydraulic conductivity and morphology of laterite soil as hydraulic barrier.* E3S Web of Conferences, International Conference on Civil & Environmental Engineering (CENVIRON 2017), 2018. **34**.

[28] Ahmad, S.Z., M.S. Ahamad, and M.S. Yusoff, *Spatial effect of new municipal solid waste landfill siting using different guidelines.* Waste Manag Res, 2014. **32**(1): p. 24-33.
[29] Ahmad, K., Y.B. Yamusa, and M.A.B. Kamisan, *Effects of Soil Recompaction on Permeability.* Science World Journal 2018. **Vol 13(No 3), 6-9**.
[30] Bello, A.A., *Hydraulic conductivity of three compacted reddish brown tropical soils.* KSCE Journal of Civil Engineering, 2013. **17**(5): p. 939-948.
[31] Amadi, A., A. Eberemu, and K. Osinubi, *Strength consideration in the use of lateritic soil stabilized with fly ash as liners and covers in waste landfills.* State-of-the-art and practice in geotechnical engineering. American Society of Civil Engineers (ASCE) geotechnical special publication (GSP), 2012. **225**: p. 3835-3844.
[32] Osinubi, K.J., et al., *Influence of Molding Water Content on Shear Strength Characteristic of Compacted Cement Kiln Dust Treated Lateritic Soils for Liners and Covers.* . Nigerian Journal of Technology, 2015. **34 (2)**: p. pp. 266 – 271.
[33] Bello, A., *Analysis of shear strength of compacted lateritic soils.* The Pacific Journal of Science and Technology, 2011. **12**(1): p. 425-434.
[34] Osinubi, K. and G. Moses. *Compacted Foundry Sand Treated with Bagasse Ash As Hydraulic Barrier Material.* in *Geo-Frontiers 2011@ sAdvances in Geotechnical Engineering*. 2011. ASCE.
[35] Ahmad, K., Y. Yamusa, and M.B. Rosly, *Laterite Soil Shear Strength Characteristics from the Use of the Standard Proctor and Compression Machine.* Arid Zone Journal of Engineering, Technology and Environment, 2018. **14**(2): p. 304-316.
[36] Yamusa, B.Y., et al., *Volumetric Shrinkage of Compacted Soil Liner for Sustainable Waste Landfill.* Chemical Engineering Transactions, 2018. **63,** 613-618.
[37] Amadi, A.A., *Improvement of Barrier Soil Properties with Fly Ash to Minimize Desiccation Shrinkage.* International Journal of Engineering Research in Africa, 2012. **7**: p. 1-11.
[38] Osinubi, K.J. and A.O. Eberemu, *Desiccation Induced Shrinkage of Compacted Lateritic Soil Treated with Blast Furnace Slag.* Geotechnical and Geological Engineering, 2010. **28**(5): p. 537-547.
[39] Osinubi, K. and A. Eberemu. *Effect of desiccation on compacted lateritic soil treated with bagasse ash.* in *Proceedings of Bi-Monthly Meetings/Workshops, Materials Society of Nigeria (MSN) Zaria Chapter, January-December*. 2008.
[40] Amadi, A.A. and A.O. Eberemu, *Delineation of compaction criteria for acceptable hydraulic conductivity of lateritic soil-bentonite mixtures designed as landfill liners.* Environmental Earth Sciences, 2012. **67**(4): p. 999-1006.
[41] Osinubi, K. and G. Moses, *Compacted foundry sand treated with bagasse ash as hydraulic barrier material,* in *Geo-Frontiers 2011: Advances in Geotechnical Engineering*. 2011. p. 915-925.
[42] Bello, A.A., *Design Parameters for Abandoned Dumpsite Soil as Liner Material.* Leonardo Journal of Sciences, 2013. **46**(23): p. 29-40.

Geomorphological Spatial Model of Risk Analysis for Natural Disasters in Heritage Railway Tunnel

A D W Rahmadana[1], S A P Rosyidi[2]*, D M Setiawan[2] and R Nuraini[3]

[1]GeoArtScience: Consultant, Planning, Research & Survey, Perum Taman Pesona Asri Kav.17, Sleman, Yogyakarta, Indonesia
[2]Department of Civil Engineering, Universitas Muhammadiyah Yogyakarta, Jalan Brawijaya, Bantul, Yogyakarta, Indonesia
[3]Department of Civil Engineering, Universitas Teknologi Yogyakarta, Indonesia
*Corresponding author: atmaja_sri@umy.ac.id

Abstract. Flood event in Ijo Tunnel led to disruption to schedules and train routes. The objectives of this research are to determine the susceptibility, vulnerability, capacity, and risk, to analyze the geophysical and geotechnical condition, and to compile the strategy of disaster mitigation of Ijo Tunnel. A spatial approach based on landform analysis was performed to establish the disaster components. Geophysical and geotechnical investigation was conducted to describe the condition of the affected region. A field survey was accomplished to collect data and to study supporting works of literature. The descriptive analysis was completed to determine strategy and recommendation in flood and potential landslide mitigation. The result depicts that Ijo Tunnel holds multi-disaster risk such as floods and landslides. The occurrence of floods is induced by extreme rainfall and inability of drainage channels to retain water. The landslide potential is influenced by morphological factors and composting materials in Ijo Tunnel's region. The existence of spring fracture on the north slope also contributes to increasing the slopes' landslide potential. The flood mitigation measure can be carried out by building a well-prepared drainage system and watershed and surface runoff management. Areas with a high level of landslide susceptibility can be overcome by building permanent retaining walls and pipes on slopes to reduce the soil saturation level.

1. Introduction

Rail transport is a mass transportation mode with more advantages that the other transportation modes. These advantages include: (1) large capacity (bulk); (2) fast; (3) safe; (4) energy-saving and environmental-friendly; and (5) requires relatively small area. Compared to land transportation modes such as buses which have energy consumption of 0.0125 litters of energy per km/Passenger or private cars which have energy consumption of 0.02 litters per KM/Passenger [1], the train energy consumption is considered the most efficient, which is 0.002 litters per km/Passenger. Therefore, trains become the backbone of freight and passenger transportation in Indonesia.

The smoothness of trains in freight and passenger transportation is also affected by natural conditions. One of the disruptions to rail traffic is caused by the disaster aspect [2]. Various natural disasters such as flood, landslide, earthquake, volcanic eruption, and tsunami can affect the rail transportation services negatively [3]. Disaster risk needs to be minimized by identifying the level of vulnerability, susceptibility, and capacity of the railway operator. Such action is needed since the cost of developing railway infrastructure is very expensive [1].

IOP Conf. Series: Earth and Environmental Science **971** (2022) 012024 doi:10.1088/1755-1315/971/1/012024

One of the natural disasters that disrupted rail traffic occurred on June 17-18th, 2016 around Ijo Tunnel, Bumiagung Village, Rowokole District, Kebumen Regency. Flood and landslide occurred in Ijo Tunnel have disrupted the railway traffic in the southern part of Central Java. Therefore, several trains (KA) were delayed in departure and arrival.

The effect of the flood and landslide occurred in Ijo Tunnel is material losses. The affected train route certainly needs to be improved regarding its facilities and infrastructure so that similar incidents will not happen again [4]. Therefore, efforts to identify and formulate strategies for handling floods and landslides in the Ijo Tunnel need to be made to maintain the safety of the train travel. Hence, this study has three objectives, those are: (1) determining the level of susceptibility, vulnerability, and capacity in Ijo Tunnel; (2) analyzing the geophysical and geotechnical conditions of Ijo Tunnel; and (3) developing a strategy of flood mitigation strategy in Ijo Tunnel.

2. Method

This research applied a spatial approach based on an analysis of the landform approach in establishing disaster components. Landform units were obtained from the data processing of slope class, lithology, geological structures, and river flow patterns. Slope identification was obtained from each cell of the earth's surface in digital format (raster) which contains information of location point. Raster data is used to build a 3D model into a DEM. The output of the slope raster can be calculated in degrees or percentages with the following triangulation basic equation [5].

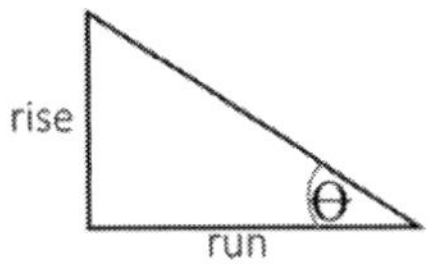

$$\text{Degree of slope} = \tan \Theta = \text{rise/run} \quad (1)$$

$$\text{Percent of slope} = \text{rise/run} * 100 \quad (2)$$

Geophysical and geotechnical analysis are the main keys in describing the condition of the area affected by the disaster. Furthermore, the strategies and recommendations in handling disasters in Ijo Tunnel was established descriptively.

2.1. Tools and Materials

This research required certain tools and materials. The tools used included a computer set, ArcGIS 10.1 software, and survey equipment (camera, GPS, compass, and geological hammer). Meanwhile, the research materials used are presented in table 1.

Table 1. Research materials.

Parameter	Research Variable	Data
Susceptibility of landslide disaster	Slope Class	Digital Indonesian Topographical Base Map (RBI) in the scale of 1:25,000 by the Geospatial Information Agency
	Level of incision	Shuttle Radar Topography Mission (SRTM) of 30 m resolution by the United States Geological Survey (USGS)
Subsubsection	Slope Class	Digital Indonesian Topographical Base Map (RBI) in the scale of 1:25,000 by the Geospatial Information Agency
	Landform	Slope Class (Digital Indonesian Topographical Base Map (RBI) in the scale of 1:25,000 by the Geospatial Information Agency)
		Lithology and geological structure (Digital Geological Map shape file (.shp) by Geological Research and Development Center)

	River flow patterns (Digital Indonesian Topographical Base Map (RBI) in the scale of 1:25,000 by the Geospatial Information Agency)
Rainfalls	Monthly rainfall (Meteorology, Climatology and Geophysics Agency – BMKG)

2.2. Assessment of Multi-hazard

2.2.1. Landslide. The susceptibility of landslides was assessed using slope class and level of incision variables. The landslide susceptibility class was determined by summing up the slope class scores and the level of incision. Classification of slope class, level of incision, and landslide susceptibility is presented in table 2. and table 3.

Table 2. Classification and score of slope class and level of incision.

Skor	Slope	Morphology	Level of Incision
0	0-2 %	Flat	Low (Score 1)
1	2-7 %	Choppy	
2	7-13 %	Wavy	Medium (Score 2)
3	13-20 %	Hills	High (Score 3)
4	20-55 %	Mountains	
5	>55%	Steep Mountains	

Table 3. Classification of landslide susceptibility.

Scores of Slope Class and Incision	Class of Landslide Susceptibility
0-2	Low
3-5	Medium
6-8	High

2.2.2. Flood. The flood susceptibility was identified and assessed using three variables, those are slope class, landform, and rainfall. The susceptibility class was determined by summing up the scores of each variable. Classification of slope classes, landforms, rainfall, and flood susceptibility are presented in table 4. and table 5.

Table 4. Classification and score of slope class, landform, and rainfall.

Skor	Slope	Landform	Monthly Rainfall (mm/month)
0	5-6%	Hills, mountains, scarp	0-100
1	4%	Volcanic fluvio fans, colluvial cones, foot of slopes of volcano, debris flow of the foot of slopes, the foot of volcano	100-200
2	2-3%	Foot plains, alluvial plains,	200-400

		fluviomarin plains, intervolcanic plains, valleys between hills	
3	1%	River valley, floodplain	>400

Table 5. Classification of flood susceptibility.

Scores of Slope Class, Landforms, and Monthly Rainfalls	Class of Flood Susceptibility
0-3	Low
4-6	Medium
7-9	High

2.2.3. Multi-Hazard. Multi-hazard is information on the level of vulnerability of an area to various disaster threats [6]. The multi-hazard level was determined by summing up the flood and landslide susceptibility scores. Classification and classes of multi-disaster are presented in table 6.

Table 6. Multi-hazard class.

Multi-hazard scores	Multi-hazard Class	Score
0-5.7	Low	1
5.8-11.5	Medium	2
11.6-17	High	3

2.3. Assessment of Vulnerability

Disaster vulnerability is a condition of a community or society that leads to an inability to deal with disaster threat. Such vulnerability means an assessment of the magnitude of the element at risk of being affected by a disaster related to facilities, infrastructure, staff, and use of services. There are three assessment parameters used, including the completeness of supporting infrastructure, the quality of the train route, and the density of rail traffic (see table 7).

Table 7. Class of disaster vulnerability.

Class and Score	Completeness of supporting infrastructure	Quality of train route	Density of rail traffic
Low (1)	The existence of buildings such as gabions, embankments, and others to reduce the effect of disasters in almost all vulnerable zones, as well as complete disaster emergency response tools	Good	Not dense
Medium (2)	The existence of buildings such as gabions, embankments, and others to reduce the effect of disasters in parts of all vulnerable zones, yet the disaster emergency response tools is not complete	Quite good	Quite dense
High (3)	No gabions, embankments, and other buildings to reduce the effect of disasters in all vulnerable zones, and the disaster emergency response tools is not complete	Poor	Very dense

2.4. *Assessment of Capacity*
Disaster capacity is the ability of a region and communities to take action to reduce the level of threat and the level of losses due to disasters. Such capacity means the preparedness of the railway operator in this case is PT. Kereta Api Indonesia (Persero) and related agencies in responding to disasters along the train route. There are three assessment parameters used, including Standard Operating Procedures (SOP) for disaster emergency response, studies, and information on disaster risk in the Operational Area (DAOP), as well as communication and coordination of PT. Kereta Api Indonesia (Persero) and related agencies (table 8).

Table 8. Class of disaster capacity.

Class and Score	SOP of disaster and emergency response	Study and information regarding the disaster risk in DAOP	Communication and Coordination
Low (1)	None	Few	Less correlated
Medium (2)	Presents but not in accordance with the standard yet	Quite complete	Correlated but not quite intensive
High (3)	Presents and in accordance with the standard	Complete	Correlated and intensive

2.5. *Assessment of Risk*
Disaster risk is the potential loss caused by a disaster in an area within a certain period. Disaster risk is Mathematically formulated in equation (3). The level of disaster risk will increase as the hazard and vulnerability increase. However, the value of risk can be small if the level of capacity is high. The disaster risk class is divided into three classes, which are low, medium, and high (see table 9).

$$R = (H \times V) / C \qquad (3)$$

In which:

R = Risk
H = Hazard
V = Vulnerability
C = Capacity

Table 9. Class of disaster risk.

Class of Risk	Score of Disaster Risk
Low	0.00-3.22
Medium	3.23-6.11
High	6.12-9.00

3. Results and Discussion

3.1. *Multi-Hazard*
Multi-hazard is information regarding the vulnerability level of an area to various natural disaster threats. Multi-hazard refers the vulnerability of Ijo Tunnel area to landslide and earthquake. Multi-hazard mapping is important considering that almost no area on earth has a single vulnerability [7].

Landslide is the mass movement of soil or rock in the potential landslide area. If the movement of the soil mass is excessive, it is called a landslide [8]. Landslide in Ijo Tunnel is triggered by high rainfall so that the water content in the soil increases. Hence, then the resistance of the shear will decrease and there will be mass movement of the soil.

Hardiyatmo [8] claimed that the behavior of annual rain can affect the frequency of landslides. There are two factors that cause such occurrence, including: (1) an increase in groundwater content

IOP Conf. Series: Earth and Environmental Science **971** (2022) 012024 doi:10.1088/1755-1315/971/1/012024

due to softening of the slope-forming material; and (2) an increase in the groundwater level which reduces the average shear strength of the soil. When it rains, the soil that was originally not saturated with water gradually becomes saturated. Along with the increase in water content in the soil, the shear strength of the soil gradually decreases, especially in fine-grained soils. This is because the pore pressure which was initially negative will become positive.

In further investigation, there are several causes of landslides in Ijo Tunnel, those are: (1) landforms; (2) slope geometry; and (3) the presence of springs. Ijo Tunnel is located in a strongly eroded to light denudational hill form composed of sandy limestone (calcarenite) located at N 060°E/20° (figure 1a.). When the calcarenite rock undergoes weathering, it will form a soil with a sandy clay fraction that has a moderate permeability level. The soil will then become easily saturated with water so that the shear resistance in the soil decreases. Hardiyatmo [8] added that the clay minerals contained in the soil greatly determine the value of the Atterberg boundaries which affect the shear strength empirically.

Soil saturation is also affected by the presence of springs on the northern slopes (figure 1b.). Soil or rock material that has a high-water content will have a high mass weight so that it will tend to move down the slope if it is located on a large slope angle and/or a layered structure that is tilted in the direction of the slope [9]. Thus, the landslide vulnerability in Ijo Tunnel is categorized as high. According to Hardiyatmo [8], springs can also be an indication that the location is the foot or tip of the landslide.

Figure 1. (a) Outcrop of sandy limestone (calcarenite); (b) Spring seepage of North side slope.

In addition to landform, slope geometry also determines the level of landslide susceptibility. The steeper the slope, the greater the gravitational force that pulls the soil mass down the slope. The slope geometry of Ijo Tunnel is as follows: slope direction N 170°E; slope continuity direction N 080°E; and the tilt of slope is 60°. A slope value of ($\angle 60^o$) indicates high landslide susceptibility in Ijo Tunnel.

Landslides that occur in Ijo Tunnel are of topples type, which is the movement of collapsed material on rock slopes which is very steep to upright. The discontinuity area on a collapsed slope is relatively vertical. The main factor that causes the topples type landslides is the presence of water that fills the cracks [8]. Cracks in cliffs can be caused by clay mineral content in the soil which has the characteristics of swelling, cutting slopes, and vibration due to rail traffic.

Ijo Tunnel area is a flood-prone area. Flood is an incident or condition in which part, or all of the land is submerged because the volume of water increases [10]. Aspects that determine flood susceptibility include landform and rainfall. The train route in Ijo Tunnel was formed by cutting slope activities on the denudational hilly landform so that the train route occupies the valley part. According to Sartohadi [9], slope cutting to obtain a flat area used as a place for human activities tends to increase the rate and capacity of rainwater infiltration and irrigation. If so, then the surface runoff will be higher and the potential for flood incidence will also be greater. Based on this description, it can be concluded that the flood vulnerability of Ijo Tunnel is categorized as high.

IOP Conf. Series: Earth and Environmental Science **971** (2022) 012024 doi:10.1088/1755-1315/971/1/012024

Figure 2. Flood indicator point.

The flood occurred on June 17th, 2016, inundated Ijo Tunnel up to 20 cm (figure 2.). Due to the incident, the train route was eroded, and 13 trains (KA) were delayed. The trains included Pasundan, Progo, Gajahwong, Senja Utama Solo, Kahuripan, Jaka Tingkit, Taksaka Malam, Lodaya, Malabar, Argo Dwipangga, Turangga, Gajayana and Bima trains.

The flood in Ijo Tunnel is an overflow river caused by the inability of river and drainage channels to flow the flood discharge or the existing flood discharge is greater than the existing drainage capacity (figure 3). Therefore, it is necessary to re-evaluate the construction of drainage channels that also considers the return period of rain. Changes in rainfall patterns and intensity as well as changes in land from time to time are possible causes of the mismatch between drainage capacity and the amount of runoff.

Figure 3. (a) Drainage channel; (d) river channel in Ijo Tunnel.

The multi-hazard in Ijo Tunnel was assessed by summing up the flood susceptibility score and the landslide susceptibility score. Based on the calculations, the multi-hazard of Ijo Tunnel to landslides and floods was high (table 10.). Multi-hazard mapping using a landform approach is a product that facilitates the socialization process of disaster to the community because its boundaries are easy to find in the field [7]. In addition, multi-hazard mapping also makes it easier for PT. Kereta Api Indonesia (Persero) as the organizer of railways to determine disaster-prone areas, update disaster information, and formulate policies and procedures to monitor and maintain the railway facilities and infrastructure to ensure the safety of rail transport users. It should be noted that safety is one of the pillars used as the ba1sis for the operation of the railway to realize the principle of zero fatality and zero accident [11].

Table 10. Assessment of Multi-hazard.

	Parameter	Score	Total Score
Landslide Susceptibility	Slope	5	8
	Level of incision	3	
Flood	Landform	3	9

Susceptibility	Rainfall	3	
	Slope	3	
	Total Multi-hazard		17 (High)

3.2. *Vulnerability*

Vulnerability is a condition of risk elements that lead to an inability to deal with the disaster threat. The higher the level of disaster vulnerability, the easier it is for elements at risk to be affected by disasters. Disaster vulnerability is measured by three assessment components, including the completeness of supporting infrastructure, the quality of the train route, and the density of rail traffic.

Ijo Tunnel is located in a denudational hilly landform with a hilly morphology. The train route is built by cutting the slope so that it is prone to disasters. Therefore, good infrastructure is needed to prevent the occurrence of disasters. The train route in Ijo Tunnel is basically equipped with drainage channels considering its location is a valley where water flows. However, the capacity of the drainage channel is not sufficient to accommodate the water flow so that it is susceptible of flood. In addition, the slopes on the right and left sides of the train route are not entirely built with permanent cliffs so they are vulnerable to landslides (figure 4.). This condition causes Ijo Tunnel's vulnerability to landslides and floods to be high.

Figure 4. Drainage channels along rails and slopes that are not entirely permanent cliffs.

The train route condition around Ijo Tunnel was identified from the type and type of rail, the type and condition of the bearing, the distance of the rail to the edge of the pavement, and the use of the surrounding land. The type of rail used on the train route around Ijo Tunnel is R54 with good condition. The type of rail bearing used is concrete, while the bridge is wooden (figure 5). The condition of the rail bearings is classified as slightly damaged. The distance between the rails and the edge of the pavement is the distance between the rails and the nearest road access. Ijo Tunnel can be accessed by car via a rocky road with rough terrain for 3 km from the main access (arterial road). It is very important to know the road access to Ijo Tunnel to facilitate the process of monitoring rail conditions and the evacuation process during the disaster incident. The use of land around the train route is also important because if there are people living around the train route, areas vulnerable or damaged rails can be easily informed to officers. In addition, the community can also assist officers in the evacuation process when a disaster occurs. The land uses around the Ijo Tunnel are farm, plantations, and forests. By paying attention to the three aspects above, the vulnerability of Ijo Tunnel from the aspect of rail conditions is high.

The determination of disaster vulnerability along the train route can also be done based on the density of rail traffic. The more trains that pass, the higher the potential for vibration. Ijo Tunnel is the main train route across the southern island of Java. This route is traversed by passenger and goods from and to Jakarta, Bandung, Purwokerto, Yogyakarta, Solo, Kediri, to Surabaya. Vibrations caused by rail traffic can cause a decrease in shear strength in some sensitive clays which will increase the potential of landslides [8]. Such conditions in the technical assessment in categorized as high vulnerability.

Based on the descriptions above, the disaster vulnerability around Ijo Tunnel is high. More monitoring and supervision are needed to the risk elements. Periodic checks in accordance with regulations and procedures can minimize the risk of disasters that occur.

Figure 5. (a) Rail condition and (b) Rail bearing.

3.3. *Capacity*

Capacity is the ability of a region and its community to take action to reduce the level of threat and level of loss due to disasters. Improving the capacity factor is an important aspect since it will accelerate the recovery process due to disasters. Disaster capacity was measured by the presence or absence of disaster SOPs, whether database management related to disasters is established, as well as communication and coordination between agencies in dealing with disasters [12].

Disaster SOPs are systematic stages in disaster mitigation and management efforts. If disaster SOPs are carried out systematically, the disaster mitigation and management process will run quickly and efficiently in terms of the costs and energy required. Ijo Tunnel is included in the working area of DAOP 5 Purwokerto which already has standard operating management in dealing with disasters so that when a disaster occurs, the officers responsible for dealing with disasters are alert. However, SOPs related to disaster emergency response have not yet been formally drafted and ratified to be applied in all DAOPs. This condition is categorized as medium capacity.

According to Central Java Technical Center officers, most of the disaster-prone areas have been recorded and there are procedures for 1×24 hour security at the disaster-prone areas. However, there has been no activity for updating the data for disaster-prone areas [13]. In fact, periodic data updating is very important to monitor disaster-prone areas. This condition is categorized as medium capacity.

During disaster management process, good communication and coordination between agencies engaged in disaster aspects is also required. Based on the statement from the Central Java Technical Center officer, communication and coordination were only carried out internally by DAOP. However, internal coordination is sufficient in dealing with disasters. Based on the assessment parameters (scoring), such conditions are categorized as low capacity. By considering the three aspects of disaster capacity assessment above, the disaster capacity in DAOP 5 is categorized as medium (figure 6).

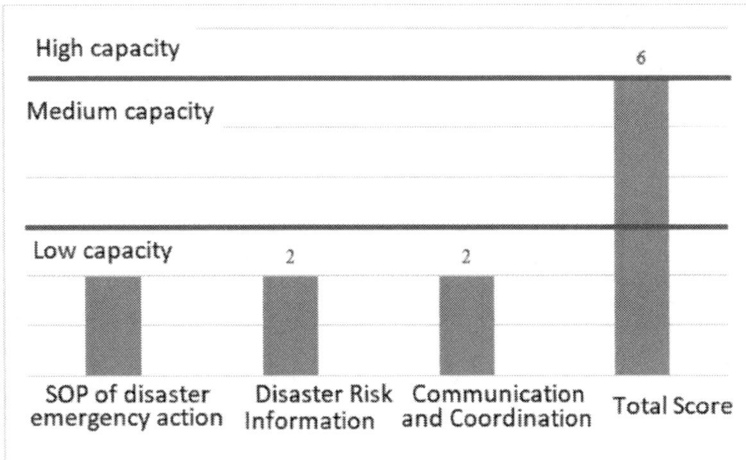

Figure 6. Disaster capacity based on the parameters.

3.4. *Risk of Disaster*

Disaster risk is produced from a systematic calculation of susceptibility, vulnerability, and disaster capacity. Disaster risk in an area was assessed as an integrated step that can be used as a reference in managing disaster aspects as a whole. For example, an area that has a high disaster risk does not necessarily have a low disaster capacity but can be caused by a low susceptibility or vulnerability value. A careful study needs to be done on the efforts to reduce disaster risk in an area so that the decisions made can resolve existing problems.

Disaster risk in Ijo Tunnel area is categorize as medium (figure 7.). This category was obtained from the high value of multi-hazard and disaster vulnerability but the capacity to deal with disasters is categorized as medium (table 11.). Identification of disaster risk provides a basis for decision making to accept how much risk can be faced or borne and design plans and efforts to mitigate it [14].

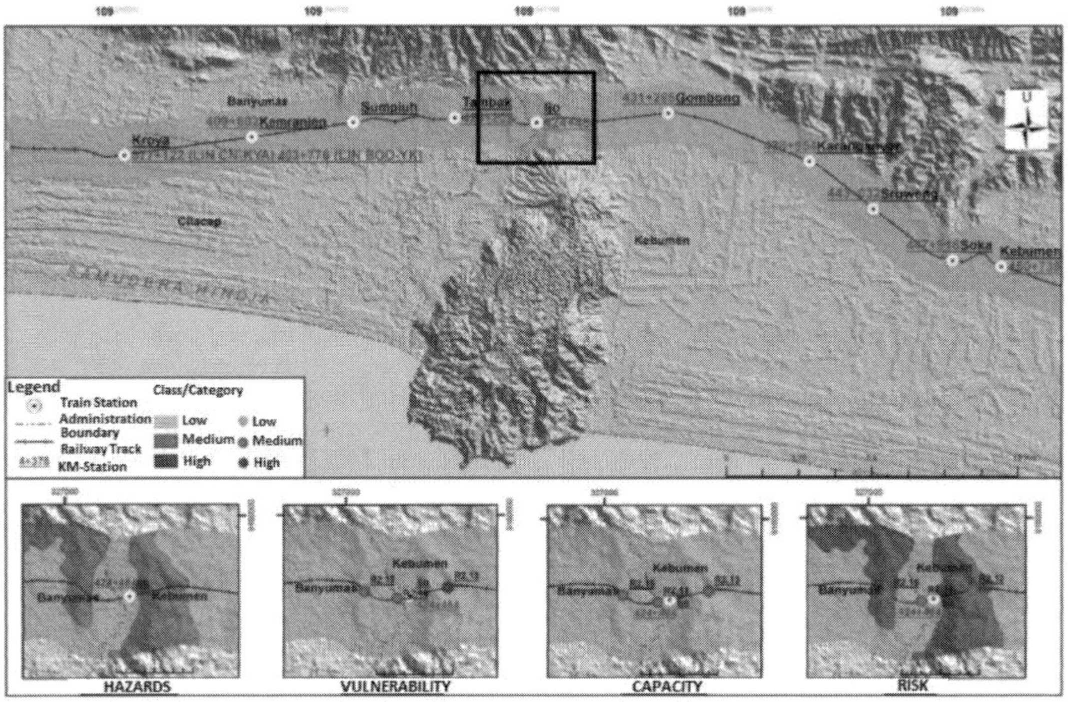

Figure 7. Ijo Tunnel Multi-hazard Risk.

Table 11. Multi-hazard Assessment.

Parameter	Score
Multi-hazard	3
Susceptibility	3
Capacity	3
Risk	3 (Medium)

3.5. *Strategy in Reducing Disaster Risk*

The reduction of disaster risk is a series of strategies and activities performed to anticipate and overcome disaster incidents. The reduction of disaster risk includes three main aspects of its constituents, those are multi-hazard, vulnerability, and capacity. Risk reduction stages are conducted to reduce the effect of disasters [15].

Two things that must be used as the foundation to formulate a disaster risk reduction strategy are by paying attention to the characteristics of the existing landforms and the current field conditions (Figure 7). Landforms are composed of four main aspects, those are [16,17]: (1) morphology; (2) morphogenesis; (3) morpho-chronology; and (4) morpho-arrangement. These four aspects summarize geomorphological processes that cause changes in the configuration of the earth's surface from time to time [18]. If a scientist has the ability to find, identify, and analyze the working geomorphological processes, the factors causing the disasters and their solutions can be formulated properly.

Disaster risk reduction strategies in Ijo Tunnel can be carried out structurally and non-structurally [19]. Structural strategy is carried out by handling in physical form in the form of construction of landslide-retaining buildings and waterways (figure 8.). Meanwhile, non-structural strategies are carried out by increasing the human resource capacity of officers and residents around Ijo Tunnel for disaster management.

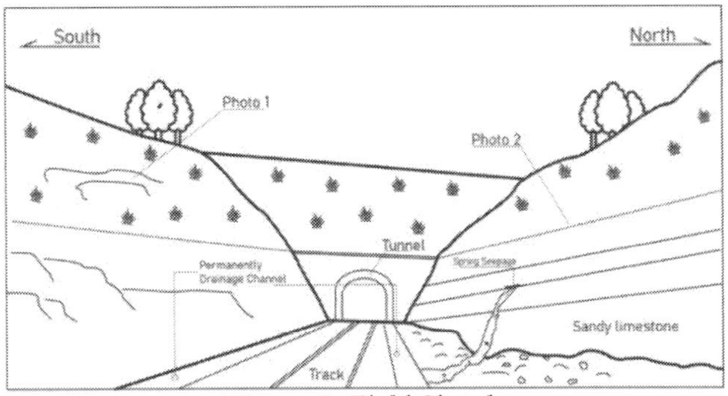

Figure 8. Field Sketch.

Disasters will occur if there is a meeting between multi-hazard and vulnerabilities so that it must be avoided [14]. Stages that can be taken are: (1) strengthening the slope stability to reduce landslide incidents by making permanent cliffs; (2) installing pipes on cliffs to channel excessive water when it rains, especially on cliffs where there are seepage springs; (3) evaluating drainage channels that have been made; (4) implementing the culture of not throwing garbage in rivers which can reduce the capacity of river distribution; and (5) repairing and strengthening of the railway bearing structure which is indicated to be slightly damaged.

Disaster risk can be reduced by increasing capacity. Stages that can be done are: (1) making an SOP of integrated disaster emergency response that can be applied to all DAOPs; (2) periodic updating of disaster information; (3) establishing coordination and cooperation with relevant agencies engaged in the field of disaster; and (4) strengthening knowledge of train operators on disaster risk assessment along train route. The stages above are urgent to take considering that risk management examines all strategic and implementation options both in the context of structural and non-structural handling [14]. Non-structural handling in the form of SOPs and so on needs to be legally legalized in the form of laws and regulations as well as government and institutional regulations.

4. Conclusion

Based on the research results, it can be concluded that Ijo Tunnel has a high level of disaster susceptibility, medium level of disaster vulnerability, medium level of disaster capacity, and medium level of disaster risk. The surrounding condition of Ijo Tunnel is a minor fault area with loose material. Furthermore, there is water seepage that triggers floods and landslides so that volume, construction, and age of buildings around the walls need to be recalculated. Therefore, the disaster management strategy in Ijo Tunnel is carried out structurally through physical and non-structural handling by increasing the capacity.

5. References

[1] Kementerian Perhubungan 2011 *Rencana Induk Perkeretaapian Nasional* (Jakarta: Kementerian Perhubungan, Ditjen Perkeretaapian)

[2] Soemardi K, Prasidi A and Desrianty A 2010 Kajian penyebab laten kecelakaan kereta api, menggunakan kerangka HFACS dengan pendekatan QFD *Proc. Seminar Nasional IV Manajemen & Rekayasa Kualitas 2010* (Itenas, Bandung)

[3] Direktorat Jenderal Perkeretaapian 2014 *Buku Informasi Perkertaapian 2014* (Jakarta: Direktorat Jenderal Perkertaapian, Kementerian Perhubungan Republik Indonesia)

[4] Yusrizal Y, Saleh S M and Fadhly N 2013 Studi revitalisasi jalur kereta api Banda Aceh – Batas Sumatera Utara, *Jurnal Teknik Sipil.* **2(1)** 26-38

[5] Esri n.d. Resources https://pro.arcgis.com/en/pro-app/2.7/tool-reference/spatial-analyst/how-slope-works.htm

[6] Maulana E and Wulan T R 2015 Pemetaan multi-rawan kabupaten malang bagian selatan dengan menggunakan pendekatan bentanglahan *Proc. Simposium Nasional Sains Geoinformasi IV/2015* (Yogyakarta)

[7] Wulan T R, Maulana E, Maulia N, Anggara I H, Putra A S, Putra M D, Ibrahim F, Wahyuningsih D S and Siswanti S 2016 Analysis of land capability in alluvial plain and volcanic slope of rembang district using landforms approach *Proc. The 2nd International Conference of Indonesian Society for Remote Sensing* (Yogyakarta)

[8] Hardiyatmo H C 2012 *Tanah Longsor dan Erosi: Kejadian dan Penanganan* (Yogyakarta: Gadjah Mada University Press)

[9] Sartohadi J 2005 *Pemanfaatan Informasi Kerawanan Gerakan Massa Untuk Penilaian Kemampuan Lahan di Sub-DAS Maetan, Daerah Aliran Sungai Luk Ula Jawa Tengah* Majalah Geografi, **19** pp 21-39 (March 2005)

[10] Kodoatie R J and Sugiyanto S 2002 *Banjir: Beberapa Penyebab dan Metode Pengendaliannya dalam Perspektif Lingkungan* (Yogyakarta: Pustaka Pelajar)

[11] PT. Kereta Api Indonesia (Persero) 2016 *Laporan Tahunan 2015: Transformasi dan Inovasi Pelayanan Berkelanjutan Persembahan Untuk Negeri* (Bandung: PT. Kereta Api Indonesia (Persero))

[12] Iridiastadi H dan Izazaya E 2012 Kajian Taksonomi Kecelakaan Kereta Api Di Indonesia Menggunakan Human Factors Analysis and Classification System (HFACS). Prosiding InSINas 2012. Bandung: Asdep Relevansi Program Riptek, Deputi Bidang Relevansi dan Produktivitas Iptek, Kementerian Riset Dan Teknologi;Hal 51-55.

[13] Nugrahini Y 2012 Analisis kinerja pelaksanaan kewajiban pelayanan publik bidang angkutan kereta api penumpang kelas ekonomi, *Jurnal Perencanaan Wilayah dan Kota*. **23**(1) 19-36

[14] Latief H 2011 *Siaga Tsunami di Pesisir Sumatra* in Abdurahman O P 2011 *Hidup di Atas Tiga Lempeng: Gunungapi dan Bencana Geologi* (Bandung: Badan Geologi, Kementerian Energi dan Sumberdaya Mineral)

[15] Novianta M A and Setyaningsih E 2015 Sistem informasi monitoring kereta api berbasis web server menggunakan layanan GPRS, *Jurnal Momentum*. **17(2)**

[16] Zuidam and Cancelado, 1985 Terrain Analysis and Classification Using Aerial Photograph; a Geomorphologic Approach. ITC Enchede. The Netherlands

[17] Santosa, 2015 Keistimewaan Yogyakarta Dari Sudut Pandang Geomorfologi. Gadjah Mada University Press. Yogyakarta.

[18] Sartohadi J 2007 Terapan geomorfologi dalam pengelolaan sumberdaya air, *Jurnal Alami*. **12(1)** 16-21

[19] Muhtadi A 2011 Kinerja lalu lintas jalan nasional kabupaten sidoarjo, *Neutron Jurnal Teknik Sipil*. **11(2)** 17-25

Acknowledgments
I would like to express my gratitude to the Director of Railway Safety, the Director General of Railways, and the Ministry of Transportation of the Republic of Indonesia as the provider of disaster data updating work along the southern Java Island train route who also provided assistance during the data collection in the field and licensing arrangements. The authors also express their gratitude to PT. Danureksa Sarana Cipta as the party that accommodates the implementation of activities. This research also could not be separated from the assistance of IDRB Team.

IOP Conf. Series: Earth and Environmental Science **971** (2022) 012025 doi:10.1088/1755-1315/971/1/012025

Correlation between Uniaxial Compressive Strength and Point Load Strength of Penang Island Granites

J D Mohamad[1], D Z Abang Hasbollah[2*], A Mohd Taib[3], M F Md Dan[4], S N Jusoh[1], K N Mat Said[1],

[1] School of Civil Engineering, Faculty of Engineering Universiti Teknologi Malaysia, 81310 UTM Skudai, Johor, Malaysia
[2] Centre of Tropical Geo-engineering, School of Civil Engineering, Faculty of Engineering Universiti Teknologi Malaysia, 81310 UTM Skudai, Johor, Malaysia.
[3] Department of Civil Engineering, Faculty of Engineering and Built Environment, Universiti Kebangsaan Malaysia
[4]Department of Infrastructure & Geomatic Engineering, Faculty of Civil and Built Environment, Universiti Tun Hussein Onn Malaysia, 86400 Batu Pahat, Johor
*Corresponding author: dzulaika@utm.my

Abstract. Uniaxial compressive strength (UCS) is one of the most widely used strength parameters in geotechnical design. The value is obtained through direct testing method, carried out through careful procedures of testing in a laboratory setting. As such, it is relatively more expensive, more tedious and requires longer time to complete. Considering that UCS is highly important, Engineers often utilize indirect methods, namely the Point Load Test, to obtain the Point Load Index Strength (Is(50)). From this test, the results are correlated in order to predict the UCS value. One of the general correlations is provided by ISRM where UCS=20-25 Is(50). However, it is highly contested as it is not universal for all types of rocks. The main focus of this research is to find the correlation of UCS and Is(50) in the Penang Island area. This research would also attempt to classify the granites according to strength. This study showed that the Penang Island Granites are "weak to very strong" granites with UCS values between 24.76 to 156.82MPa. The granites from the North Pluton are classified as "medium strong to very strong" with UCS values between 37.11 to 156.82MPa. The granites from the South Pluton are classified as "weak to very strong" with UCS values between 24.76 to 141.59 MPa. The recommended correlation between UCS and Is(50) are (1) Overall: 8.385 Is(50) + 30.016 (2) North Pluton: 7.93 Is(50)+ 35.606 (3) South Pluton: 9.03Is(50)+20.138.

1. Introduction

Reliable estimates of the strength and deformation characteristics of rock masses are required for almost any form of analysis used for the design of slopes, foundations and underground excavations. As such, a Geotechnical Engineer should have a proper understanding of the rock mass behavior and strength. In order to achieve this, qualitative measurements of the intact rock strength parameters are commonly carried out both in the laboratory and in-situ. Uniaxial Compressive Strength (UCS) and Point Load Index Strength (Is (50)) are among the most important intact rock strength properties. A reliable and good testing program and results could help Engineers and Geologists to classify the rock properly as well as predicting the behaviour of the rock mass and the performance of our design. This paper is aimed to classify Penang Island Granites based on the UCS and Is(50) as well as to establish the correlation between UCS and Is(50) of the Penang Island granite samples.

2. Problem background

In order to obtain the UCS value, the test involved is expensive, time-consuming, and requires high-quality samples [1-4]. Because of these drawbacks, UCS is a bit more hassle to carry out at the site. Not only that, in sites where the rock is highly weathered, the rocks are harder to prepare as the rock would crumble, resulting in failure to comply with the testing standard size as per ASTM and ISRM Standards. In order to overcome this, Engineers often utilize indirect testing methods to obtain rock strength parameters as they are simpler, cheaper, quicker and easily adaptable to the field [5-6]. One of the popular indirect methods used for this purpose is the Point Load Test (PLT). From this, Point Load Index Strength (Is(50)) is obtained. Previously multiple researchers attempted to predict the UCS values by correlating the Is(50) values with UCS values. From this, the correlation factor, (k) is obtained. Many studies had produced correlation between these two parameters for various types of rocks. Among the most renowned studies was by Broch and Franklin [7] giving a correlation factor of 24 x Is(50), for samples referred to a standard size of 50mm. According to the International Society of Rock Mechanics [8], the correlation factor recommended is between 20 Is(50) to 25 Is(50). Subsequent studies found that k=24 is not applicable for all types of rocks, instead, there appeared to be a broad range of correlation factors.

3. Literature review

3.1 Uniaxial Compressive Strength

Generally intact rock strength is divided into two, compressive strength and tensile strength. Compressive strength is the ability of a rock sample to withstand a compressive force applied axially on the rock. It is the maximum load applied to the rock before being fractured. When a rock fails in compression, it means that the compressive stress has exceeded the compressive strength of the rock. On the other hand, tensile strength is the maximum tensile stress a rock material can bear. Rocks is often assumed to have zero tensile strength since the test results for tensile strength are usually extremely low. One of the most important intact rock strength parameters is the uniaxial compressive strength (UCS). It is one of the important parameters when determining rock mass strength [9]. UCS gives the Engineers a rough index and some indication on the likely issues to occur during construction [10]. At the very basic level, the UCS is used as part of the Basic Geotechnical Description [11-13]. Other than that, UCS is also as one of the important parameters in rock mass classifications such as the Rock Mass Rating [14], Rock Mass Quality (Q) [15], Mining Rock Mass Rating (MRMR) [16] Geological Strength Index (GSI) [17] and Rock Mass Index (RMI) [18]. UCS could also be used to predict the shaft friction resistance in rock sockets [19-27]. Based on the various applications of the UCS mentioned above, it can be concluded that the UCS of a rock is undoubtedly very important. It is indeed versatile as it could help Engineers to obtain a lot of parameters for geotechnical design.

3.2 Uniaxial Compression Test (UCT)

The UCS value of a rock is obtainable via a direct strength test called the Unconfined Compression Test (UCT). The testing procedures are standardized by the International Society of Rock Mechanics (ISRM) as well as the American Society for Testing and Materials (ASTM). Rock samples are cut into the standard sizes of right circular cylinders before being crushed between two steel platens, with the load applied axially until the sample fails. The diameter preferably not lesser than NX core size, about 54mm [28-29]. Thuro *et al.* [30] noted that there are no significant effects by the sample diameter size on the UCS value. The length-to-diameter ratio (L:D) varies depending on which testing standard we follow as the ISRM recommends L:D between 2.5:1 and 3:1, while ASTM recommends L:D between 2.0:1 and 2.5:1. It is also noted that the measured strength would increase if the ratio is lesser than 2. Chiu and Johnston[31] pointed out that increased strength with a shorter specimen is due to lateral restraint at the ends of the specimen, caused by the platens, which may result in non-uniform stress distributions under compression [32]. Considering this, it is common practice that the ratio is typically 2:1. UCS value can be calculated using the following simple equation:

IOP Conf. Series: Earth and Environmental Science **971** (2022) 012025 doi:10.1088/1755-1315/971/1/012025

$$UCS = F/A \quad (1)$$

where F and A are maximum applied load and specimen cross sectional area, respectively.

3.3 Point Load Test (PLT)

PLT is carried out to obtain the point load strength index (Is (50)) of an intact rock. PLT is a popular method, used together with UCS in rock classification [33]. PLT is carried out by compressing rock samples (applying point load) between two conical steel platens (radius of 5mm, conical angle of 60°) until failure occurs. It is noted that only failures where one or more extensional planes developed through the whole body of the sample, below the conical platens are considered as valid tests. Samples which chipped to the sides or where the failure plane did not go through directly below the conical platens are considered invalid. Typical failure patterns, as well as the valid and invalid tests are shown in Figure 1.

ASTM has established the basic procedure for conducting and calculation of point load strength index (ASTM D5731-08). In practice, there are three main types of PLT; axial, diametral and lump or block. Axial and diametral types are performed on rock core samples, while lump or block tests are carried out on rock samples with irregular shapes. The point load strength index determined by PLT must be corrected to the standard equivalent diameter (De) of 50mm [34], and is denoted by Is(50). The suggested equation for Is(50), by ASTM is as the following:

$$Is(50) = (De / 50)\ 0.45 \times Pu / De^2 \quad (2)$$

where Pu and De are the failure load and the equivalent diameter which is the core diameter. The load configurations and specimen shape requirement for PLT is as shown in Figure 2.

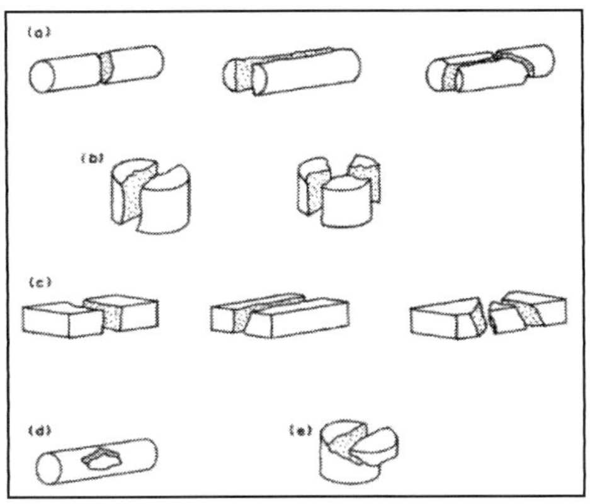

Figure 1. Typical Modes of Failure for Valid and Invalid Tests (a) Valid diametral tests; (b) valid axial tests; (c) valid block tests; (d) invalid core test; and (e) Invalid axial test.

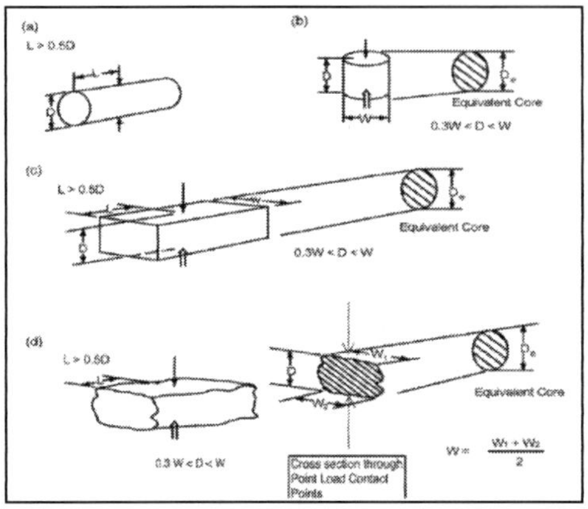

Figure 2. Load Configurations and Specimen Shape Requirement for (a) the Diametral Test, (b) the Axial Test, (c) the Block Test, and (d) the Irregular Lump Test Legend: L = distance between contact points and nearest free face, and De = equivalent core diameter.

Point Load Index Strength is considered as an index property of the rock. Tan and Chan[35] defined index properties as the physical or mechanical property of a rock, which are determined through a method that is relatively easy but still bears a significant relationship to its engineering properties.

3.4 UCS vs Is(50)

This method is highly popular, as there are literally hundreds of this method carried out by multiple researchers in the past, with various (k) values. As such, only past studies related to granites are included as shown in Table 1.

Table 1. Suggested correlations between UCS and Is(50) for granites.

Reference	Correlation	No.s of sample	R or R^2
Basu and Aydin [36]	UCS =18 Is(50) MPa	40	R^2=0.97
Ghosh and Srivasta[37]	UCS=16 Is(50) MPa	22	-
Tuğrul and Zarif [38]	UCS=15.25Is(50) MPa	19	R=0.98
Tan [39]	UCS= 7 to 17 Is(50) MPa	-	-

3.5 Previous studies in Penang Island area

Based on the Geology Map Sheet 28 by published by the Director General of the Malaysian Geological Society in 1994, the main island of Penang is underlain by two granite plutons of different magma origins namely the North Pinang Pluton and the South Pinang Pluton with sedimentary rocks of Mahang Formation and recent Quaternary deposits [40].

The North Penang Pluton is subdivided into three units which are the Tanjung Bunga Granite, Feringgi Granite and Muka Head microgranite. The South Penang Pluton is further divided into two, namely the Batu Maung granite and Sungai Ara Granite.

The Batu Maung Granite is characterized by medium to coarse-grained biotite granite which occupies about 80% of microcline granite while the Sungai Ara Granite has the characteristics of a fine-grained biotite granite that occupies about 20% of the area underlain by the microcline granite. A fairly large area of altered granite occurs at the contact between these two granite plutons. Based on Lee *et al.* [41], the Batu Maung Granite subsoil mainly consists of silty SAND and sandy SILT.

The recent Quaternary deposits which consist of relatively recent unconsolidated sedimentary of Gula Formation and Beruas Formation are observed. According to Ong [42] the top one meter of recent alluvium deposits consists of light yellowish-brown clay with brownish spots. Below this is a thicker layer of brownish grey soft clay with abundant plant remains of greyish fine to medium grained sand. At the base of alluvium, a layer of coarse sand with some greyish clay or a layer of peat may occur. Figure 3 shows the Geological Map of the Penang Island while previous relevant studies carried out in the Penang Island Granites are compiled in Table 2.

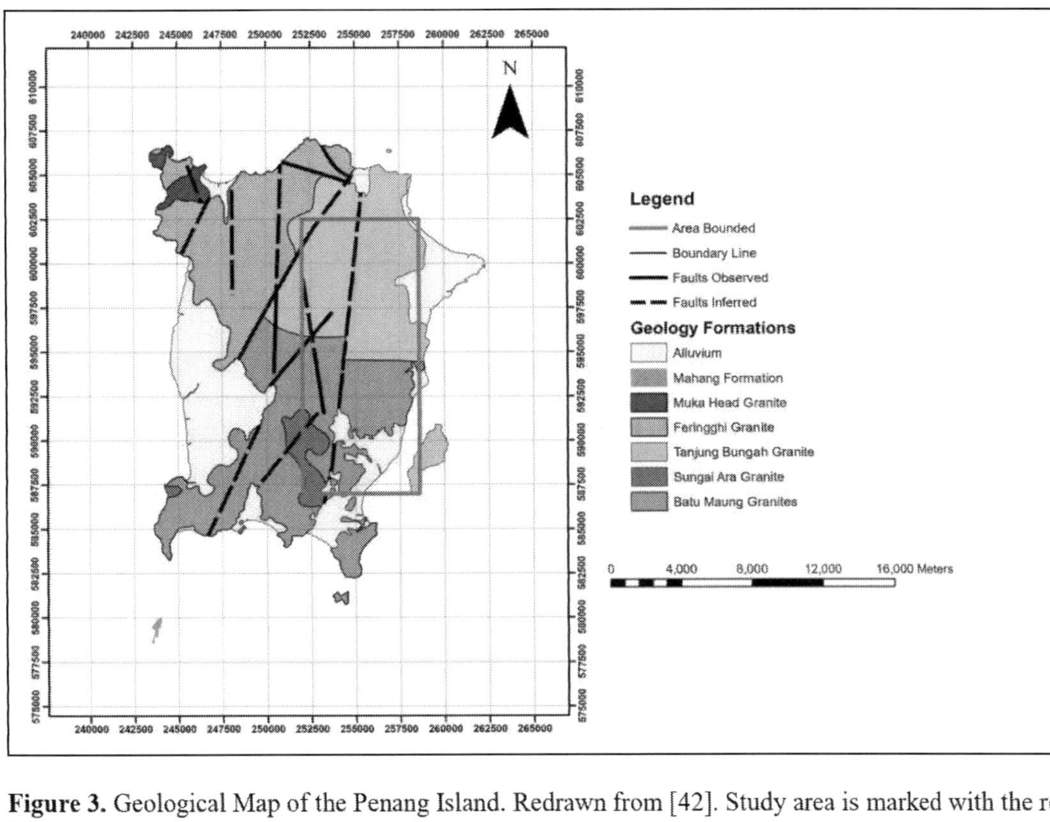

Figure 3. Geological Map of the Penang Island. Redrawn from [42]. Study area is marked with the red box.

Table 2. Previous relevant studies carried out in Penang Island area.

Reference	Geo-material studied	Properties studied	Gaps
Ong [42]	Geology and Engineering Geology of Penang Island	General Geology Engineering Geology (Aggregates)	Data on Engineering Geology not typically used in Geotechnical Engineering
Tan [39]	Soils Rocks and slope stability in Penang Island urban area. The study covered both North and South Pluton	i) Engineering properties of the granitic soils ii) Engineering properties of the granites iii) Slope stability. iv) Produced a correlation of UCS=7 to 17 Is(50)	Did not differentiate the two granites (The correlation is combined for the two plutonic groups)
Ahmad *et al.* [43]	North Pluton: (Batu Feringghi, Tg Bunga & Muka Head Microgranite) South Pluton: (Batu Maung Granite and Sungai Ara Granite.)	Soil Index Properties SPT-N Values Stiffness and Compressibility	No Rock Strength

4. Methodology

Given that the study takes place during the COVID-19 Pandemic, the data used for this study is secondary data. A total of fifty (50) Soil Investigation (SI) points from around the Penang Island area are gathered and analysed. The rock samples are distributed all around the Penang Island, mainly on the east coast of the Penang Island. Since there is a restriction on the exact locations of the boreholes by the data contributor, the authors will only include the approximate area bounded by SI points, which is summarized as shown in Table 3.

Table 3. Area Bounded by the SI Points.

Axis	RSO Coordinates (meters)	
	From	To
Northing	587 000	602 500
Easting	252 000	258 600

5. Data Analysis

From the data sets, UCS and Is(50) results are extracted. A set of rock sample is when UCS and Is(50) is available for a given depth. Only rock samples that comply with the test requirements outlined by ISRM and ASTM were selected for this study. A total of forty-two (42) sets of rock samples are used for this study. Any inconsistencies between the records shown on the results sheet and the recorded dimensions of the rock sample is checked, and corrections are made accordingly. It must be noted that the laboratory tests were done at the respective laboratories, and the accuracy and reliability of the results are borne by the respective Soil Investigation Contractors. Table 4 summarized the available data for this study:

Table 4. Summary of the data available for this study.

No	Pluton Member	Geological Formation	Sample No.s
1	North	Tg, Bungah	19
2	South	Batu Maung	22
3	South	Sg. Ara	1

5.1 Strength Classification of the rock samples

The rock samples are classified based on its strength. The strength classification shall be based on the BS 5930. BS5930:2015 were used to classify the strength of the rock samples. Clause 36.2.1 is referred from BS5930:2015. The rock strength descriptions are summarized as Table 5 below:

Table 5. Terms for description of rock strength. Adapted from [12].

Term for use in field or based on measurement	Definition for field use	Definition on basis of Unconfined Compressive Strength measurements (MPa)
Extremely weak	Can be indented by thumbnail. Gravel sized lumps crush between finger and thumb.	0.6 – 1.0
Very weak	Crumbles under firm blows with point of geological hammer. Can peeled by a pocketknife.	1 – 5
Weak	Can be peeled by a pocketknife with difficulty. Shallow indentations made by firm blow with the point of geological hammer	5 – 25
Medium strong	Cannot be scraped with pocketknife. Can be fractured with a single blow of geological hammer.	25 – 50
Strong	Requires more than one blow of geological hammer to fracture.	50 – 100
Very strong	Requires many blows of geological hammer to fracture.	100 – 250
Extremely strong	Can only be chipped with geological hammer.	>250

From this, it is learned that in overall, the granites from Penang Island are categorized as "weak to very strong" rocks, with the UCS values between 24.76 to 156.82 MPa. The granites from the North Pluton are classified as "medium strong to very strong", with UCS values between 37.11 to 156.82

MPa. The granites from South Pluton are classified as "weak to very strong" with UCS values between 24.76 to 151.59 MPa. The result from this stage of analysis is summarized in Table 6:

Table 6. Summary of the UCS value range and strength descriptions based on BS5930:2015.

Geology	UCS (MPa)	Strength description
North Pluton	37.11-156.82	Medium strong to very strong
South Pluton	24.76-141.59	Weak to very strong
Overall	24.76-156.82	Weak to very strong

5.2 Correlation between UCS values and Is(50) values

This stage of analysis aims to find out the correlation between the UCS value and the Is(50) values for the rock samples. In order to achieve this, a simple mathematical approach is employed. The UCS and Is(50) values are plotted into a graph using Microsoft Excel. An upper boundary line and a lower boundary line were drawn to limit the highest and the lowest value. Then, gradient for these two boundary lines shall be determined in order to find the formula. The formula UCS= x Is(50) will be derived from this exercise, where x is the gradient. Next, a line of simple linear regression is also obtained from the graph. The line is the best fit line. The equation of the line shall be the correlation between UCS and Is(50). The analysis is done for overall, North Pluton and South Pluton samples. The graph obtained for overall samples is as shown in Figure 4.

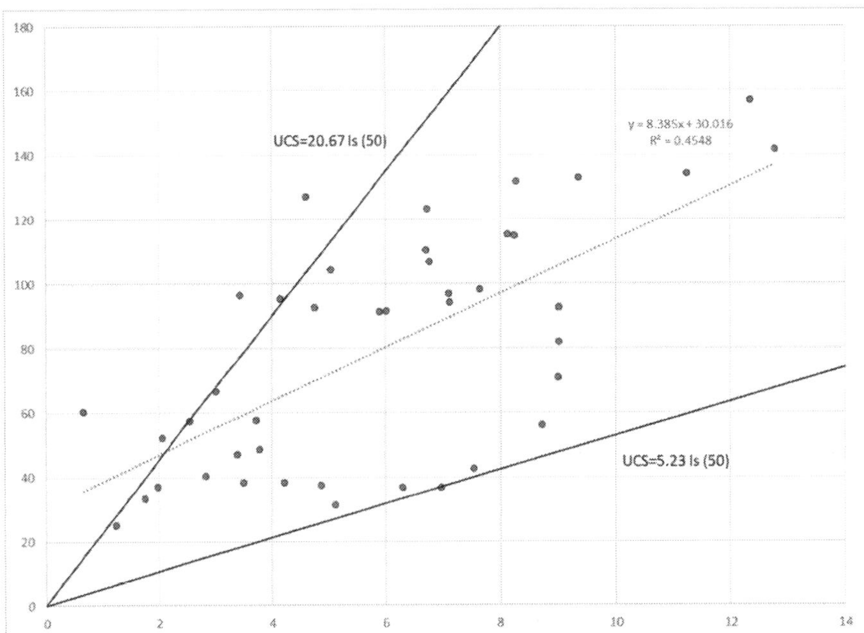

Figure 4. UCS vs Is(50) graph for Overall Samples.

Based on Figure 4, the correlation between UCS and Is(50) for overall Penang Island Granites is 8.385 Is(50) + 30.016, with a correlation coefficient (R^2) of 0.455 and R value of 0.674. This shows that the data is scattered around the fitted regression line, with a positive correlation. This also tells us that 45.5% of the variability of UCS value is affected by the Is(50) values.

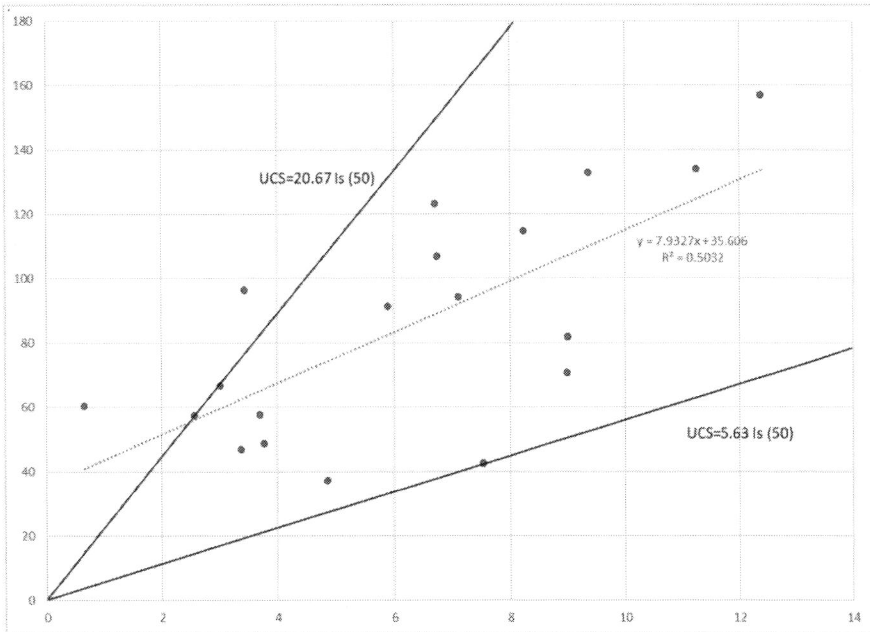

Figure 5. UCS vs Is (50) graph for North Pluton Samples.

Based on Figure 5, the correlation between UCS and Is(50) for North Pluton Granites is 7.93 $Is_{(50)}$+ 35.606, with a correlation coefficient (R^2) of 0.503 and R value of 0.709. This shows that the data is scattered around the fitted regression line, with a positive correlation. This also tells us that 50.3% of the variability of UCS value is affected by the Is(50) values.

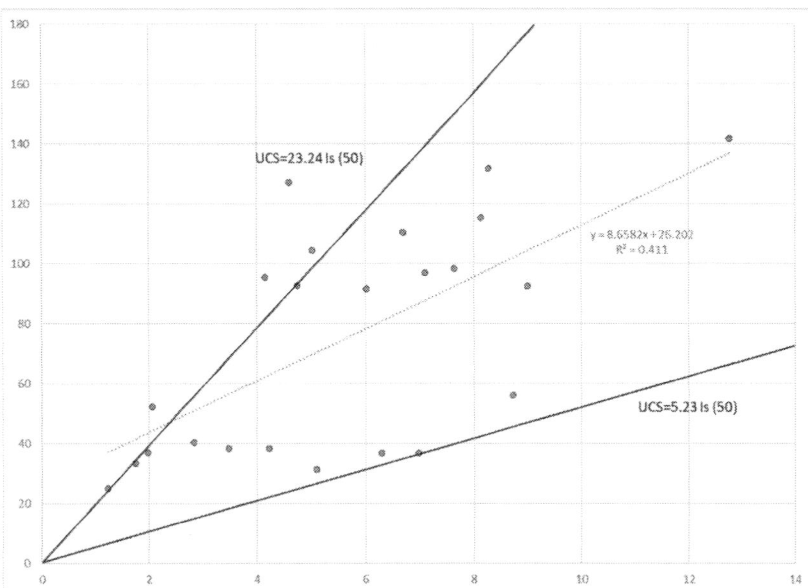

Figure 6. UCS vs Is (50) graph for South Pluton Samples

Based on Figure 6, the correlation between UCS and Is(50) for the South Pluton Granites is 9.03Is(50)+20.138, with a correlation coefficient (R^2) of 0.411 and R value of 0.641. This shows that the data is scattered around the fitted regression line, with a positive correlation. This also tells us that 41.1% of the variability of UCS value is affected by the Is(50) values. From graphs shown in Figure 4, Figure 5 and Figure 6, the following can be summarized, shown in Table 7.

Table 7. Summary of the recommended correlations.

Data Set	Lower Boundary	Upper Boundary	Recommended Correlation	R^2	R	Equation No
Overall	5.23 Is(50)	20.67 Is(50)	8.385 Is(50) + 30.016	0.455	0.674	(3)
North Pluton	5.63 Is(50)	20.67 Is(50)	7.93 Is(50)+ 35.606	0.503	0.709	(4)
South Pluton	5.23 Is(50)	23.24 Is(50)	9.03Is(50)+20.138	0.411	0.641	(5)

5.3 Comparison with previous studies

One of the main goals of this study is to build upon the work by [39] Through this, existing knowledge on the Penang Islan Granites can be improved. [39] proposed the upper and lower boundary lines for Penang Island Granites. As such, the upper boundary and lower boundary is compared for the overall samples. Figure 7 shows the plotting from [39] and this study.

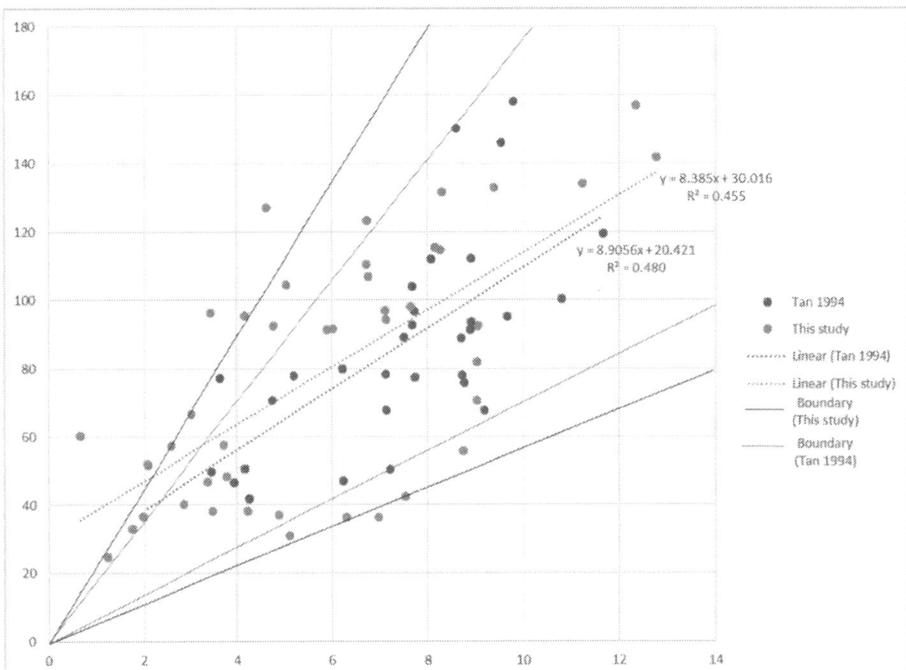

Figure 7. UCS and Is(50) values from Tan [39] and this study plotted together.

From Figure 7, it is learned that this study expanded the range of values of the lower boundary and the upper boundary. More data are added into the range proposed by Tan [39].This is something positive as more data is available, increasing the reliability of the proposed value by Tan [39]. Twenty-four (24) more data sets fall be between the 7-17 Is(50) proposed by Tan[39]. A similar result to [4] was also obtained where the R^2 value decreased when the sample number increases. The R^2 value for North Pluton (19 samples) are higher than the South Pluton (23 samples).

From the laboratory results, it is calculated that the average value for UCS for North Pluton Granites is higher than the South Pluton Granites. This was also observed for Is(50) values as well. This is thought to be as a result of the difference of (1) rock weathering condition and (2) difference of grain sizes of granite. This is in line with previous findings by previous granitic rock researchers which found that the petrological properties, texture, grain size, mineral composition and weathering grades affects the mechanical properties of a granite [44-46].

Equations 3, 4 and 5 were then compared with other correlations published. It is learned that the equations tend to underpredict the UCS values of the granites. Twenty-four (24) samples were underpredicted when the UCS is predicted using the proposed equations from this study, while the remaining eighteen (18) were overpredicted. This is expected since the R^2 values are relatively low.

6. Conclusion and recommendations

This study accomplished in classifying the Penang Island Granites based on UCS value and Is(50). Strength classifications were successfully made based on BS5930:2015. This study also proposed the correlation between UCS and Is(50) for the Penang Island Granites, as well as the upper boundary and the lower boundary lines. The correlations proposed as well as the strength classification were made for overall, North Pluton and South Pluton members of the Penang Island Granites. This study showed that the Penang Island Granites are "weak to very strong" granites with UCS values between 24.76 to 156.82MPa. The granites from the North Pluton are classified as "medium strong to very strong" with UCS values between 37.11 to 156.82MPa. The granites from the South Pluton are classified as "weak to very strong" with UCS values between 24.76 to 141.59 MPa. The recommended correlation between UCS and Is(50) are (1) Overall: 8.385 Is(50) + 30.016 (2) North Pluton: 7.93 Is(50)+ 35.606 (3) South Pluton: 9.03Is(50)+20.138.

There are a few suggestions that can be considered for future works such as to use primary data for analysis. Even though the majority of the fieldworks were supervised by the authors, as well as by experienced Geotechnical Engineers, ultimately the sampling and lab tests were carried out by a third party i.e., SI Contractors. The skill in drilling may affect the sample in which coring-induced fracture may occur, which in return would affect the strength. We could also better monitor the sample preservation. The samples in this study were stored in a wooden box, handled carefully. However, there could be some unknown accidents from the site to the laboratory. A poor sample preservation could also cause, loss of moisture over time may affect the rock quality eventually affecting the UCS results. The second suggestion is to increase the number of samples, in order to obtain better correlation and the boundary lines of the data. More data would give a more accurate simple linear regression line, and as a result, a more accurate correlation could be proposed. The third suggestion is to make the data uniform in terms of the testing standards followed. In this study, samples from ASTM and ISRM are combined. It would be beneficial if future study would be made up exclusively from one standard of testing. Not only that, there is also identified research gap in this area where the major difference in results between ASTM and ISRM standards can be explored.

7. References

[1] Aliyu M M, J. Shang, W. Murphy, J. A. Lawrence, R. Collier, F. Kong and Z. Zhao 2019 Assessing the Uniaxial Compressive Strength of Extremely Hard Cryptocrystalline Flint, *Int. J. of Rock Mech. Min. Sci.* **113** 310–321

[2] Basu A, N Ghosh and M Das 2012 Categorizing Weathering Grades of Quartzitic Materials and Assessing Brazilian Tensile Strength with Reference to Assigned Grades, *Int. J. of Rock Mech. Min. Sci.* **49** 148–155

[3] Cai M 2010 Practical Estimates of Tensile Strength and Hoek–Brown Strength Parameter m_i of Brittle Rocks, *Rock Mech. Rock Eng.* **43** (2) 167–184
[4] Momeni E, Nazir R, Armaghani D J and Mohamad E T 2015 Prediction of Unconfined Compressive Strength of Rocks: A Review Paper, *J. Tekno.* **77**(11)
[5] Aksoy C O, Ozacar V, Demirel N, Ozer S C and Safak S 2011 Determination of Instantaneous Breaking Rate by Geological Strength Index, Block Punch Index, and Power of Impact Hammer For Various Rock Mass Conditions, *Tunnel Underground Space Tech.* **26**(4) 534–540
[6] Gunsallus K L and Kulhawy F H 1984 A Comparative Evaluation of Rock Strength Measures, *Int. J. of Rock Mecha. and Min. Sci. and Geomech.* **21** 233-248.
[7] Broch E and Franklin J A 1972 The Point Load Strength Test, *Int. J. of Rock Mech. and Min. Sci. and Geomech.* **9** 669-697
[8] ISRM. International Society of Rock Mechanics Commission on Testing Methods 1985 Suggested Method for Determining Point Load Strength, *Int. J. of Rock Mech. and Min. Sci. and Geomech.* **22** 51-60
[9] Edelbro C 2004 *Evaluation of Rock Mass Strength Criteria* PhD Thesis (Sweden: Luleå University of Technology) pp 11
[10] Jahanger Z K and Ahmed A A 2013 Correlation Between Point Load Index And Very Low Uniaxial Compressive Strength Of Some Iraqi Rocks, *Australian J. of Basic and Appl. Sci.* **7**(7) 216-229
[11] Deere D U and Miller R P 1966 *Engineering Classification and Index Properties For Intact Rock*, Technical Report No. AFWL-TR-65-116 (New Mexico: Air Force Weapons Laboratory) pp 136
[12] British Standards Institute (BSI) 2015 *BS5930 Code of Practice For Site Investigations* (London: British Standard) pp 134
[13] Geotechnical Engineering Office (GEO) 2017 *Guide to Rock and Soil Descriptions* (Hong Kong: Geotechnical Engineering Office) pp 12
[14] Bieniawski Z T 1974 Geomechanics Classification Of Rock Masses And Its Application In Tunneling, *Proc. 3rd Int. Congress on Rock Mechanics* (Denver) 27-32
[15] Barton, N 2002 Some New Q-Value Correlations To Assist In Site Characterization And Tunnel Design, *Int. J. of Rock Mech. and Min. Sci.* **39** 185-216
[16] Laubscher D H and Jakubec J 2001 The MRMR rock mass classification for jointed rock masses. Underground mining methods: engineering fundamentals and international case studies eds. Hustrulid WA, Bullock RL (Littleton, CO: Society for Mining, Metallurgy, and Exploration, Inc) pp 475–481
[17] Hoek E (1994) Strength of Rock and Rock Masses ISRM News J. **2** (2): 4–16
[18] Palmstrøm A 1996 Characterizing Rock Masses by The RMi For Use in Practical Rock Engineering: Part 1: The development of the Rock Mass index (RMi), *Tunnel Underground Space Tech.* **11**(2), 175-188
[19] Horvath R G and Kenney T C 1979 Shaft Resistance of Rock Socketed Drilled Piers, *Proc. of a Symposium on Deep Foundations* (Atlanta, Georgia, USA) pp. 182–214
[20] Rosenberg P and Journeaux N L 1976 Friction and End Bearing Tests on Bedrock for High-Capacity Socket Design, *Canadian Geotech. J.* **13**(3): 324–333
[21] Rowe R K and Armitage H H 1987 A Design Method for Drilled Piers in Soft Rock, *Canadian Geotechnical Journal* **24**(1) 126–142
[22] Carter J P and Kulhawy F H 1988 Analysis and Design of Drilled Shaft Foundations Socketed into Rock. Final report, EPRI-EL-5918/Project 1493-4 (New York:Electric Power Research Institute Cornell University)
[23] Zhang L and Einstein H H 1998 End Bearing Capacity of Drilled Shafts in Rock, *J. of Geotech. and Geoenv. Engineering* **124**(7) 574–584
[24] Meigh A C and Wolski W 1979 Design Parameters for Weak Rock, *Proc. of the 7th Euro. Conf. on Soil Mech. and Foundation Eng.* (Brighton, UK) **5** 59–79

[25] Williams A F and Pells P J N 1981 Side Resistance Rock Sockets in Sandstone, Mudstone, And Shale, *Canadian Geotechnical Journal* **18**(4) 502–513
[26] Kulhawy F H and Phoon K K 1993 Drilled Shaft Side Resistance in Clay Soil to Rock *Proc. of Conference on Design and Performance of Deep Foundations: Piles and Piers in Soil and Soft Rock* Geotechnical Special Publication **38** 172–183
[27] Hoek E and Brown E T 1980 Empirical strength criterion for rock masses. Jour. of the Geotech. Eng. Division, ASCE **106**(GT9): 1013–1035
[28] Bieniawski Z T and Bernede M J 1979 Suggested Methods for Determining The Uniaxial Compressive Strength And Deformability Of Rock Materials: Part 1. Suggested Method for Determining Deformability of Rock Materials in Uniaxial Compression, *Int. J. of Rock Mecha. and Min. Sci. and Geomech.* Abstracts **16**(2) 138-140
[29] ISRM. International Society of Rock Mechanics Commission on Testing Methods, (2007). The Complete ISRM Suggested Methods for Rock Characterization, Testing and Monitoring: 1974–2006. (Ankara:Int. Soc. for Rock Mechanics)
[30] Thuro K, Plinninger R J, Zäh, S and Schütz S 2001 Scale Effects In Rock Strength Properties. Part 1: Unconfined Compressive Test and Brazilian Test, *ISRM regional symposium, EUROCK* pp. 169-174
[31] Chiu H K and Johnston I W 1983 The Uniaxial Properties of Melbourne Mudstone, *5th Congress of the International Society for Rock Mechanics*, 209-214
[32] Matthews M C and Clayton C R I 1993 Influence of intact porosity on the engineering properties of a weak rock, *Geotechnical Engineering of Hard Soils – Soft Rocks*, pp. 693-701
[33] Bieniawski Z T 1975 The Point-Load Test in Geotechnical Practice, *Eng. Geol.* **9** 1-11
[34] Reese L C and O'Neill M W 1988 Drilled Shaft: Construction Procedures and Design Methods, Publication No. FHWA-H1-88-042 (Washington: US Department of Transportation-Federal Highway Administration
[35] Tan B. K and Chan S F 1980 Preliminary Studies on The Correlation of Index Properties and Engineering Properties of Malaysian Rocks, *Geol. Soc. Malaysia Bulletin* **12** 31-43
[36] Basu A and Aydin, A. 2006 Predicting Uniaxial Compressive Strength by Point Load Test: Significance of Cone Penetration, *Rock Mech Rock Eng.* **39** 483–490
[37] Ghosh D K and Srivastava M 1991 Point-Load Strength: An Index for Classification of Rock Material, *Bulletin of the Int. Assoc. of Eng. Geol.*, **44**(1), 27-33
[38] Tuğrul A and Zarif I H 1999 Correlation of Mineralogical and Textural Characteristics with Engineering Properties of Selected Granitic Rocks from Turkey, *Engineering Geology*. **51**(4) 303-317
[39] Tan B K 1994 Engineering Properties of Granitic Soils and Rocks of Penang Island, Malaysia. Geol. Soc. Malaysia Bulletin **35** 69- 77
[40] Chakraborty K R 1990 Granitic Rocks of Penang Island: Minor and Trace Element Variation Patterns and Implications. *Geol. Soc. of Malaysia Annual Geological Conference '90* (Ipoh)
[41] Lee P, Kwang K, Kasim K I B and Azhar A B 2019 Hand-Dug Caisson Piles In Granitic Formation, Penang, *1st Malaysian Geotechnical Society (MGS) and Geotechnical Society of Singapore (GeoSS) Conference* (Petaling Jaya, Malaysia)
[42] Ong W S 1993 The Geology and Engineering Geology of Pulau Pinang. (Kuala Lumpur: Jabatan Penyiasatan Kajibumi Malaysia)
[43] Ahmad F, Yahaya A S and Farooqi M A 2006. Characterization and Geotechnical Properties of Penang Residual Soils with Emphasis on Landslides, *Am. J. Env. Sci*, **2**(4) 121-128
[44] Irfan T Y 1996 Mineralogy, Fabric Properties and Classification of Weathered Granites in Hong Kong, *Quarterly J. of Eng. Geol. and Hydrogeo.*, **29**(1) 5-35.
[45] Brace W F 1961 Dependence of Fracture Strength of Rocks on Grain Size, *Penn. State Univ. Mineral Expt. Sta. Bulletin* **176** 99-103
[46] Yusof, N Q A M and Zabidi H. 2016 Correlation of Mineralogical and Textural Characteristics with Engineering Properties of Granitic Rock From Hulu Langat, Selangor. *Procedia Chemistry* **19** 975-980

Acknowledgments
Authors would like to acknowledge UTM Research Grant-Tier 2 (Vot No: Q. J130000.2651. 16J84) for financial support. Authors would also like to thank Mr. Andrew Tham and Mr. Devendran Arumugam and Mr. Yeoh Gim Heng for their assistance in this research.

IOP Conf. Series: Earth and Environmental Science **971** (2022) 012026 doi:10.1088/1755-1315/971/1/012026

Soil Improvement using Limestone, Prefabricated Drained and Geotextile at Pejagan Pemalang Toll Road via Numerical Program

S Abdurrohman [1]*, K Rinda [2], R Abdul [2]

[1] Postgraduate, Civil Engineering Department, Faculty of Engineering, Universitas Islam Sultan Agung, Indonesia
[2] Lecturer, Civil Engineering Department, Faculty of Engineering, Universitas Islam Sultan Agung, Indonesia
*Corresponding author: abdurrohman.sholeh.1@std.unissula.ac.id

Abstract. Pejagan Pemalang Toll Road is the highway with toll that passes through Tegal city and Pemalang regency. This area has soft soil such as very soft clay, soft clay, etc. The soil improvement is needed to reach the suitable excess pore pressure which can make optimum moisture content to strengthen the soil. The soil improvement is using Limestone, Prefabricated Drain and Geotextile are methods available that can reach faster consolidation 90% and avoids collapse with safety factor as the guidance criteria. In this study, the numerical software of Plaxis 8.2 with soil parameters from SPT correlation was incorporated for a series of 2D model. The result of the cross section model without soil improvement shows collapse at embankment 5 m height and soil improvement without limestone is collapse at embankment 7 m height. T90 on the cross section at stage of Consolidation 5 and the long section at stage of Loads of 90.4 kPa. The settlement on cross section at Consolidation 5 stage with soil improvement is 103.5 x 10-2 m with excess pore pressure -9.52 kPa and safety factor 1.2522. The settlement on long section at Loads of 90.4 kPa stage is 907.36 x 10-3 m with excess pore pressure 16.84 kPa and safety factor 2.353. The safety factor on the cross section and the long section with soil improvement is more than 1.2 as criteria design.

1. Introduction

The Pejagan Pemalang Toll Road (PPTR) is the paid highway that connects Jakarta and Semarang city which passes through Tegal city and Pemalang regency. This highway is included in national routes and as an alternative to avoid Pantura. The PPTR passes through the Kaligangsa river. Nearby the river there are so many farmlands which have soft soil. It would be a problem if the routes of PPTR used embankments without soil improvement. It could be able to use the embankment of approaches bridge without soil improvement, but the consolidation will take longer for water pore pressure to reach out. The factor of primary consolidation needs to take more than 3 years for soft soil to reach out the excess pore pressure. After the primary consolidation, the soil still experiences a settlement but not faster than primary consolidation because it has been reaching out of 90% degree of consolidation as there is no moisture content beneath the soil and sub grade to be stiff soils.

Prefabricated drain which consists of vertical and horizontal drain can be accelerated to the consolidation process, and it can combine with preloading on top of embankment. This combination can support the consolidation process shorter than only using prefabricated drain without surcharge preloading. The moisture content can be reached out faster by combining both of two methods. Most

IOP Conf. Series: Earth and Environmental Science **971** (2022) 012026 doi:10.1088/1755-1315/971/1/012026

soft soil has been using these combination methods to reach the consolidation to be faster (to reach optimum moisture content). Refers to research ground improvement using prefabricated drain vacuum with preloading can increase 60% consolidation coefficient and it has reduced 70% for 90% consolidation time than without using prefabricated drain with surcharge preloading method (Krishnapriya, Sandeep, Jency, 2016) [1]. The prefabricated drain with surcharge preloading is using geotextile because the embankment is more than 5 meters so that it can prevent from collapse on embankment because existing soil on subgrade with soft soil and it can improve the stability of embankment.

A factor of UBC with using geotextile has been increased 2.76 to 3.37 and a factor of BCR at low strains has been increased 2.03 to 2.47. the result explains that using geotextile can improve the bearing capacity. (Panigrahi and Pradhan, 2019) [2].

The effect using limestone powder on the clayey soil has explained that plasticity index has reduced up to 32% and liquid limit has been reduced up to 32%. Swelling index has reduced up to 61%. The Cc and Cs has reduced up to 27% and 31%. (Pastor, Tomas, Cano, Riquelme, Gutiérrez, 2019) [3].

The aggregate CBR using 60% based on Amu et al that CBR value has been increasing when it combines with lime and wood ash is better for pavement (Kesharwani, Sahu and Khan, 2016) [4].

Rigid pavement is one type of pavement that uses Portland-Cement as material. On highway first class, rigid pavement using steel reinforcement with its thickness is 30 cm due to SK SNI T-15-1990-03 [5].

The soft soil problems at PPTR nearby Kaligangsa river must determine the scope of analysis by using the limestone, prefabricated drain for horizontal and vertical drain with surcharge preloading, it combines with two geotextiles on embankment in which the area of analysis for CPT that correlated to SPT at stationing 291+100 for cross section and stationing from 291+000 until 291+200 for longitudinal section. The software for simulation using Plaxis 8.2. The objectives of study to determine the excess pore pressure, safety factor and 90% of consolidation time (primary). So that, it needs to identify the settlement on the embankment approach bridge with limestone, prefabricated drain (PVD and PHD) and geotextile both of cross section without and with limestone and then simulation for long section model.

2. Numerical model development

Model for soil improvement simulation using Plaxis 8.2. Plaxis is the software which can simulate a model of soil improvement with an analysis model for each stage. The details numerical model development using Plaxis 8.2 is presented as the following:

- The model geometry consists of cross section without limestone and soil improvement (Figure 1), cross section with soil improvement and without limestone, cross section with limestone and soil improvement (Figure 2), long section with structure, limestone and soil improvement (Figure 3). After the model has created, the mesh development must be generated such as mesh generation for checking the connection of ground model and then determine the ground water level (water table) with 2 meters beneath the existing ground level.

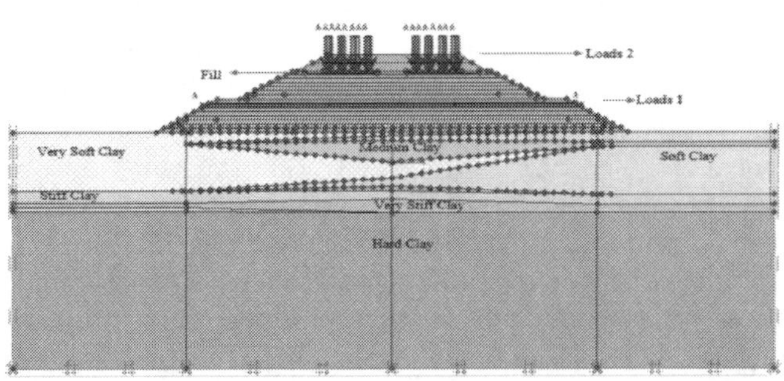

Figure 1. Cross Section of road embankment without soil improvement.

IOP Conf. Series: Earth and Environmental Science **971** (2022) 012026 doi:10.1088/1755-1315/971/1/012026

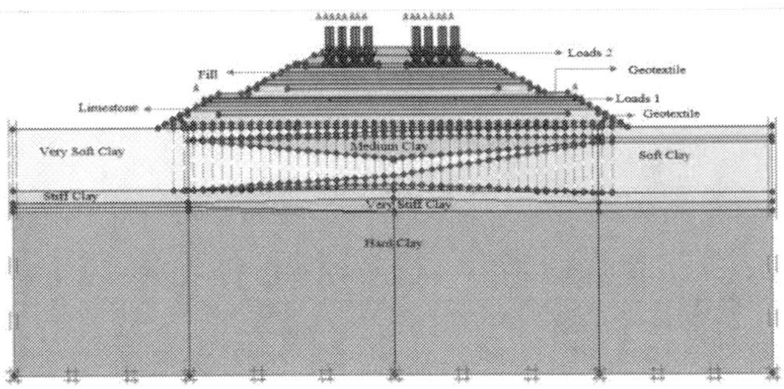

Figure 2. Cross section of road embankment with soil improvement.

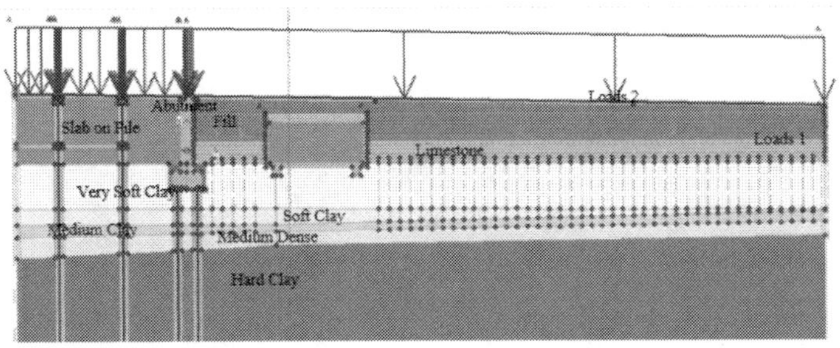

Figure 3. Longitudinal cross section of road embankment with soil improvement.

- Properties for numerical program using N-SPT (Standard Penetration Test) data. γsat using interpolation from N-SPT. The soil improvement with prefabricated drain (horizontal and vertical drain), geotextile, limestone, concrete quality, embankment soil on table 1 until table 9. γsat is the weight of soil that has been on the optimum moisture content but γunsat is the weight of soil before the soil reach the optimum moisture content. Modulus elasticity (E') is comparison between strain and stress in the soil elements. Cohesion (c') is the shear strength that is independent interparticle friction in the soil. The phi value is the Mohr's circle that explains about the angle of failure envelope tangent. The Vs is the share wave velocity (secondary, shear wave) that are related to modulus elasticity (Young's modulus), Poisson ratio (v) and Bulk modulus (k). The Np is axial force that specified in force units per unit width. The Mp is maximum bending moment that is given in force units times length per unit with.

IOP Conf. Series: Earth and Environmental Science **971** (2022) 012026 doi:10.1088/1755-1315/971/1/012026

Table 1. Table of Interpolation.

N-SPT	γ_{sat}
2	16
2	16
4	19

Table 2. Cross Section Soil Properties (S1).

No.	Consistency	Depth (m)	N-SPT Average	γ_{unsat} (kNm^{-3})	γ_{sat} (kNm^{-3})	E' (kPa)	c' (kPa)	phi' (Φ)
1	Very Soft Clay	0.2 - 7.2	2	14	16	1400	12,5	10
2	Stiff Clay 3a	7.4 - 8.6	12	17	19	8400	50	16.7
3	Very Stiff Clay	8.8 - 9.2	17	17	19	11900	75	25
4	Stiff Clay 3b	9.4 - 9.6	13	17	19	9100	50	16.7
5	Hard Clay	9.8 - 10.8	50	17	19	35000	100	33.3

Table 3. Cross Section Soil Properties (S2).

No.	Consistency	Depth (m)	N-SPT Average	γ_{unsat} (kNm^{-3})	γ_{sat} (kNm^{-3})	E' (kPa)	c' (kPa)	phi' (Φ)
1	Very Soft Clay 2a	0.2 - 0.4	2	14	16	1400	12,5	10
2	Soft Clay 2a	0.6 - 1	3	14	16	2100	12,5	10
3	Medium Clay	1.2 - 3.6	4	15	17	2800	25	15
4	Very Soft Clay 2b	3.8 - 5.8	2	14	16	1400	12,5	10
5	Soft Clay 2b	6 - 6.6	3	14	16	2100	12,5	10
6	Very Stiff Clay 2a	6.8 - 7.2	15	17	19	10500	75	25
7	Stiff Clay	7.4 - 8.2	11	17	19	7700	50	16,7
8	Very Stiff Clay 2b	8.4 - 9.8	21	17	19	14700	75	25
9	Hard Clay	10 - 10.8	50	17	19	35000	100	33,3

Table 4. Cross Section Soil Properties (S3).

No.	Consistency	Depth (m)	N-SPT Average	γ_{unsat} (kN/m3)	γ_{sat} (kN/m3)	E' (kPa)	c' (kPa)	phi' (Φ)
1	Very Soft Clay	0.2 - 7.2	2	14	16	1400	12,5	10
2	Stiff Clay 3a	7.4 - 8.6	12	17	19	8400	50	16,7
3	Very Stiff Clay	8.8 - 9.2	17	17	19	11900	75	25
4	Stiff Clay 3b	9.4 - 9.6	13	17	19	9100	50	16,7
5	Hard Clay	9.8 - 10.8	50	17	19	35000	100	33,3

Table 5. Long Section Soil Properties STA 291+000.

No.	Consistency	Depth (m)	N-SPT Average	γ_{unsat} (kNm^{-3})	γ_{sat} (kNm^{-3})	E' (kPa)	c' (kPa)	phi' (Φ)
1	Medium Clay BL1	0.2 - 0.4	8	15	17	5600	25	8,33
2	Soft Clay	0.6 - 1	3	14	16	2100	12,5	10
3	Medium Clay BL2	1.2 - 3.6	6	15	17	4200	25	8,33
4	Medium Dense	3.8 - 5.8	18	17	19	12600	11,67	35
5	Hard Clay BL1	6 - 6.6	32	17	19	22400	100	33,3
6	Medium Clay BL3	6.8 - 7.2	7	15	17	4900	50	16,7
7	Very Stiff Clay	7.4 - 8.2	24	17	19	16800	75	25
8	Hard Clay BL2	8.4 - 9.8	40	17	19	28000	100	33,3
9	Dense Sands	10 - 10.8	34	17	19	23800	13,33	40

Table 6. Long Section Soil Properties STA 291+200.

No.	Consistency	Depth (m)	N-SPT Average	γ_{unsat} (kNm^{-3})	γ_{sat} (kNm^{-3})	E' (kPa)	c' (kPa)	phi' (Φ)
1	Very Soft Clay	0.2 - 5.2	2	14	16	1400	12.5	10
2	Soft Clay	5.4 -7	3	14	16	2100	12.5	10
3	Medium Clay	7.2 - 7.6	7	15	17	4900	25	8.33
4	Medium Dense	7.8 - 9.4	20	17	19	14000	11.67	35

Table 7. Rigid Pavement, Spun Pile and Geotextile.

No.	Platform	Type	γ sat (kNm^{-3})	E' (kPa)	v	Vs
1	Embankment	Drained	19	1.10E+04	0.3	-
2	Limestone	Drained	18	1.85E+04	0.3	-
3	Concrete	Non-Porous	24	2.53E+07	0.15	2120
4	CBR 60%	Drained	21	5.00E+04	0.2	-

Table 8. Embankment, Limestone, Concrete and CBR 60%.

No.	Platform	Type	EA	w	v	Np
1	Rigid 30 cm	Elastic	1.22E+06	0,471	1.50E-01	1.00E+15
2	Spun Pile	Elastic	5.32E+06	1,257	1.50E-01	1.00E+15
3	PET Woven 100 Geotextile	Elastoplastic	688			70

- The traffic load model used is loads of trucks which is the heaviest loads of traffic. Table 9 presents the designated truck wheels loads used for this study. The traffic load is 90.4 kN/m^2. The T is loads of wheels on the front wheels, the Tx is the right and the left side distribution of loads of wheels on façade view and the Ty is the front and behind side distribution of loads of wheels. The Bxy is Tx times Ty distribution that the result is T'.

Table 9. The loads of truck's wheels.

Unit	T (kg)	Tx (m)	Ty (m)	Bxy (m^2)	T' (kPa)
Loads of Truck	5000	0.85	0.65	0.553	90.4

- The simulation of road embankment constructions was carried out phase by phase, by including the soil improvement stages starting from subgrade until the traffic load model, consolidation phase and determining of safety factor. The differences stage involved for the long section and cross section; the stage starting from structure of slab on pile and abutment. The stages are presented separately via Figure 4 and Figure 5.

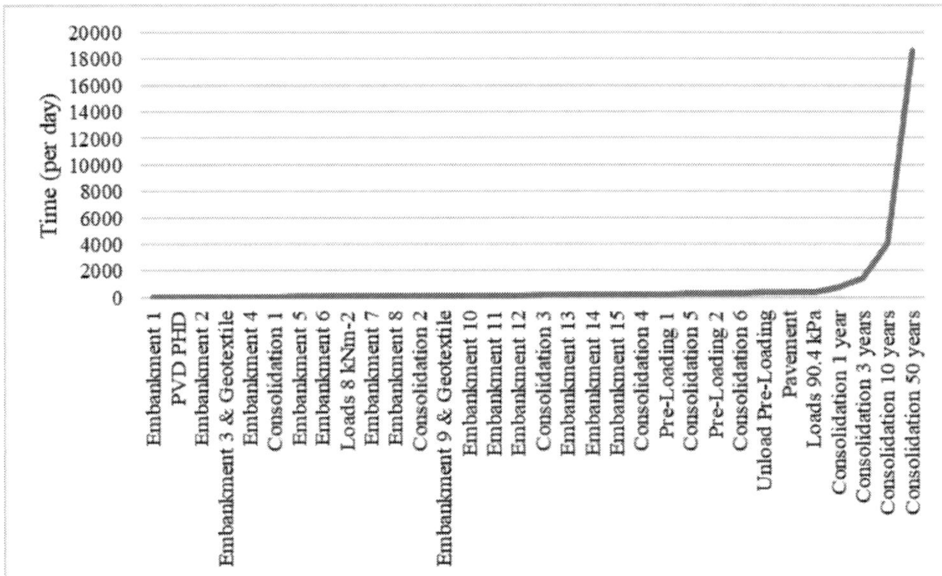

Figure 4. Time of Construction Stages (Cross Section).

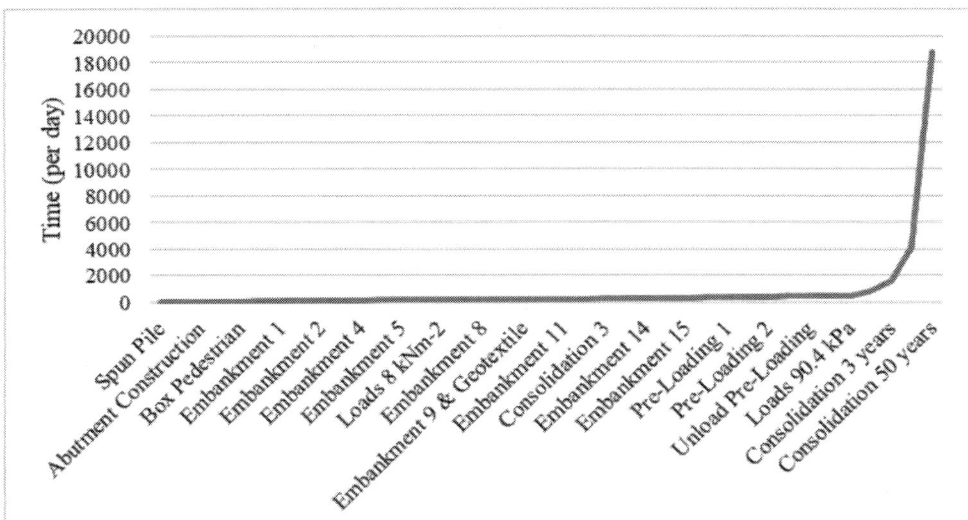

Figure 5. Time of Construction Stages (Long Section).

IOP Conf. Series: Earth and Environmental Science **971** (2022) 012026 doi:10.1088/1755-1315/971/1/012026

Based on Figure 4, the construction stages for soil improvement with limestone is on embankment 1 until stage of embankment 5 and after that construction stage as normal embankment at stage of embankment 6. Based on Figure 5, the construction stages on long section from STA 291+000 until 291+200 must be on structure stages before ground stages to avoid from collapse on embankment impact of spectrum from spun pile and slab on pile construction that model is on Figure 3.

3. Analysis and result

3.1. Consolidation 90% (T_{90})

The settlement has reached to 90% consolidation on cross section is on stage of consolidation 5 (Figure 6). For the longs ection model, the settlement can be reached to 90% consolidation onstage of Loads of 90.4 kPa (Figure 7). Due to the cross section model, there is a space for soft clay on the left side which does not balance settlement on the left side. For models without limestone and without soil improvement with prefabricated drain and geotextile collapses so that there is no information about 90% consolidation.

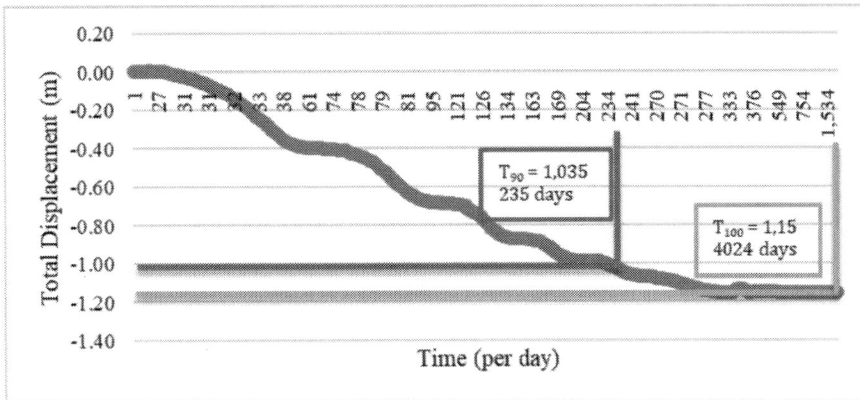

Figure 6. Graph of T_{90} (Cross Section).

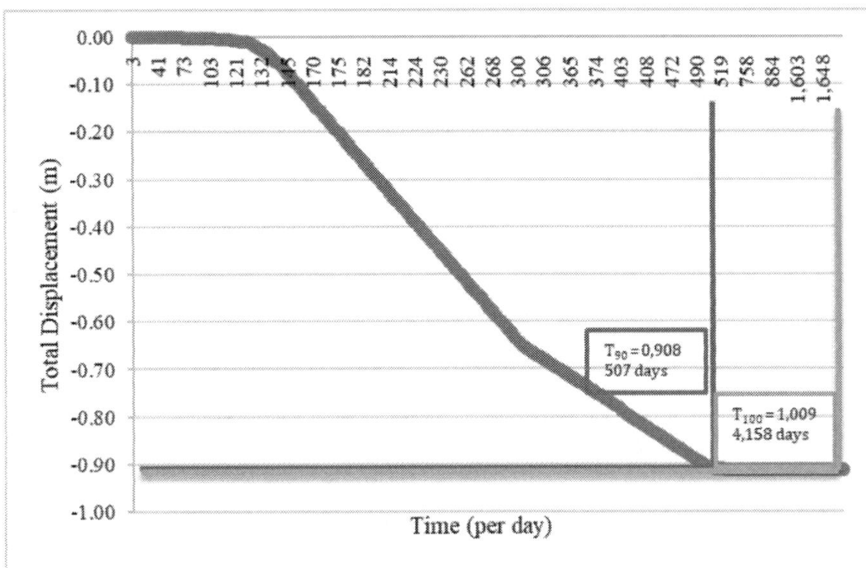

Figure 7. Graph of T_{90} (Long Section).

Table 10. Formatting sections, subsections and subsubsections.

	Font	Spacing
Section	11 point **Times bold**	1 line space before a section No additional space after a section heading
Subsection	11 point *Times Italic*	1 line space before a subsection No space after a subsubsection heading
Subsubsection	11 point *Times Italic*	Subsubsections should end with a full stop (period) and run into the text of the paragraph

3.2. Settlement

Based on Figure 8, the settlement on cross section due to 90% consolidation at stage of Consolidation 5 is 103.5 x 10^{-2} m as criteria design [6]. The settlement on the long section based due to consolidation 90% on stage of Loads of 90.4 kPa is 907.36 x 10^{-3} m (Figure 9). For the model without soil improvement, the maximum settlement predicted l is 708.04 x 10^{-3} m occurred when the stage of embankment at 5 m height. Settlement with prefabricated and geotextile without limestone model is 753.07 x 10^{-3} m at stage of embankment 7 m.

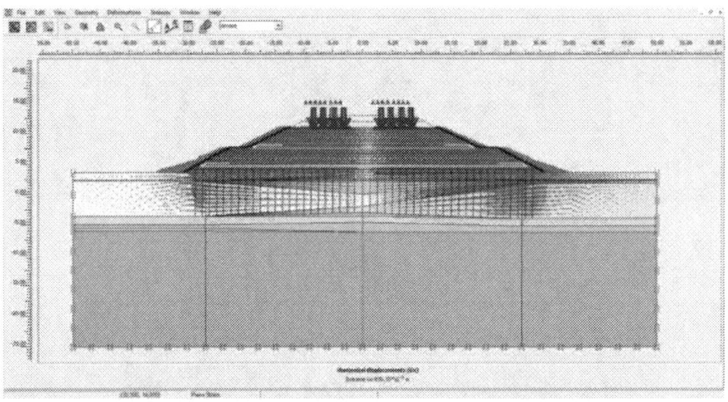

Figure 8. Total Displacement (Cross Section).

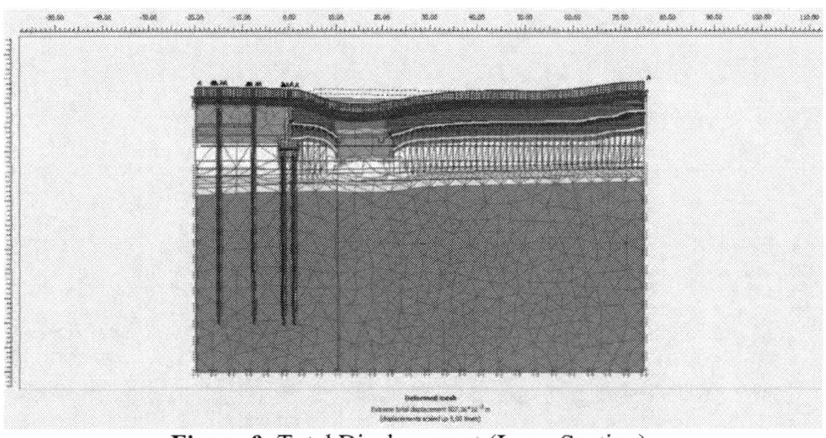

Figure 9. Total Displacement (Long Section).

3.3. Excess Pore Pressure and Safety Factor

Based on Figure 10, the excess pore pressure on the cross section at stage of Consolidation 5 is -9.52 kPa with the safety factor is 1.2522 so that the safety factor is more than 1.2 as criteria design [7]. Based on Figure 11. For the long section at stage of Loads of 90.4 kPa, the excess pore pressure is 16.84 kPa with the safety factor is 2.353, it is more than 1.2 as criteria design. For the model without soil improvement and without limestone collapses so there is no graph information.

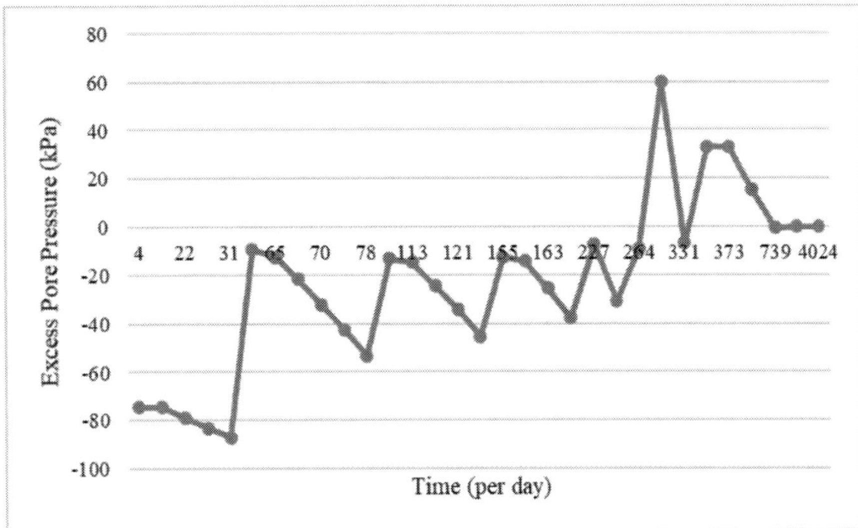

Figure 10. Excess Pore Pressure (Cross Section).

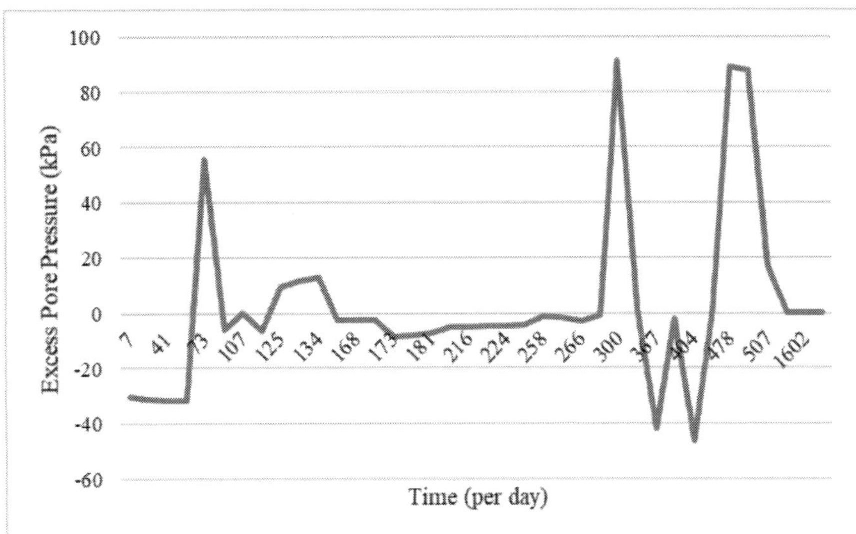

Figure 11. Excess Pore Pressure (Long Section).

4. Conclusions

The embankment on soft soil needs soil improvement which can reach the settlement faster with excess pore pressure higher than not using soil improvement. Based on the analysis that without soil improvement the model collapses on stage of embankment 5 meters and without limestone model with prefabricated drain and geotextile it collapses on stage of embankment 7 m so that two models cannot reach consolidation 90%. While using soil improvement with limestone, prefabricated drain and geotextile the stage is succeeded and the consolidation 90% on the cross section at stage of Consolidation 5 and the long section at stage of Loads of 90.4 kPa with safety factor is more than 1.2 as criteria design. The soil improvement with limestone is caused by more space for soft clay on the left side, but the soft clay decreases on the right side and then it needs to increase the soil stability with limestone to make balance for embankment.

5. References

[1] Krishnapriya P B, Sandeep M N, Antony and Jency 2016 Efficiency of Vacuum Preloading Consolidation Behaviour of Cochin Marine Clay *Procedia Technology.* **24** 256–262
[2] Panigrahi B and Pradhan P K 2019 Improvement of bearing capacity of soil by using natural geotextile *International of Geo-Engineering Springer Index.* **10**
[3] Pastor J L, Tomás R, Cano M, Riquelme A and Gutiérrez E 2019 Evaluation of the Improvement Effect of Limestone Powder Waste in the Stabilization of Swelling Clayey Soil. *MDPI Journal Sustainability.***11**
[4] Kesharwani R S, Sahu A K and Khan N U M 2016. CBR Value of Sandy Subgrade Blended with Coarse Aggregate *International Journal of Geomate.* **10** 1743-1750
[5] Prabowo A D, Setyawan A, Sumbawo K Adi 2013 Desain Beton Berpori untuk Perkerasan Jalan Yang Ramah Lingkungan *Matriks Teknik Sipile* **1** 2723-4223
[6] Pusat Litbang Prasarana Transportasi, PT Virama Karya and PT Trikarla Cipta 2001 PT Panduan Geoteknik 4, *Desain dan Konstruksi Timbunan Jalan pada Tanah Lunak. Bandung*
[7] Pertiwi D 2014 Persyaratan Perenanaan Geoteknik dan Kegempaan. ISBN: 978-602-264- 067-7 *Kementerian Pekerjaan Umum Badan Penelitian dan Pengembangan (Pusjatan)*

IOP Conf. Series: Earth and Environmental Science **971** (2022) 012027 doi:10.1088/1755-1315/971/1/012027

Tunnel Progression Effects to the Ground Surface and the Adjacent Pile

M A Zahid[1], S N Jusoh[2*], A Rahim[2], M A Hezmi[2], N Z M Yunus[2], K Mat Said[2] and M I Shahrin[2]

[1] School of Civil Engineering, Faculty of Engineering, Universiti Teknologi Malaysia, Johor Bharu, Johor
[2] Department of Geotechnics and Transportation, School of Civil Engineering, Faculty of Engineering, Universiti Teknologi Malaysia
*Corresponding author: snorafida@utm.my

Abstract. Tunnelling close to existing structure in urban area has become unavoidable. The progression tunneling activities induced ground movement and might affect the integrity of existing structure especially the one lies within the tunneling influence zone. It may cause catastrophic failures of structures and can cause losses of human lives. Therefore, considering its risk, this study focuses on the tunnel-soil-pile interaction by performing a physical model testing. By conducting a series of laboratory tests, the ground subsidence and pile behavior is presented herein. For a twice diameter distance of tunnel and pile, the pile axial settlement is 0.04% of the tunnel diameter respectively, while the maximum pile bending moment is 4928. 93kN.m. Maximum ground settlement is 0.56% respectively to tunnel diameter. To sum up, the axial displacement of pile decreases when the pile located further away from the tunneling zone. Similarly, the ground surface subsidence decreases when the pile location is more in distance during the tunneling advancement.

1. Introduction

Rapid development in the urban area led to a significant decrease of available construction land. Therefore, an alternate solution is to build either upwards or downwards. For that purpose, tunnel is one of construction method that has been used widely to overcome this problem. Tunnel excavation causes soil redistribution and settlement as a result of the ground movement, this leads to moment and movement in pile foundations. In details, tunnel excavation will induce stress relief on the soil surrounding the tunnel wall and face because the act of soil removal led to an unbalanced pressure in the soil particle. Consequently, the soil surrounding the tunnel wall and face will move inward into the tunnel cavity[1-3]. However, clay and sand have different failure mechanism

Introduced by Peck [4], analysis of the ground deformation induced by tunnelling via the empirical data suggest that the form of inverse Gaussian distribution curve as relatively the form of the settlement trough which can be seen in **Figure 1**. The Gaussian distribution curve allow the vertical settlement position at x distance from tunnel centre, S_v to be calculated based on equation (1) in which S_{max} is the maximum settlement at of the surface which likely to occur at the tunnel centre, and i_x is the horizontal distance between the tunnel centre and the point of inflection on the settlement trough.

$$S_v = S_{max} \cdot e^{\left(\frac{-x^2}{2{i_x}^2}\right)} \quad (1)$$

O'Reily [5] discovered the correlation between the parameter i and tunnel's depth and geological condition, as stated in equation (2).

$$i = Kz_0 \quad (2)$$

in which z_0 is the tunnel depth and K is the empirical constant.

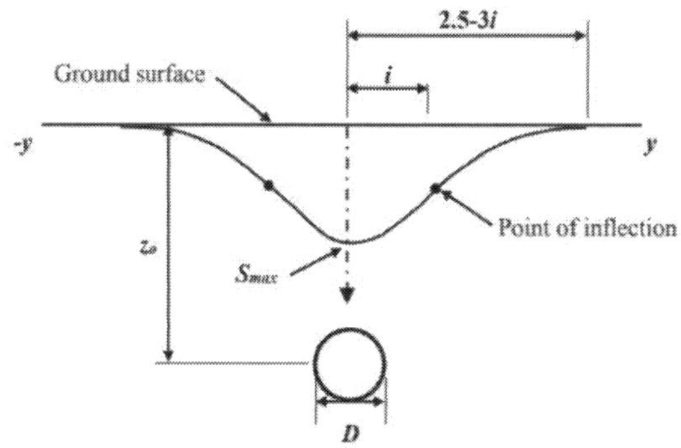

Figure 1. Settlement trough in Gaussian distribution curve after Khoo [20]

Engineers poses a major challenge in estimating the effect of tunnelling on existing pile foundation of buildings. It is particularly vital to estimate the tunnelling effects when new tunnels are to be built near an existing pile [6]. The presence of pile adjacent to the tunnel provide differ result on the ground surface settlement compared to greenfield condition. A study on tunnel-pile interaction conducted by Marshall and Haji [7] through an analytical approach, stated that the tunnelling activity affect the pile capacity due to the ground stress redistribution and displacement. Induce by tunnelling activity, pile is susceptible to reduction in end bearing capacity and shaft frictional resistance. A review of case studies from several researchers done by Dias and Bezuijen [8] conclude that the tunnelling activities have shown to cause low impact in term of damage toward the nearby existing pile. Nevertheless, it was noted due to lack of understanding in the mechanism of pile-tunnel interaction, these constructions confront a lot of uncertainty thus, future studies are needed. In addition, Jongpradist [9] reveals in his study that maximum deformation occur in influence zone. For each case, the zone of influence varies based on the tunnel depth and diameter. It is where more stress is released than its defined amount. A plenty number of previous researchers focus their studies on the existing adjacent pile to the tunnel rather than the pile above the tunnel [7][10-14]. In doing so, the affected structural integrity of a foundation has been the main interest due to increased lateral loading and bending moment from the pile induced by tunnelling [15].

It is common nowadays researcher studying geotechnical problem tend to apply centrifuge modelling technique [16]. An appealing alternative such as the physical modelling provide a better understanding upon the mechanism and reaction of complex problem relating soil structure interaction. On the contrary, due to the behaviour soil directly influence by soil stress, physical model test cannot extrapolate the result to a prototype scale. Nonetheless, this flaw can tackle by using centrifuge modelling technique. Furthermore, under the execution of a control environment, centrifuge model test can measure soil profile strength, deformation of soil, and break off time can be quantified with a reasonable preciseness in dependable test result. Therefore, this study aims to use physical modelling to estimate and analyse these effects. What make centrifuge model technique an appealing method is their ability to repeat the test with a consistency and not to mention their relatively economical.

2. Methodology

In this study, physical model development, model properties and the apparatus setup will be present for the testing. The model constructed will cover the tunnel at depth (C/D) equal to 3, with a distance 2.0D (diameter) from the tunnel centre.

2.1. Soil Model

Concerning the incorporation of sand as the soil material in this study, the basic and engineering properties of the soil that were relevant to the study was obtained and the sandy soil was a poor graded soil (SP). A controlled relative density sand was poured into the model box through a hopper that was put in position above the box until it reaches the level of necessary cover-to-diameter ratio (C/D). Variation of density was achieved by changing the flow velocity of the dry pluviator system [17]. The system itself comprise of sand hopper, shutter, fixing device, diffuser system and soil collector, whereby the hopper is manually moved within the range of the soil box area. Applying trial and error calculation to obtain the accurate distance between the sieve and the surface of the sand to incorporate a constant of 50% relative density.

2.2. Model Box, Tunnel, and Pile

In reference to the research done by Sohaei [18], the physical model was prepared. The model box shown in **Figure** with a dimension of 60 cm length x 60 cm width x 50 cm height will applied to act as a medium to replicate the underground condition for the tunnelling work. Previous researcher has also adopted similar concept in a ground loss prediction that also using box [19]. The size of the model was larged enough to minimize the influence of the rigid boundary on the measured stress. Several researchers such as Potts and Zdravkovic [20] and Tan *et al.*[21] recommend that the measurement of the model box should be at least four times of the tunnel diameter from tunnel circumference.

Figure 2. Model box used in physical modelling test

Taking into account the comparison of the actual condition and the physical laboratory condition, scaling factor are implemented to scale the physical model test. C/D ratio and tunnel diameter will be the main concern while the other parameter was decided accordingly. The simplified reduction scale and the tunnel and pile diameter was presented in **Table 1**. Note that, both tunnel and pile was fabricated with respective geometry. A casing method introduced by Meguid and Mattar [22], was used for tunnel model here in.

2.3. Tunnel Excavation Rate

Error! Reference source not found. depicts the revolution per minute (rpm) of similar test setup conducted by different researchers. The data in the table shows that variant motor speed is applied during the test to simulate the field excavation rate. The motor speed for this experiment is set to be higher than the previous researcher since the offset pile with two diameters of tunnel theoretically has less inducement by the tunnel excavation. Moreover Shehata *et al.*[23] in his study emphasize on the unavailability of a guideline for tunnel excavation rate in soft ground in adjacent to existing structure. Based on Table 2, Siti Ai'dah [24] have the lowest motor speed while Feras [25] and Sohaei [18] have similar motor speed.

IOP Conf. Series: Earth and Environmental Science **971** (2022) 012027 doi:10.1088/1755-1315/971/1/012027

Table 1. Reduction scale from the prototype to model.

Tunnel and Pile Parameters	Prototype (m)	Model (mm)
Tunnel Depth (C)	19.7	147
Tunnel Diameter (D)	6.57	49
Pile Length (Lp)	24.32	181.5
Pile Diameter (dp)	1.2	9
Pile Thickness	0.1	0.8

Table 2. Physical Model Motor rpm from previous researcher.

	Motor Speed (cm/sec)	Motor circumference(meter)	Rpm
Sohaei (35)	0.130	0.2316	0.337
A'idah (43)	0.097	0.2316	0.273
Feras (44)	0.137	0.2316	0.354
This study	0.194	0.2316	0.531

2.4. Testing configurations

The test for physical model system were run three times or/and more to achieve valid and verified data that is enough to be used in the data analysis. **Error! Reference source not found.** shows schematic diagram for 3D view of physical model test shows. In details, Figure 4(a) is the cross-sectional schematic diagram that illustrates the position of tunnel and pile set for the physical modelling testing while Figure 4(b) is the parallel-sectional schematic diagram. For data measurement purpose, LVDTs and strain gauge (SGs) was placed on respective locations here in. Upon the completion of the instrument set up, electric motor will pull out the barrel (tunnel shield) to stimulate the construction of the tunnel with an average speed of 0.13 cm/sec over 6 consecutive minutes that is being provided. The average speed applied is reference to the actual excavation rate at about 0.12 m/hr. The tunnelling construction will be progressively carried out until the shield is completely out of the box.

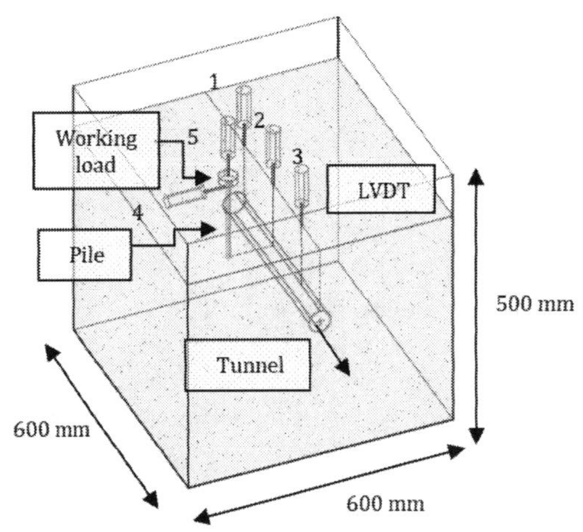

Figure 3. Schematic diagram for 3D view of physical model test.

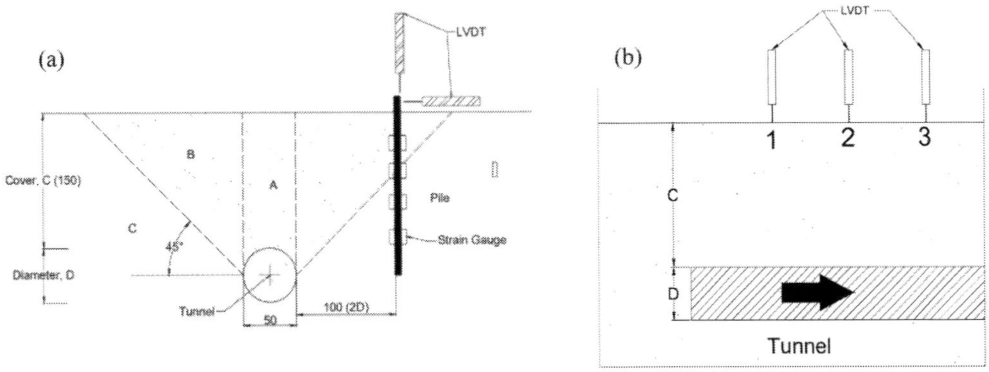

Figure 4. Schematic diagram for 2D view of physical model test (a) Cross sectional view that illustrate the test setup for pile movement and stress induce by pile from tunnel progression (b) Side view that illustrate the test setup for ground movement induced by tunnel progression.

3. Result and Discussion

The laboratory physical model test relies on other basic properties as a secondary data obtained from the previous researcher. The main focus of this study is to determine the inducement of tunnel progression on the existing pile. Therefore, it was anticipated from this study, there will be movement on the pile either vertically or laterally. The outcome from this physical modelling test will set to determine the relation of tunnel-soil and tunnel pile based on the set of data obtained. Figure 5 shows that the result presented afterwards based on tunnel influence zone by Selementas's influence line [26]. The results of this study will be compared to previous researcher which include ground surface and pile head movement, and axial force along the pile during tunnel excavation.

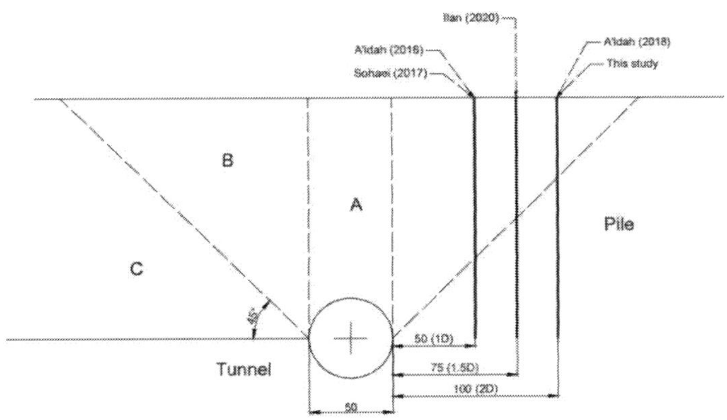

Figure 5. Previous research with variant pile distance from tunnel axis.

3.1. *Ground Surface Settlement*

The tunnel induced ground settlement presented here in.

3.1.1. Transverse Ground Surface Settlement

The transverse settlement troughs shown in Figure 6 were obtained through calculation using Peck's [4] formula as shown previously in equation (1) and selecting a value of 0.42 for the K in accordance to Sohaei's[18] work. **Figure** show the comparison between the transverse ground surface settlements against the transverse distance from tunnel centre. It is noticeable that the value of the maximum settlement can be seen increase as the advancement of the tunnel from the point 1(50mm) to point 3(300mm). However, there is a slight contradiction on the behaviour of the maximum settlement between point 2 and 3. Maximum settlement at point 2 can be observe to be slightly higher value than point 3. Uneven sand distribution and the rapid tunnel excavation through the simulation of the barrel pull out could cause a build up at the face pressure that would contribute a small build-up of sand on the surface at point 3. Therefore, surface at point 3 would have slight lesser settlement than point 2.

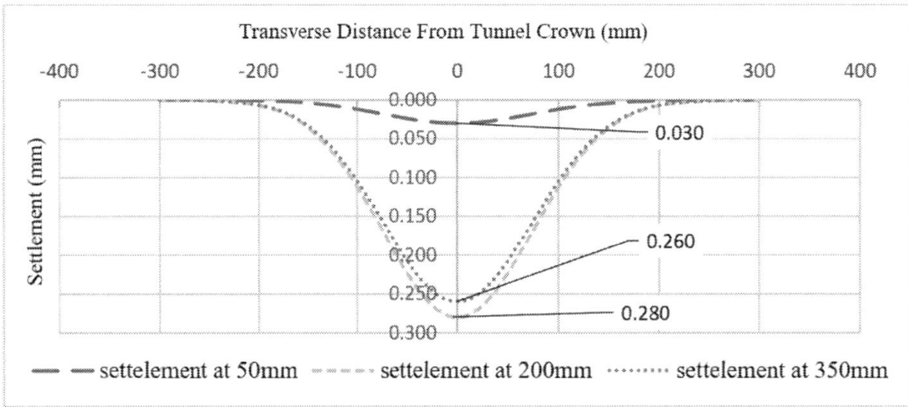

Figure 6. Comparison between transverse ground surface settlement troughs at 50mm,200mm and 350mm from tunnel face.

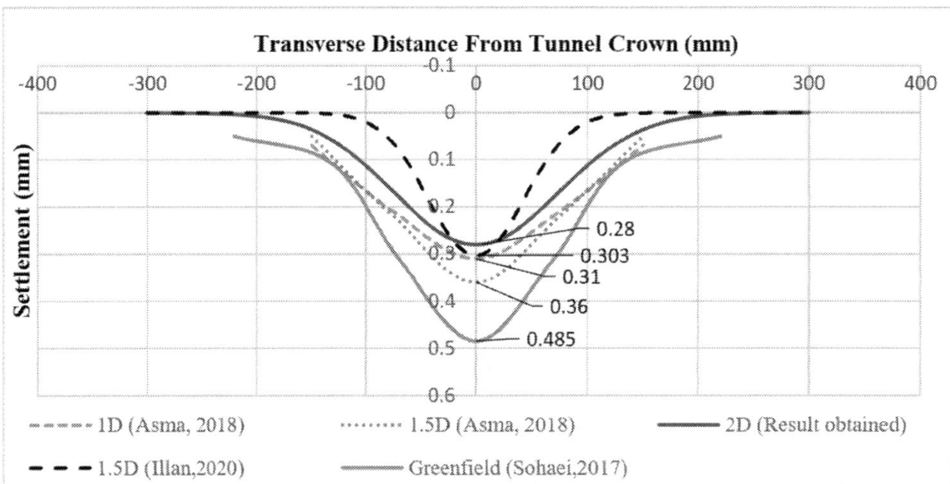

Figure 7. Comparison between transverse surface settlement troughs from previous research.

Based on **Figure** 7, the obtained result shown to have a wider trough width with a value of maximum settlement that is the lowest compare to the previous result. However, Ilan [27]'s 1.5D distance have narrower trough width while Siti A'idah's [24] 1D and 2D distance have similar wider trough width with the obtained result. Pinto and Whitttle [28] in their study revealed that the narrow settlement through measured for surface settlement of tunnels in sand was due to the influence of dilation of the free or partially draining soils. They also stated that an approximate analytical solution is well suited to describe this soil settlement behaviour. Base on Peck's equation, Kolivand [29] stated that only two parameter that influence the shape and magnitude of the trough that is the point of inflection and volume loss. However, since volume loss are similar with the previous researcher thus considered to be a constant variable, the difference of inflection point plays a major role in determining the magnitude of the settlement trough width in this result. To sum up, the trend from the previous result differs with the obtain maximum surface settlement trough, which shows pile with 2D distance from tunnel crown to have lower value than 1D and 1.5D distance. Prior discussion has stated on the excavation rate for this experiment to be higher than the excavation rate from the previous experiment. Therefore, higher excavation rate has produced a lower magnitude of maximum settlement and a wider through width of settlement profile.

3.1.2. Longitudinal Ground Surface Settlement

The longitudinal surface settlement result was obtained from three similar point from transverse settlement trough, located across the tunnel axis. Based on Hajjar's[30] calculation, the longitudinal settlement trough for each point is shown in 8, exhibit the comparison between longitudinal ground surface settlement trough and the longitudinal distance of tunnel face from each point.

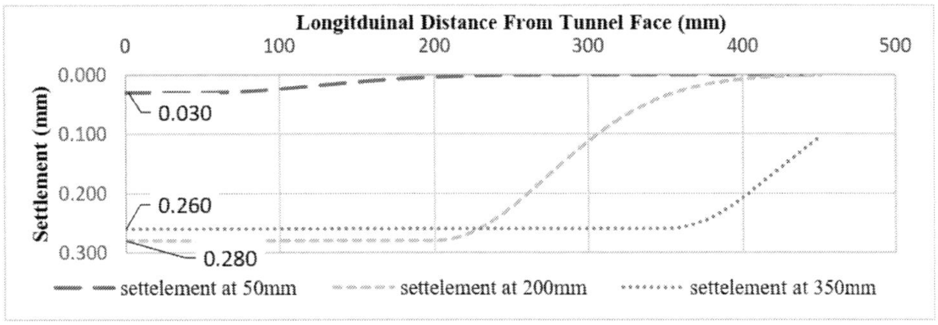

Figure 8. Comparison between longitudinal settlement trough at 50mm,200mm and 350mm from tunnel face.

As shown in Figure 8, the longitudinal settlement trough manages to showcase how the cumulative disturbance of tunnel excavation can increase the maximum settlement. Similar discussion in transverse settlement, the value of maximum settlement can be seen increase as the advancement of the tunnel from the point 1(50mm) to point 3(300mm) and there is a slight contradiction on the behaviour of the settlement between point 2 and 3.

The x-axis in **Figure** is the length of the tunnel whereby the positive direction of the x-axis is the direction of the tunnelling. It can be seen from the graph for the result obtained have a steady settlement until Y/D at 1.0 which can be seen progressively start to settle from Y/D at 2 until 5 in relatively linear trend. Meanwhile for distance pile at 1D,2D A'idah [24] and 1.5D Ilan [27], the settlement that occurred between Y/D at 1 and 2 shows a sudden increment then as it approach and after Y/D at 3 the settlement become steady. The existing pile provide reduction on the surface settlement. In can be notice that the maximum ground settlement in greenfield condition was 0.97%, whereas it gives 0.58% ground settlement in 1D. The increasing distance between a pile and a tunnel centre has less influence on the surface settlement as the pile withstand the soil movement while the tunnel excavates close to the existing pile [18][31]. However, the trend from the previous result differs with the obtain maximum surface settlement, which shows pile with 2D distance from tunnel crown to have lower value than 1D and 1.5D distance.

3.2. Pile Induced by Tunneling

Pile behavior induced by soil stress redistribution due to tunnel excavation is presented here in.

3.2.1. Pile Head Settlement

Figure 9 and Figure 10 shows the measured pile displacement (S_p/D) against normalized pile depth during the tunnel advancement. The obtained pile head settlement has a close magnitude and trend to

IOP Conf. Series: Earth and Environmental Science **971** (2022) 012027 doi:10.1088/1755-1315/971/1/012027

Ilan [27]. It can be observed in the Figure 11 that the settlement of the pile remained constant and only to occur minimal increment throughout the tunnelling progression as the settlement only gained 0.04% while 1.5D gained 0.02%. These findings indicate a good agreement with the recent studies of Ayasrah et al. [32] that have shown that an inducement of pile head settlement fades as the distance between the pile and the tunnel axis is twice the tunnel diameter. However, these behaviours are much differ from Sohaie's result which can be compared to 1D and 2D that experience an increment in their settlement gradually and began rapidly drop from -0.5D from the tunnel face.

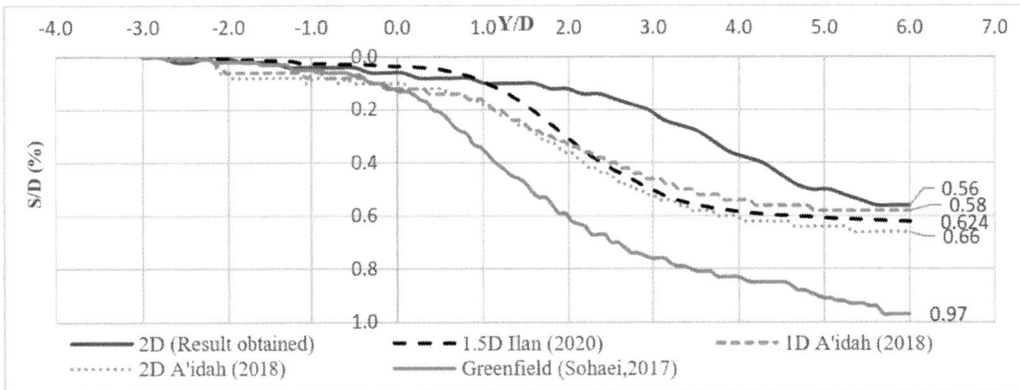

Figure 9. Comparison of longitudinal surface settlement in greenfield with a presence of pile.

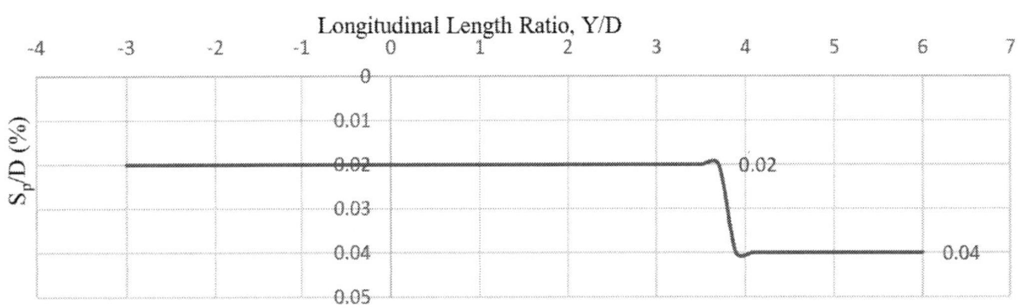

Figure 10. Pile head settlement with respect to different distance of tunnel progression.

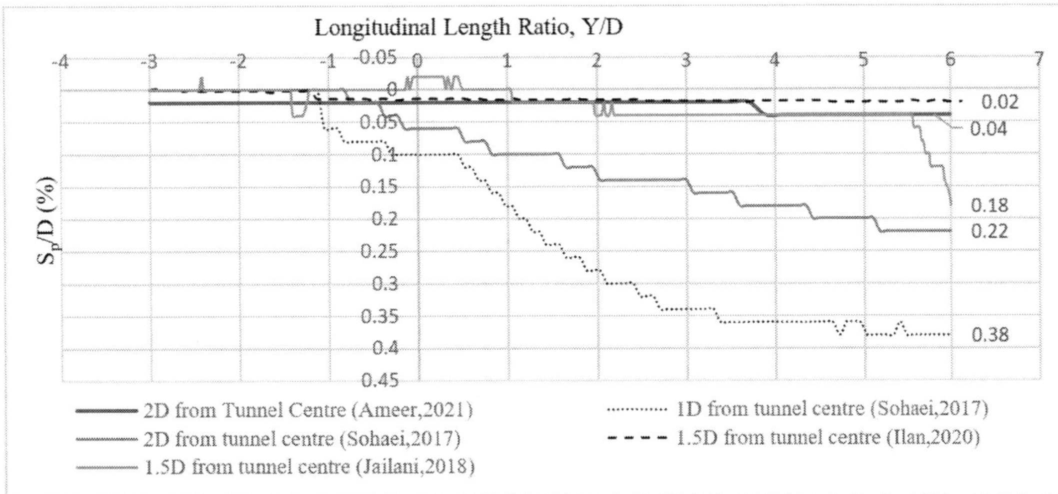

Figure 11. Comparison of pile head settlement with previous research.

3.2.2. Pile Axial Force

Figure 12 shows the plots of induced axial force against the normalized pile depth (Z/D) after tunnelling test. Mobilization of positive and negative skin friction attributed the axial force result which are induced between the movement of surrounding soil and pile during the adjacent tunnelling. Hun [33] stated that settlement by tunnelling generally caused negative skin friction on the pile shaft but the degree to such inducement depends on the distance between the pile on the tunnel. It can be observed in Figure 13 the pile with 1D distance have developed an axial compressive force as the magnitude increases to the tip of the pile as large portion of the pile are in Zone B. Meanwhile the pile with 2D distance developed a similar trend with the pile with 1D but developed an axial tensile force with a small portion of axial compressive force as small portion of the upper pile are placed in Zone B. Ayasrah [32] stated that the middle and the lower part of the pile have significant effect by the axial force during tunnel excavation. Thus, the findings are in good agreement with Jongpradist [9] and Selemetas [26] zone of influence.

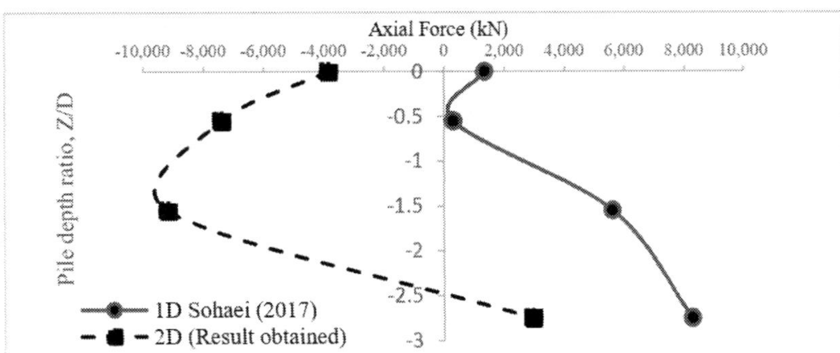

Figure 12. Comparison of axial force profile with respect to depth for pile from previous research.

4. Conclusion

The results obtained and the observation made in this study draw some conclusions. These are:

i. The tunnel progression significantly induced the surface settlement (transverse and longitudinal) but the impact is minimized as the excavation rate of the tunnel higher than the normal excavation rate.
ii. The pile exposes less in the influence zone as the horizontal distance of the pile between the tunnel axis approaches twice the tunnel diameter. Therefore less significant impact on the pile

settlement as shown that the ground surface subsidence is lesser when the pile is further during tunnelling advancement

iii. Pile with 2D distance developed a similar trend of axis tensile force when compare with the pile with 1D distance but mainly developed axial tensile force with a small portion of axial compressive force near the pile toe.

5. References

[1] Chambon P, Corté JF 1994 Shallow tunnels in cohesionless soil: Stability of tunnel face *J Geotech Eng.* **7** 1148

[2] Mair R J, Williamson M G 2014 The influence of tunnelling and deep excavation on piled foundations. In: Geotechnical Aspects of Underground Construction in Soft Ground *Proceedings of the 8th Int Symposium on Geotechnical Aspects of Underground Construction in Soft Ground* 21-30

[3] Mair R and Taylor R 1999 Bored Tunnelling in the Urban Environments **53** 6–7

[4] Peck R B 1969Deep Excavations and Tunneling in Soft Ground *7th International Conference on Soil Mechanics and Foundation Engineering* **2** 25–90

[5] O'Reilly M P, New B M 1982 Settlements above tunnels in the United Kingdom - their magnitude and prediction *Tunn '82 Pap Present 3rd Int Symp* **1** 73–81

[6] Ng C W W, Boonyarak T and Mašín D 2013 Three-dimensional centrifuge and numerical modeling of the interaction between perpendicularly crossing tunnels *Can Geotech J* **50** 35-46

[7] Marshall A M and Haji T 2015An analytical study of tunnel-pile interaction. Tunn Undergr Sp Technol. **45** 43–51

[8] Dias T G S and Bezuijen A 2014 Pile Tunnel Interaction: Literature review and data analysis. ITA World Tunn Congr 2014 - Tunnels a better life

[9] Jongpradist P, Kaewsri T, Sawatparnich A, Suwansawat S, Youwai S and Kongkitkul W 2013 Development of tunneling influence zones for adjacent pile foundations by numerical analyses *Tunn Undergr Sp Technol* **34** 96-109

[10] Basile F 2012 Pile-group response due to tunnelling *Geotech Asp Undergr Constr Soft Gr Proc 7th Int Symp Geotech Asp Undergr Constr Soft Gr.* **7** 81–90

[11] Huang M, Zhang C and Li Z 2009 A simplified analysis method for the influence of tunneling on grouped piles *Tunn Undergr Sp Technol* **24**

[12] Kitiyodom P, Matsumoto T and Kawaguchi K 2005 A simplified analysis method for piled raft foundations subjected to ground movements induced by tunnelling *Int J Numer Anal Methods Geomech.* **507** 25-29

[13] Loganathan N, Poulos H G and Xu K J 2001 Ground and pile-group responses due to tunnelling. *Soils Found.* **41** 57–67

[14] Zhang R-J, Zheng J, Zhang L and Pu H 2011 An analysis method for the influence of tunneling on adjacent loaded pile groups with rigid elevated caps *Int J Numer Anal Methods Geomech.* **71**

[15] Williamson M G, Elshafie M Z E B Mair R J Devriendt M D 2017 Open-face tunnelling effects on non-displacement piles in clay - part 1: Centrifuge modelling techniques *Geotechnique* **67** 983-1000

[16] Ng C W, Leung C F, Yong K Y and Chow Y K 2007 Performance of pile due to tunneling-induced soil movements **1** 619-624

[17] Madabushi S P G, Houghton N E and Zhang L M 2006 A new automatic sand pourer for model preparation at University of Cambridge *London, Taylor: Francis* pp 6

[18] Sohaei H 2017 *Contolling Tunnel Induced Ground Surface and Pile Movements Using Micropiles* Universiti Teknologi Malaysia **1**

[19] Juneja A and Dutta S 2008 Ground loss due to circular tunnel deformation in sands *12th Int Conf Comput Methods Adv Geomech* **5**

[20] David M P and Lidija Z 1999 Finite Element Analysis in Geotechnical Engineering: Volume One Theory *Finite Element Analysis in Geotechnical Engineering: Volume One – Theory* Thomas Telford Ltd
[21] Tan T S, Setiaji R R. and Hight D W 2005 Numerical analyses using commercial software– A black box 250-258
[22] Meguid M A and Mattar J 2009Investigation of Tunnel-Soil-Pile Interaction in Cohesive Soils *J Geotech Geoenvironmental Eng.* **135**
[23] Shehata A S, El-Kelesh A M, El-kasaby A-S E and Mansour M 2018 Rates of Soft Ground Tunneling in Vicinity of Existing Structures *Int J Adv Eng Manag Sci.* **4** 35-45
[24] Siti Ai'dah M A, Aminaton M, Siti Norafida J and Sohaie H 2018 Effect of Tunnel Construction On Existing Pile In Sand **1** 1-5
[25] Feras Mousa M A 2019 *The Effects of Tunneling on Single Piles* Universiti Teknologi Malaysia;
[26] Selemetas D 2005 *The response of full-scale piles and piled structures to tunnelling* PhD Thesis, Cambridge Univ.
[27] Polanippan I 2020 Tunnel-Pile-Soil Interaction **13** 1-15
[28] Pinto F and Whitttle A 2014 Ground Movements due to Shallow Tunnels in Soft Ground. *Analytical Solutions. J Geotech Geoenvironmental Eng.* **1**
[29] Kolivand F and Rahmannejad R 2018 Determination of settlement trough width and optimization of soil behavior parameters based on the design of experiment method (DOE) *Int J Min Geo-Engineering* **52** 7–15
[30] Hajjar M, Hayati A N, Ahmadi M M and Sadrnejad S A 2015 Longitudinal settlement profile in shallow tunnels in drained conditions *Int J Geomech.* **15**
[31] Khoo C M, Idris N I S I, Mohamad H and Rashid A S A 2018 Numerical Evaluation of Settlement trough Width Parameter *MATEC Web Conf.* 203 1-9
[32] Ayasrah M, Qiu H, Zhang X 2021 Influence of cairo metro tunnel excavation on pile deep foundation of the adjacent underground structures: Numerical study *Symmetry (Basel)* **13**
[33] Hun G K, Authority L T 2016 Case Histories of Bored Tunnelling Below Buildings in Singapore Downtown Line **3** 49–61

Acknowledgments

The authors would like to acknowledge the support of the Universiti Teknologi Malaysia and the Ministry of Higher Education Malaysia through the awarded Q.J130000.2651.17J59 UTM ER grant and Collaborative Research Grant (CRG) vote number R. J130000. 7351. 48509.

IOP Conf. Series: Earth and Environmental Science **971** (2022) 012028 doi:10.1088/1755-1315/971/1/012028

Study Restraint Pressure of Membranes on Expansive Soil

L Fitriyana[1*], A G Dian [2] and F Darwis[3]

[1,2] Sultan Agung Islamic University, Department of Civil Engineering
Jl. Raya Kaligawe Km.04, Semarang, Jawa Tengah, Indonesia
[3] Pasifik Morotai University, Department of Civil Engineering
Kec. Morotai Selatan - Kab. Kepulauan Morotai - Prov. Maluku Utara
*Corresponding author: lisa.fitriyana@unissula.ac.id

Abstract. Type of deformation in expansive soil is lateral and vertical Laboratory expansion tests are often performed on rings with restrained, lateral movement so that horizontal, swelling deformation will not occur. The road damage in the highway between Purwodadi – Semarang (Begel village, Purwodadi) was identified because the soil is expansive. Therefore, this study was conducted to identify characteristics of its soil expansion, specifically the influence of pressure of membrane restraints on expansion of expansive soil. The expansion test was conducted on modified test tools. The axial expansion strain was measured using the dial gauge which was installed in parallel, vertical position while the lateral strain was measured using the caliper digital in order to identify any change in the diameter of sample with the variations of initial water content (w_o), vertical pressure (p_v) and membrane's thickness (t). The test result show that the greater the (horizontal) lateral strain is, the greater the restraint pressure in the membranes will be. The restraint pressure in the membranes (p_h) is closely related to factors that influence the expansion possibility such as vertical pressure (p_v), initial height of test object (h_o), initial water content (w_o), and membrane's thickness (t).

1. Introduction

Expansive soil contains montmorillonite mineral which has highly negative charge and can absorb water entering the pore space, making the soil swell and reducing its strength drastically [1]. The most prominent characteristics of expansive soil are its very low load capacity in wet conditions and ability to swell and shrink in a relatively short time, causing severe damage during a change in water content (cracks in a dry condition and expansion in a moist condition) [3].

According to a study by Transportation Infrastructure R&D Center, Balitbang Kimpraswil (1992) [4], the distribution of expansive soils in Java Island comprises several areas as follows:

- The northern coast area, along the highways of Jakarta - Cikampek (West Java), Demak - Kudus, Dempet - Godong, and Semarang - Purwodadi (Central Java). They are also in the highways of Lamongan - Gersik and Surabaya - Gersik (Jawa Timur).
- The low hilly areas, in the highways of Bojonegoro - Babat (Jawa Tengah) and Ngawi - Caruban (Jawa Timur).
- The volcanic sediment area temporarily, where the expansive soils are along the highways of Yogya - Wates.

IOP Conf. Series: Earth and Environmental Science **971** (2022) 012028 doi:10.1088/1755-1315/971/1/012028

There are two directional expansion movements: lateral and vertical [1]. Laboratory expansion test is often conducted to rings with lateral restraints so that horizontal, deformational expansion will not occur. This study aims to identify the influence of restraint pressure (lateral pressure) of membranes on expansion of expansive soil (the case study: Begel village, Purwodadi).

2. Methodology

The expansion test was conducted on modified test tools. It is theoretically similar to the expansion test on the consolidation tool, Oedometer, but with a modification on the measurement tool for lateral strain. In order to create lateral strain, rubber membranes made of latex gel were used as restraints (see Figure 1). To acquire the measurement result, caliper digital was used to observe any change in the diameter of the test object. The tool scheme is shown in Figure 2. whereas the variation of test objects is shown in Table 1.

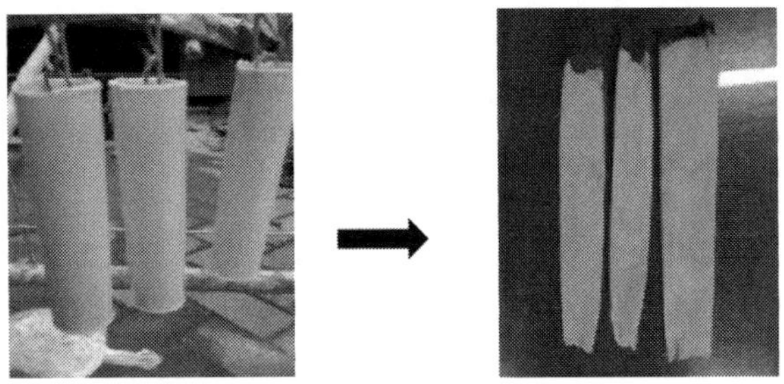

Figure 1. Rubber membranes as restrains.

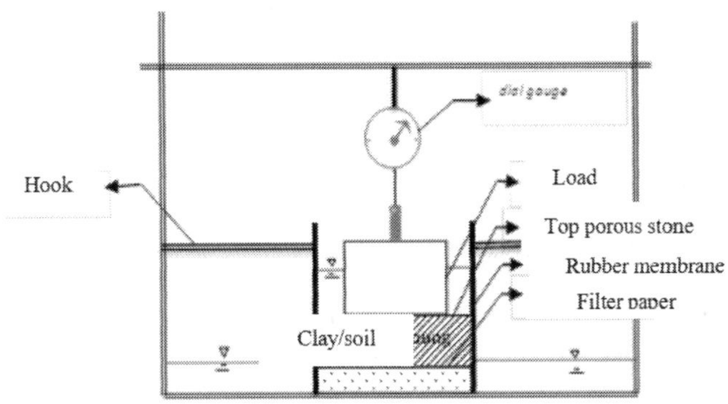

Figure 2. The scheme of expansion tools [5].

Table 1. The variation of number of test objects.

Diameter of test object D, (cm)	Sample's height, H_o (cm)	Membrane's thickness t, (mm)	Initial water content w_o, (%)	Vertical pressure, p_v (kPa)
4,5	2	0,5	18	3
			31	3
		0,7	18	3
			31	3

IOP Conf. Series: Earth and Environmental Science **971** (2022) 012028 doi:10.1088/1755-1315/971/1/012028

2.1. Expansion Potential

Expansion potential is a percentage of expansion towards lateral direction where a sample must be soaked or saturated with a pressure of 1 psi (6,9 kPa) for compacted, disturbed soil in order to achieve a maximum of density and optimum water content by using the AASHTO density [6]. Sample soil expanded in vertical direction and its changed height divided by its initial height was axial (vertical) strain, which is stated in Equation (1):

$$\varepsilon_Z = \frac{\Delta H}{H} x100\% \quad (1)$$

Lateral strain is stated in Equation (2):

$$\varepsilon_X = \frac{\Delta D}{D} x100\% \quad (2)$$

Volumetric strain is stated in Equation (3):

$$\varepsilon_V = \frac{\Delta V}{V} x100\% \quad (3)$$

where

ε_z = axial expansive strain (%)
ΔH = change in sample's height (mm)
H = sample's initial height (mm)
ε_x = lateral expansive strain (%)
ΔD = change in sample's diameter (mm)
D = sample's initial diameter (mm)
ε_v = volumetric exspansive strain (%)
ΔV = change in sample's volume (mm)
V = sample's initial volume (mm)

2.2. Internal Pressure

To stimulate lateral strain, the thin-walled pressure vessels theory was used with pressure from pipe's interior. According to Fletcher [7], thin, elastic tubes or vessels have the ability to deform to all direction if they receive an internal pressure p (see Figure 3).

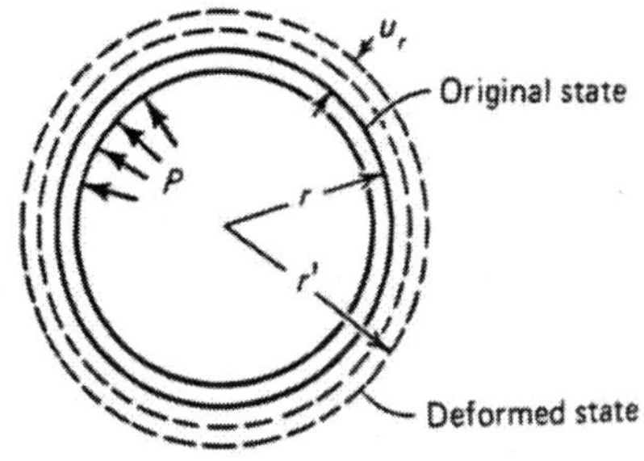

Figure 3. A deformation caused by internal pressure p [7].

Deformation is caused by change in radius, from *r* to *r'* for u_r as presented in Figure 3. It is also formulated in Equation (4):

$$\varepsilon_\theta = \frac{r' - r}{r} \tag{4}$$

where r' = r + u_r. From Hooke's Law, a deformation u_r which is caused by pressure *p* can be obtained so that Equation (6) can be formulated. To determine the value of *p*, use Equation (5):

$$u_r = \frac{pR^2}{2Et}(2-\mu) \tag{5}$$

$$p = \frac{u_r(2Et)}{R^2(2-\mu)} \tag{6}$$

p : lateral pressure (kPa)
r : radius before deformation (m)
r' : radius after deformation (m)
R : (*r* + *r'*)/2 = pipe's average radius(m)
t : pipe's thickness (m)
u_r : *r'* – *r* = change in diameter (m)
μ : *Poisson's* ratio
For next vertical direction pressure *p* is called p_h to difference with vertical pressure (p_v)

3. Discussion

For the initial test of expansive soil, soil's properties were tested (see Table 2). The test results show that the type of the tested soil is CH (Clay High).

Table 2. The results of the soil's properties test.

Properties	Value	Unit
Specific gravity, *Gs*	2,78	
Liquid limit, *LL*	90,75	%
Plastic limit, *PL*	24,14	%
Plasticity index, *PI*	66,61	%
Shrinkage limit, *SL*	26,00	%
Percentage of particles passing the sieve no. 200	89,79	%
Percentage of particles stuck in the sieve no. 4	-	
Percentage of clay fraction of 0,002 mm, *C%*	25,5	%
Activation, *A=PI/C*	2,61	%
Optimum water content, w_{opt}	31,46	%
Maximum dry unit volume weight, $\gamma_{d\,max}$	13,89	kN/m^3
Unified soil classification system	*CH*	

IOP Conf. Series: Earth and Environmental Science **971** (2022) 012028 doi:10.1088/1755-1315/971/1/012028

3.1. Membrane Elasticity Modulus Test (E)

Elasticity modulus test (E) on membranes made of latex was tested by stretching it using ASTM D-68. The test was performed in the laboratory of *Mekanika Bahan PAU UGM*. The initial length of the membrane was (H_0) = 50 mm and its thickness was (t) = 0,65 mm. The test result is presented in Figure 4.

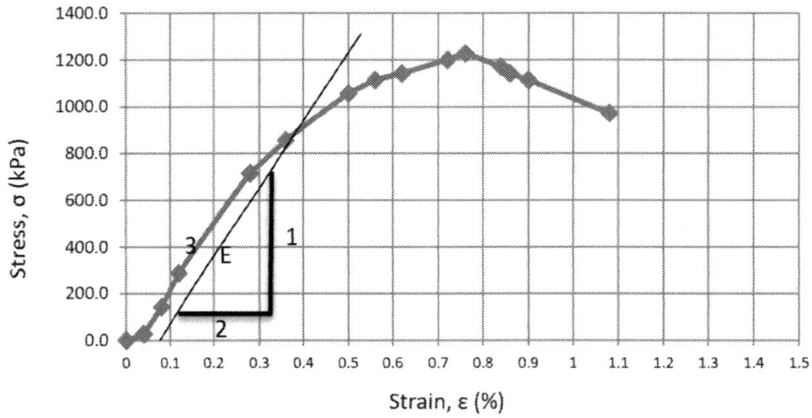

Figure 4. The result of membranes stretching test.

From the result of the rubber membranes stretching test, the strain (ε) and tension (σ) values which were substituted in the equation were obtained, resulting in the value of rubber elasticity modulus of 2239.2 kPa.

3.2. The Results of Soil Expansion

The results of expansion (strain) of the expansive soil can be viewed in Table 3.

Table 3. The recapitulation of the results of expansion (strain) and restraint pressure of membrane (p_h) with D = 4.5 cm and γ_d=13.89 kN/m^3.

Initial water content, w_0 (%)	Sample's initial height h_0 (cm)	Membrane's thickness t (mm)	Vertical pressure p_v (kPa)	Initial radius number, e_0 (%)	Axial strain ε_z (%)	Lateral strain ε_x (%)	Membrane strain pressure, p_h (kPa)
18	2	0.5	1	1.06	15.700	10.911	7.120
			3.5	1.09	11.550	13.822	8.776
			6.9	1.10	9.250	15.844	9.872
		0.7	1	1.09	17.000	10.467	9.602
			3.5	1.11	12.050	13.333	11.906
			6.9	1.10	9.400	15.800	13.787
31	2	0.5	1	1.09	14.400	8.844	5.886
			3.5	1.08	10.000	10.667	6.977
			6.9	1.12	8.900	11.333	7.366
		0.7	1	1.09	16.000	8.622	8.051
			3.5	1.09	12.000	9.400	8.712
			6.9	1.09	9.250	11.733	10.636

3.3. The Influence of Membrane's Thickness (t)

Variation in membrane's thickness was performed in this test. They were 0.5 mm and 0.7 mm. Its aim was to identify its influence on membrane's pressure on the test object when expansion strain emerged. The result of the influence of membrane's thickness on expansion strain is presented in Figure 5.

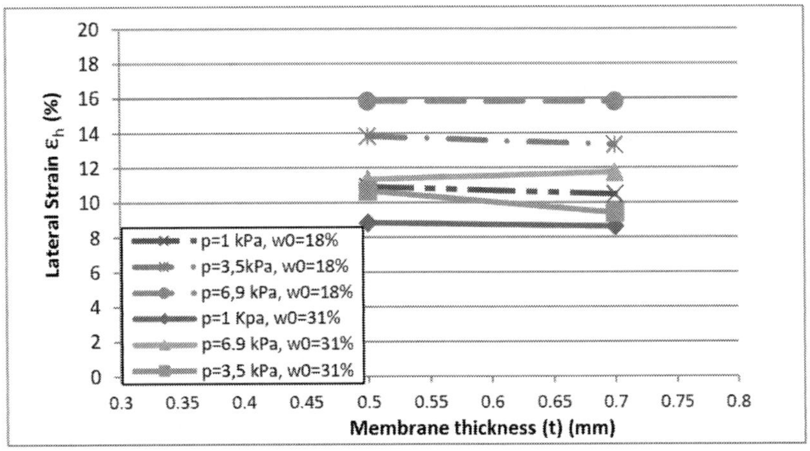

Figure 5. The influence of variation in membrane's thickness (t) on lateral strain.

Figure 5 shows that membrane's thickness also influences lateral strain. The thicker a membrane in a test object is, the smaller the lateral strain will be and the membrane's restraint pressure (p_h) will be greater. Conversely, the thinner a membrane in a test object is, the greater lateral strain will be and membrane's restrain pressure (p_h) will be greater. A maximum lateral strain can reach 16 % in t = 0,5 mm.

3.4. The Influence of Vertical Pressure (p_v) on Membrane's Restraint Pressure (p_h)

In this study, three variations of vertical pressure (p_v) of 1 kPa, 3,5 kPa, and 6,9 kPa were used. The aim of this variation in vertical pressure (p_v) was to identify any influence of vertical pressure (p_v) on membrane's restraint pressure (p_h)

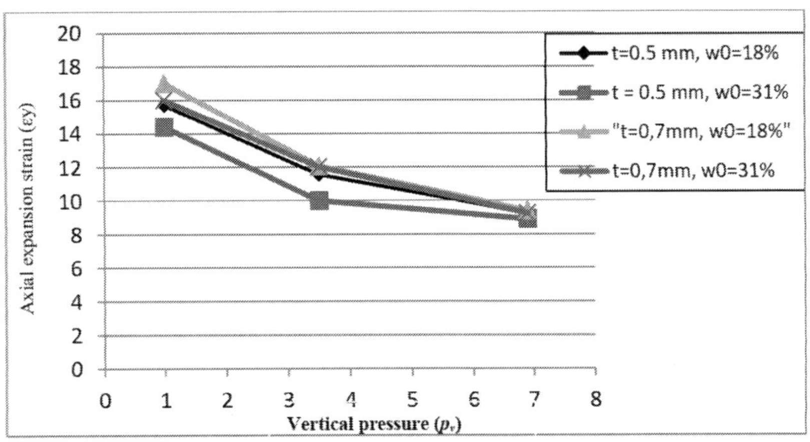

Figure 6. The influence of vertical pressure (pv) on the axial expansion strain (εy).

Figure 6 shows a relationship between vertical strain (p_v) and expansion strain that the smaller the vertical pressure (p_v) is, the greater the axial strain will be while membrane's restraint pressure will be smaller. Conversely, the greater the vertical pressure (p_v) is, the smaller the axial strain will be and membrane's restraint pressure will be greater (see Table 3).

3.5. The Influence of Vertical Pressure (p_v) on the Membrane's Restraint Pressure (P_h)

The influence of membrane's restraint pressure (P_h) on the soil expansion is related to the restraint's material. The smaller the elasticity modulus (E) value of a material is, the more elastic its nature will be. Rubber membranes made of latex belong to materials with small (elastic) rigidity. To simulate the restraint pressure (p_h) of membranes, Fletcher's equation was used [7]. The value of Poisson's ratio of rubber ranges from 0.48 to 5 [8]. It was substituted into Equation (6) so that a value of membrane's restraint pressure (p_h) was acquired (see Table 3).

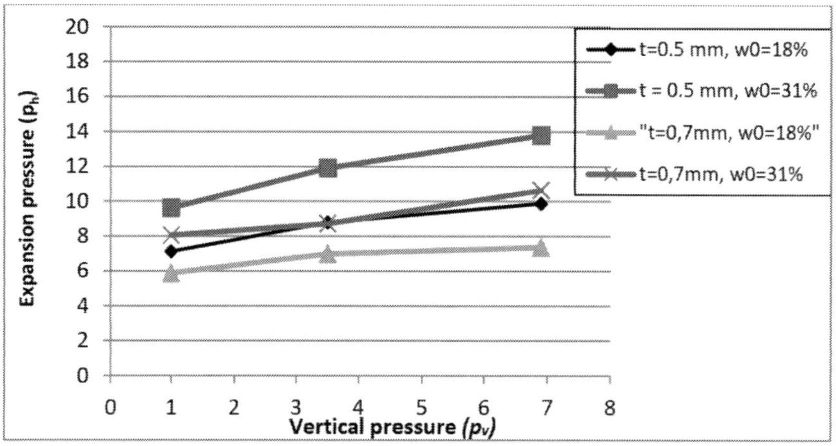

Figure 7. The influence of membrane's restraint pressure (p_h) on lateral strain (εx) with w_o = 21 %, h_o = 2 cm and D= 4.5 cm.

The relationship between membrane's restraint pressure (p_h) and vertical pressure (p_v) is shown in Figure 7. The greater the vertical pressure (p_v) is, membrane's restraint pressure (p_h) will also be greater. This is because the vertical, compressive strength causes lateral expansion to become greater.

4. Conclusion

Membrane's restraint pressure (p_h) is closely related to factors that influence expansion potential such as vertical pressure (p_v), test object's initial height (h_o), initial water content (w_o) and membrane's thickness (t).

5. References

[1] Hardiyatmo H C 2014 *Tanah Ekspansif*, Edisi 1, Gadjah Mada University Press, Yogyakarta
[2] Permadi, Wisnu,et.al 2016 Stabilisasi Tanah Lempung Ekspansif Godongpurwodadi Km 50 Menggunakan Proses Elektrokinetik Dengan Stabilisator Accu Zuur Dan Kapur, Semarang: Jurusan Teknik Sipil, Fakultas Teknik, Universitas Diponegoro
[3] Hardiyatmo Hary C 2012 Mekanika Tanah I Edisi Enam. Yogyakarta: Gadjah Mada University Press
[4] KONSTRUKSIMANIA Green Construction Today, *Hati-Hati Dengan Tanah Ekspansif* Retrieved on 8th January of 2017 from Http://Konstruksimania.Blogspot.Com/2012/06/Hati-Hati-Dengan-Tanah-Ekspansif.Html

[5] Fitriyana Lisa 2015 *Pengaruh Tekanan Kekang Membran Terhadap Pengembangan Tanah* PhD Thesis. Universitas Gadjah Mada, Inodnesia.

[6] Hangge E E 2009 *Pengaruh Kadar Air Terhadap Potensi Pengembangan Bebas Pada Lempung Ekspansif* PhD Thesis .Universitas Gajah Mada, Indonesia

[7] Fletcher Q David 1985 *Mechanics Of Materials* Holt Rinehart And Winston Holt-Saunders, Japan

[8] J A Rinde 1970 Poisson's Ratio For Rigid Plastic Foams, *Journal Of Applied Polymer Science.* **14(8)** 1913-1926

Physiochemical characterization of lateritic bauxite mining soil

N I N Ismail[1], **H Awang**[2*], **N Z Mohd Yunus**[3] and **S Saleh**[4]

[1] Faculty of Civil Engineering, Universiti Teknologi MARA Pahang, Jengka Campus, 26400 Bandar Tun Abdul Razak, Pahang, Malaysia
[2] Department of Civil Engineering, College of Engineering, Universiti Malaysia Pahang, Lebuhraya Tun Razak, 26300 Gambang, Kuantan, Pahang, Malaysia
[3] School of Civil Engineering, Faculty of Engineering, Universiti Teknologi Malaysia, 81310 Johor Bahru, Malaysia
[4] Hassan Usman Katsina Polytechnic Nigeria
*Corresponding author: haryatiawang@ump.edu.my

Abstract. Physiochemical characteristics play a significant role in evaluating the engineering behaviour of soil material and its suitability for foundation. Investigation upon the physical and chemical characteristics of lateritic bauxite soils were done on samples collected from three ex-mining bauxite sites (Bukit Goh, Semambu and Indera Mahkota) in Kuantan District, Pahang, Malaysia. The presence of chemical element was evaluated based on the elemental mineral composition content identified through X-Ray Fluorescence (XRF). The results revealed that Semambu lateritic bauxite soil has the highest content of alumina (Al_2O_3), 25.54%. The alumina content enrichment is one of the effects from the laterization of bauxite process. In addition, the physical testing included are moisture content, specific gravity and Atterberg Limit. Further investigation on the physical properties of the soil has found that Semambu has the highest MC, 33.27%, but at the same time PI is less than 12%. This is as a key indicator that lateritic bauxite is prone to surface erosion and unsuitable for construction purposes in its natural condition. The risk of the surface erosion and settlement of the ground causes it requires stabilizer that can rapid the curing time. Additionally, the high moisture content is likely to have higher chance to experience liquefaction and causes foundation problem to future infrastructures that may be built in the studied area.

1. Introduction

Bauxite deposits form when source rocks rich in alumino-silicate minerals are altered and chemically weathered. A number of authors have looked into the genetic modelling, changes in mass, element mobility, and the textures of ore in the bauxite deposits all around the world. Bauxite are classified into three categories based on its mineralogy, geochemistry, and bedrock lithology: (1) Lateritic-type bauxites, (2) Tikhvin-type bauxites, and (3) Tikhvin and Karstic-type. Lateritic bauxites are formed by in-situ lateritization from the underlying alumosilicate rocks, where mobile minerals been leached and immobile element such as Al_2O_3 and Fe_2O_3 remained to form bauxites [1]–[3]. The most critical elements in determining the extent and grade of bauxites are parent rock composition, climate, drainage, topography, groundwater chemistry, microbial activity, water table location, and the duration of weathering processes [1].

In Kuantan, Pahang, Malaysia, bauxite ex-mining land is in a critical stage, with eroded soil causing enormous erosion. According to reports, the earth that was disturbed using the open pit method was left naked, with no revegetation protection in place for rehabilitation purposes [4]. Either common mining heritage or inheritance, both contribute to landslide occurrences and problematic soils for the geotechnical engineer to contend with in subsurface and surface areas [5]. The bare ground surfaces are highly subjected to wind and water forces which exposed the soil structures to micro-aggregation instability. However, physiochemical properties of the lateritic bauxite soil are not popular to be discussed in the previous literatures compared to bauxite residue. Among all previous research, a single study that characterize lateritic bauxite soil from ex-mining sites is unclear. Therefore, characterization of the lateritic bauxite soil at the bauxite ex-mining sites is necessary to investigate the soil characteristics as well as to predict its engineering behaviour.

Besides that, as the population and construction sector develop at a quick pace, demand for bauxite output products has increased. The most common post-mining land use purposes include agriculture, forestry, recreation, construction, conservation and lakes [6, 7]. Effort studying on problematic soils has also expanded considerably [8, 9], since the natural resources business has long been recognised as the cornerstone in the design and building of a variety of creative constructions, infrastructures, and facilities aimed at improving global society [10]. Thus, physiochemical soil characterization at ex-mining areas is required to be studied, to predict its potential for foundation design and construction.

2. Material and methods

The lateritic bauxite samples were tested for physical and chemical characterization. The physical properties, including moisture content, Atterberg Limit, and specific gravity of the lateritic bauxite were tested based on the British Standard procedures as shown Table 1. Meanwhile, for the chemical elements in the lateritic bauxite were tested using XRF. Following figures show the ground conditions of the sites. Most of the eroded area are bare and no vegetation grown on top.

Table 1. Physical properties lateritic bauxite of each site

Properties	Standards
Natural moisture content	BS EN ISO 17892-1:2018
Specific gravity	BS EN ISO 17891-3:2018
Atterberg Limit	BS EN ISO 17892-12:2018

The lateritic bauxite samples were collected from Bukit Goh, Semambu and Indera Mahkota. About 0.5 m of topsoil been removed before samples are taken. The sampling areas were chosen based on display trace of bauxite open pit mining activities. These areas display uncovered disturbed surface which exposed to potential soil surface erosion as per shown in Figure 1, Figure 2 and Figure 3.

Figure 1. Ground condition at site Bukit Goh.

Figure 2. Ground condition at Indera Mahkota.

Figure 3. Ground condition at Semambu.

3. Results and discussion

The results of the physiochemical properties of the lateritic bauxite soil are presented and discussed in this section.

3.1. Physical properties of the lateritic bauxite

The results of the natural moisture content (MC), specific gravity (SG) and Atterberg Limit for each site are presented in following table. Results of Atterberg Limit are indicated as LL (liquid limit), PL (plastic limit), PI (plasticity index) and LS (linear shrinkage) in the Table 2.

Table 2. Physical properties lateritic bauxite of each site

	MC (%)	SG	LL (%)	PL (%)	PI (%)	LS (%)
Bukit Goh	21.74	2.93	46.25	37.80	8.45	7.81
Semambu	33.27	2.79	54.66	46.90	7.76	7.19
Indera						8.08
Mahkota	13.07	2.61	31.62	25.59	6.03	

Summary of the physical properties are shown in Table 2. It shows that the soil sample from Semambu has the highest LL value compared to Bukit Goh and Indera Mahkota. As referred to standard specifications of materials for roads by Federal Ministry of Works (1970), the absence or very low contents of montmorillonite clay minerals in the soils are shown from the LL value [11]. Even though the moisture content is highest, the LL for Semambu show its significantly low content of clay minerals. The value of the PI of all samples are less than 12%, and thus may be classified as having low swelling potentials [12].

3.2. Chemical properties based on XRF Analysis

The mineral composition of the bauxite is verified using the XRF data received from the soil samples collected at each site. Bauxite is a deposit that contains aluminium minerals in general [13]. Oxide concentrations, Fe concentration variations in weathering profiles, mineral leaching degree, and element geochemistry have all played a part in the distribution of trace and rare earth elements during lateritic bauxite weathering [14]. Hence, results from each sites are presented and discussed in the following section.

Based on Table 3, alumina is largely prominent in the bauxite soil from Semambu. Alumina is present in greater than 20% of all locations. Iron (Fe) is the second most abundant element in the samples. As a result of the dominance of alumina and iron in these soil samples, it may be determined that they are all bauxite [13], [14].

Table 3. Major element in soil sample

Samples	Major Element (%)			
	Alumina, Al_2O_3	Iron, Fe_2O_3	Titanium, Ti	Silica oxide, SiO_2
Bukit Goh	22.79	18.26	4.31	9.52
Semambu	25.54	15.31	3.36	15.65
Indera Mahkota	20.58	21.71	5.14	9.07

4. Conclusion

Several laboratory experiments were performed on lateritic bauxite samples collected from Kuantan, Pahang. The following conclusions are drawn from the findings of the study, characterising the physiochemical characteristics of the lateritic bauxite. The lateritic bauxite has poor to moderate physical properties that need improvement before used for construction purposes. The lateritic bauxite has alumina content of 20% to 26%, iron content 15% to 22% and silica oxide 9% to 16%. The alumina and iron content in the lateritic bauxite sample remained in the soil as immobile element, whilst silica oxide has been leached during the laterization and thus lead to bauxisation process. The MC is more than 10% which shown that the lateritic bauxite from these sites is not suitable for construction as in its natural condition. But may require soil stabilizer especially for soil performance improvement. Moreover, more extensive research shall be done on characterizing the lateritic bauxite soil prior proceed for the foundation design and construction.

5. References

[1] G Bardossy and G J J Aleva 1990 Lateritic bauxites *Clays and Clay Minerals* **39** 108-109

[2] G Bardossy, I Bogardi and L Duckstein 1982 Decision Models for Bauxite Mining Under Water Hazard in Organ by Hung Min and Metall Soc *Budapest and Cent Inst for Min Dev of Hung*

[3] H Zamanian F Ahmadnejad and A Zarasvandi 2016 Mineralogical and geochemical investigations of the Mombi bauxite deposit, Zagros Mountains, Iran *Chemie der Erde* **76** 13-37

[4] Akademi Sains Malaysia 2017 Sustainable Mining: Case Study for Bauxite Mining in Pahang

[5] M N Abd Rashid, H Ahmad, S J T Jamil, N A Yahaya and M H Othman 2015 A Review of Ex-Mining Land Reclamation as Construction Project Activities : Focusing in City of Ipoh A Review of Ex-Mining Land Reclamation as Construction Project Activities : Focusing in City of Ipoh *Proc. Postgrad. Conf. Glob. Green Issues (Go Green) UiTM (Perak), Malaysia* 7 pp 7-8

[6] H Soltanmohammadi, M Osanloo and A Aghajani Bazzazi 2010 An analytical approach with a reliable logic and a ranking policy for post-mining land-use determination *Land use policy* **27** 364-372

[7] C. Mborah, K. J. Bansah, and M. K. Boateng 2015 Evaluating Alternate Post-Mining Land-Uses: A Review *Environ. Pollut.* **5** 14-22

[8] H Moayedi and R Nazir 2018 Malaysian Experiences of Peat Stabilization, State of the Art *Geotechnical and Geological Engineering* **36** 1-11

[9] A S A Rashid, S A Tabatabaei, S Horpibulsuk, N Z Mohd Yunus and W H W Hassan 2019 Shear Strength Improvement of Lateritic Soil Stabilized by Biopolymer Based Stabilizer *Geotech. Geol. Eng.* **37** 5533-5541

[10] E Goh and S Effendi 2017 Overview of an effective governance policy for mineral resource sustainability in Malaysia *Resour. Policy* **52** 1-6

[11] I B Duruojinnaka, O C Okeke and C C Amadi 2016 Geotechnical and geochemical characterization of lateritic soil deposit derived from ajali sandstone in ihube- okigwe, southeastern nigeria for road construction *Int. J. Res. Mech. Civ. Eng.* **3** 30-47

[12] A A Bello, J O Owoseni and I O Fatoyinbo 2019 Evaluation of plasticity and consolidation characteristics of migmatite–gneiss-derived laterite soils *SN Appl. Sci.* **1** 934

[13] J Bower 2017 Bauxite Technical Note pp. 1-4

[14] J Gu, Z Huang, H Fan, Z Jin, Z Yan and J Zhang 2013 Mineralogy, geochemistry, and genesis of lateritic bauxite deposits in the Wuchuan-Zheng'an-Daozhen area, Northern Guizhou Province, China *J. Geochemical Explor* **130** 44-59,

Acknowledgments
This paper was written based on the research carried out using the Internal University Grant (RDU1803158) from Universiti Malaysia Pahang (UMP). The authors would like to express their gratitude to UMP for the generous support.

IOP Conf. Series: Earth and Environmental Science **971** (2022) 012030 doi:10.1088/1755-1315/971/1/012030

Effect of Surfactant in Molasses Based Soil Solidification Material to Tensile Strength in Seven Day Material Age

T Tsutsumi[1*] and S Uchida[2]

[1] Department of Urban Environmental Design and Engineering, National Institute of Technology, Kagoshima College, Kirishima, Kagoshima 899-5193, Japan
[2] Department of Applied Chemistry and Biochemistry, National Institute of Technology, Fukushima College, Iwaki, Fukushima 970-8034, Japan
*Corresponding author: tsutsumi@kagoshima-ct.ac.jp

Abstract. The method of ground improvement using lime or cement is popular because of the speed, adaptability, and profitability of this method. However, pollution due to alkaline water from lime or hexavalent chromium from cement may occur as a result. To avoid this issue, molasses has been used for the improvement of clay in several studies. However, a satisfactory value indicating ground improvement has not been achieved yet. In this study, surfactant has been added to molasses based soil solidification material previously studied by the authors. The Brazilian test was used to obtain tensile strength value for evaluation of the mechanical properties of the specimens. The tensile strength at seven days of this material obtained through this procedure is compared with that of specimens using distilled water, 10% molasses solution or the specimens used in the authors' former study. Four types of curing condition combined with air drying and moist curing were adopted. The highest average value of the tensile strength appeared during 7-day air curing and was 1.3 times that in the authors' former study.

1. Introduction

Soil stabilization can be described as a process which incorporates the various methods used for improving the strength and stability of a given soil mass as well as other engineering and physical properties. It is usually employed when the soil available for construction work is not suitable for its intended use. By increasing the shear strength of the soil, improving its load bearing capacity and controlling the shrink-swell properties, a better engineering performance can be ensured through stabilization with a reduced chance of engineering failures such as washing collapse [1].

The methods of soil stabilization are classified into two categories, one is mechanical and the other use a stabilizer. The rearrangement of soil particles and soil solidification materials can be observed during the latter. Lime or cement has been the most widely used soil solidification material for clay. However, the pH value in ground improved using these materials is very high. Therefore, some damage due to the surface stream or percolating water may occur for the vegetation around the site of the earth works. In addition, it is reported that hexavalent chromium flows out from ground improved with cement or cement-based agents. The soil solidification material should not only induce mechanical strength but also have little impact on the environment.

Sugar cane molasses is a viscous by-product during sugar processing from sugar cane. Increased demand for sugar increases production of cane molasses which is constituted of about 30 - 40% of sugar. Some attempts have been conducted in India and a few African countries to improve ground containing clay using molasses. For example, Taye et al. [2] or M'Ndegwa et al. [3] have used molasses for expansive clay, Ravi et al. [4] for compressive clay, and Kiran et al. [5] for Black cotton clay. Issues due to lime or cement as mentioned above could be avoided and unconfined compressive strength and CBR were seen to increase in their studies. Furthermore, in the author's former work it is shown that tensile strength of a Kaolin clay specimen with solution of 10% molasses, 0.9% acetic acid and 1.0 g/l calcium oxide is 1.2 times that using 10% molasses solution [6]. In this study, specimens were prepared by adding surfactant into the molasses-based soil solidification material dealt with in the author's former work. Four types of curing condition including air drying and moist curing were adopted. To estimate the effect on strength, tensile strength in a 7-day material adopted as the standard [7] for roadbeds was selected and Brazilian tests carried out. At the same time, the dry density of the specimen was also obtained. Tensile strength, water content and dry density attained through this procedure were compared with those of specimens using molasses-based material in the author's former work, 10% molasses solution and distilled water.

2. Methodology and materials

2.1. Material

2.1.1 Kaolin clay

Kaolin clay used in this study was mined in Utsunomiya City, Japan. Its specific gravity and natural water content are 2.65 g/cm^3 and 0.48 %, respectively. The mineral components include SiO_2 63.3%, Al_2O_3 18.9%, Fe_2O_3 0.19% and K_2O 2.9%. Its grain size accumulation curve is shown in Figure 1. Its liquid limit w_L and plastic limit w_P are 25.61% and 17.88%, respectively [6].

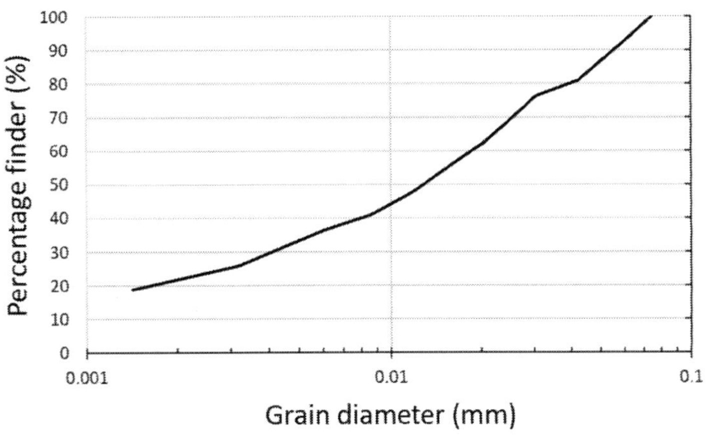

Figure 1. Grain size accumulation curve of Kaolin clay in this study [6].

2.1.2. Solution

Molasses used in this study was produced by Nippon Garlic Corporation. pH value of 10% water solution is 5.12, moisture content is 26.80% and mineral content is shown in Table 1.

Table 1. Minerals in molasses used in the present study.

Mineral in molasses (%)				
Ca	P	Mg	K	Na
1.19	0.11	0.47	3.17	0.37
Mineral in molasses (mg/kg)				
Fe	Cu	Zn	Mn	Mo
300.00	79.40	43.00	56.00	0.10

Four types of solution shown in Table 2 were prepared for this study. D indicates a specimen with distilled water, M denotes a 10% molasses solution, and MCAD and MCAA were used in the authors' former and present study, respectively. Acetic acid, Calcium oxide and surfactant used in this study are reagents. Three sets of specimens were prepared for each type. 48 g of each solution was used for 370 g of Kaolin clay to prepare the four kinds of specimens as the largest value of tensile strength appears at this water content [6].

Table 2. Type of solution used in present study.

Solution	Molasses	Calcium oxide	Acetic acid	Surfactant
D	0	0	0	0
M	10%	0	0	0
MCAD	10%	1 g/l	0.90%	0
MCAA	10%	1 g/l	0.90%	0.10%

2.2. Methodology

To estimate the binding force between each particle of clay, tensile strength was determined by the Brazilian test. As shown in Figure 2, The Brazilian test can be performed using the apparatus for an unconfined compression test. The advantage of the Brazilian test is that the procedure is not that complicated and only a small amount of clay is required for the specimens. The shape of the specimen is columnar, and its radius and thickness are 25 mm and approximately 25 mm, respectively. The tensile strength of the specimen σ_t is shown;

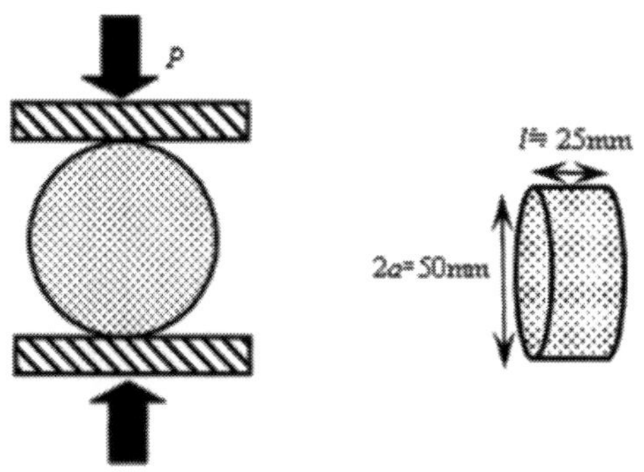

Figure 2. Schematic diagram of Brazilian test [6].

$$\sigma_t = \frac{P}{\pi a l}. \quad (1)$$

where, P, a, and l represent applied force to specimen, radius and thickness of specimen, respectively. Compactive effort E_c is;

$$E_c = \frac{W_R H N_B N_L}{V}. \quad (2)$$

where, W_R, H, N_B, N_L and V represent weight, height of the rammer, number of compaction per layer, number of layer and volume of specimen, respectively. In the author's former work, maximum tensile strength was obtained when E_c was almost 600N/cm^2 [6]. In this study, W_R is 14.7 N (1.5 kg), H is 20 cm and N_L is 1. Therefore, N_B obtained as 99.

The mechanical strength of Kaolin clay seems to be influenced by temperature or humidity during the curing period. In this study, specimens were kept in an incubator during the moist curing period and in a circulable dryer during the air drying period. Temperature was set at 25 °C for both situation.

Four types of specimens were prepared for each solution. The periods of moist curing and air drying respectively were 7 and 0, 6 and 1, 4 and 3, and 0 and 7 days for the four specimens as shown in Table 3. 3 sets were prepared for each type.

Table 3. Curing condition.

	Period after specimen preparation (day)						
Air drying period (day)	1	2	3	4	5	6	7
0			Wet curing (25°C)				
1							
3							
7					Air drying (25°C)		

3. Results and discussion

Figure 3 shows the relation between tensile strength and air drying period. In all specimens, tensile strength is almost 10N/cm^2 when period of air drying is 0. In addition, tensile strength values increase upto a 3 day air drying period except for specimens with distilled water. For 3 to 7 day air drying, the increase in tensile strength is small for all solutions. Average tensile strengths for 7 day air drying of MCAA, MCAD, M and D are 31.06, 23.82, 21.74 and 12.57 N/cm^2, respectively.

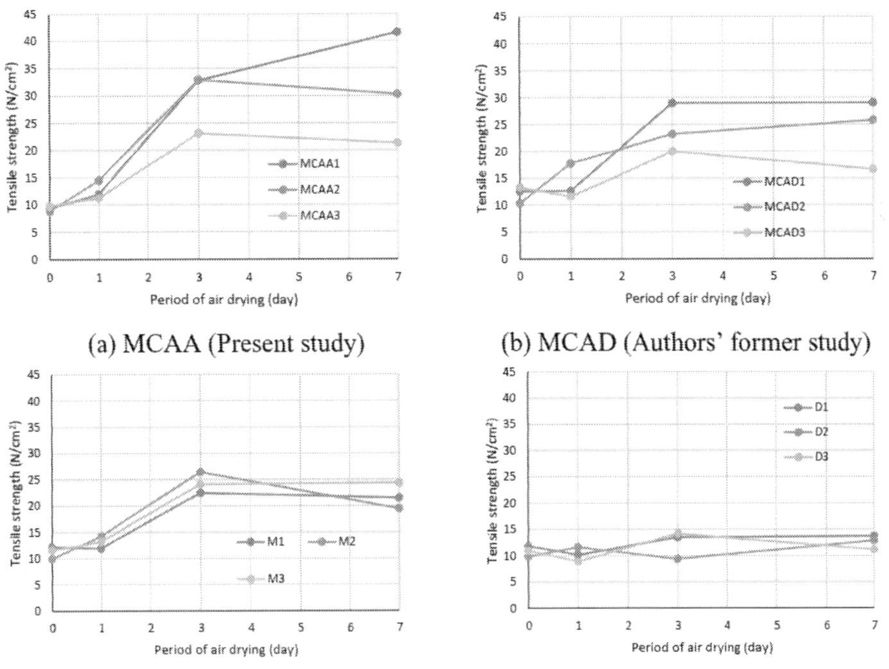

(a) MCAA (Present study) (b) MCAD (Authors' former study)

(c) M (10% molasses-solution) (d) D (Distilled water)

Figure 3. Relation between tensile strength and air drying period.

Figure 4 shows the relation between the ratio of water content to the initial value and air drying period. Almost the same value of water content, 90%, appears in all specimens when air drying period is 0. The tensile strengths of all specimens which underwent 0 day air drying are very small because the specimens are not dry enough. When air drying period is 1 day, water more than 10% of the initial water content remains in all specimens containing 10% molasses. On the other hand, about 5% of the initial water content remains in specimens with distilled water. Therefore, a hydration reaction in clay is expected to strengthen specimens since the authors' former or present soil solidification materials included calcium oxide and acetic acid. In the authors' present and former studies, large variation

IOP Conf. Series: Earth and Environmental Science **971** (2022) 012030 doi:10.1088/1755-1315/971/1/012030

among tensile strength on 7 day air drying and water content on 1 day air drying was observed. Also, it is difficult to find any correlation between them. However, the variation in tensile strength or water content becomes larger as air drying period increases. It seems that the difference in atmospheric

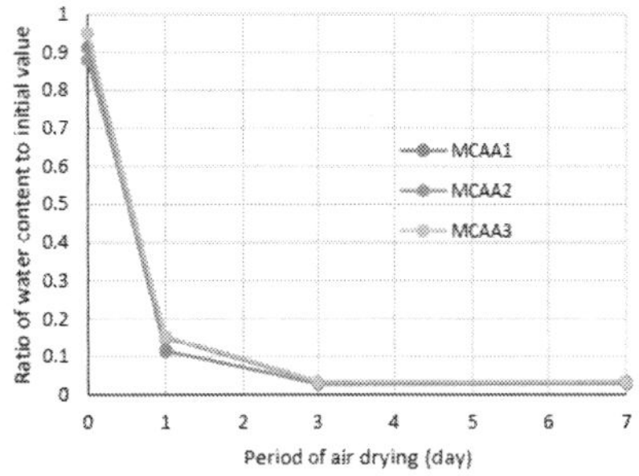

(a) MCAA (Present study)

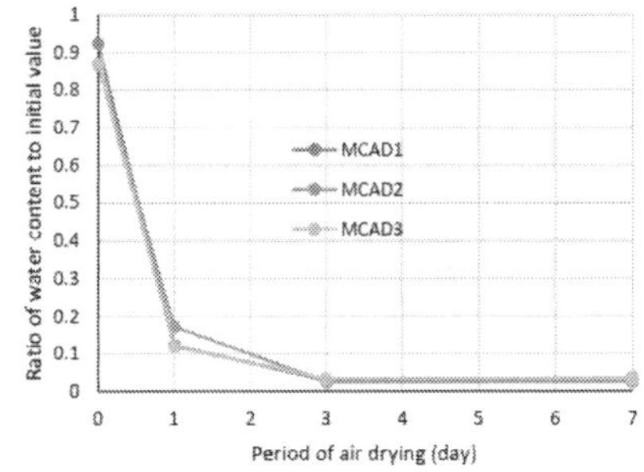

(b) MCAD (Authors' former study)

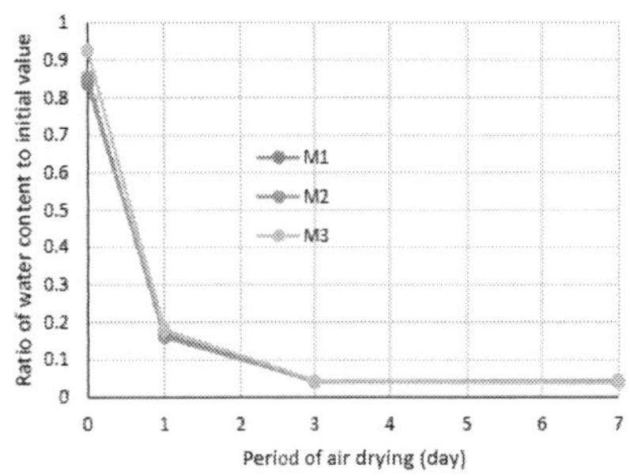

(c) M (10% molasses-solution)

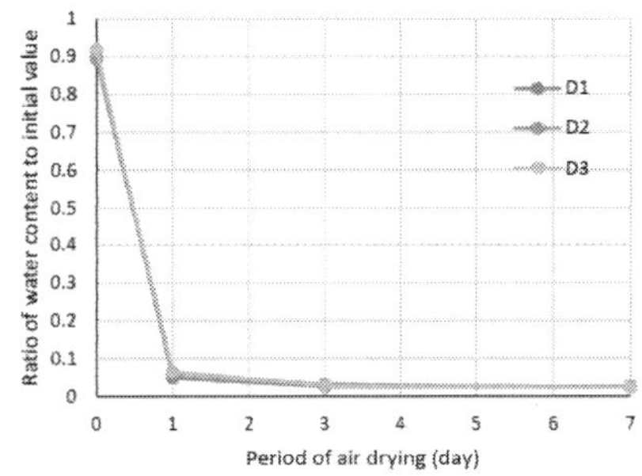

(d) D (Distilled water)

Figure 4. Relation between the water content and theair drying period.

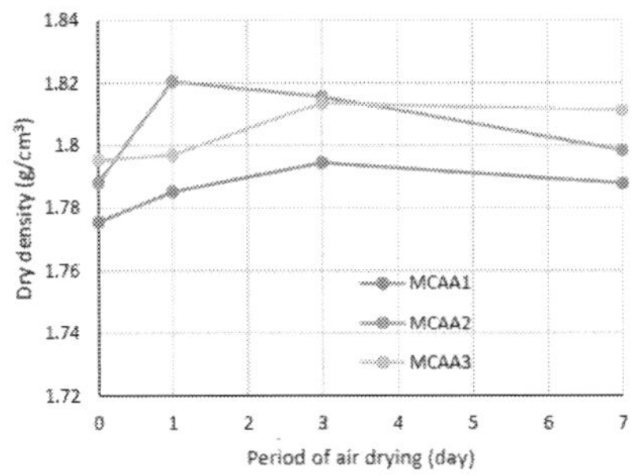

(a) MCAA (Present study)

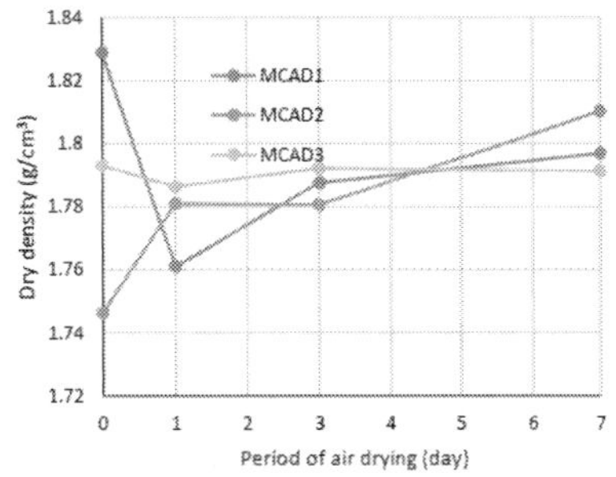

(b) MCAD (Authors' former study)

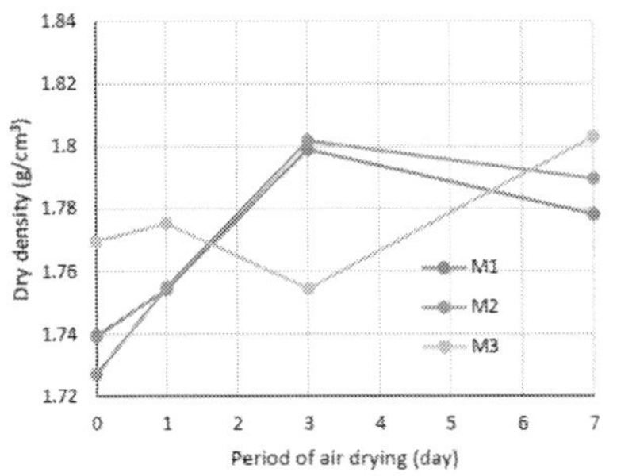

(c) M (10% molasses-solution)

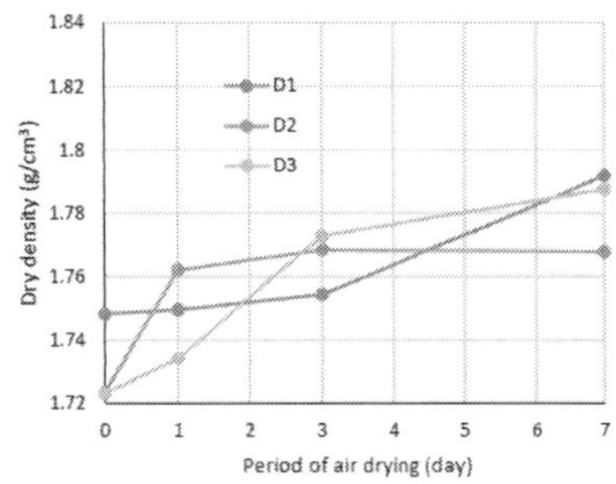

(d) D (Distilled water)

Figure 5. Relation between the dry density and the air drying period.

humidity during air drying causes this variation as the atmospheric temperature during moist curing and air drying is the same 25 °C.

Figure 5 shows the relation between the dry density of a specimen and the air drying period. Large values can be seen in the present study, especially for 0 and 1 day air drying. Generally, particles of clay are covered with adsorbed water which exists as a membrane on their surface. Surfactant can remove this adsorbed water on the soil particles. As a result, the distance between each particle decreases and sometimes the surfaces of particles may come into contact directly because of the surfactant [8]. In addition, molasses also has a weak surface-active effect. This is the reason why the dry density of a specimen using surfactant or molasses is larger than that using distilled water. Furthermore, it seems to be easy for a chemical reaction to occur connecting each clay particle as the distance between the particles decreases. This is why the tensile strength of specimen using surfactant or molasses is larger than that using distilled water.

4. Conclusion

In this study, 0.1% surfactant was added into the molasses based soil solidification material consisting of 10% molasses, 1.0 g/l Calcium oxide and 0.9% acetic acid which is used as a solidification material for Kaolin clay. Four types of curing condition were chosen to find tensile strength of a 7 day aged material using the Brazilian test. Not only tensile strength but also water content and dry density were obtained and compared with the results of the authors' former studies or results from specimens using 10% molasses solution or distilled water. The knowledge obtained from this study is as follows.

- Tensile strength increases for 0 to 3 day air drying and the increase of tensile strength is small between 3 and 7 day air drying. This is similar qualitatively to the results for specimens containing molasses.
- The highest average value of tensile strength appears for the specimen during 7 day air curing and is 1.3 times that observed in the authors' former study.

- Water content obtained from the present study is almost the same as that found in the authors' former work or in specimens using 10% molasses solution and higher than that found in the specimen using distilled water.
- The variation in tensile strength or water content becomes larger as the dry curing period increases.
- Dry density of a 7 day aged material in the present study is larger than that of a specimen using 10% molasses solution when air drying period is 0 and 1 day.

The authors' future endeavor is to find the relation between atmospheric humidity during air drying and tensile strength or water content of the specimen.

5. References

[1] Archibong G A, Sunday E U, Akudike J C, Okeke O C and Amadi C 2020 *Int. J. Advanced Achademic Research* **6** pp 89-114
[2] Taye B and Araya A A 2015 Proc.2nd Int. Conf. on Transportation of Africa pp 1-12
[3] M'Ndegwa J K, J. Eng. Trends in Eng. and Appl. Sci. 2 pp 1034-1041
[4] Ravi E E, Manikandan E A T, Karthick G and Jameel A A 2015 Int. J. Advanced Research Trends in Eng. and Tech. **2** pp 57-61
[5] Kiran, Harsha S, Raju K V S B and Kumar N M 2018 Int. J. Appl. Eng. and Research. **13** pp 219-223
[6] Tsutsumi T, Hidaka J and Uchida S 2020 Proc. 14th Symp. Ground Improvement, Japan pp 233-238
[7] Fukaya W and Sakakibara T 2009 Research Data of National Institute for Land and Infrastructure Management Japan, **531** pp 33-61
[8] Niiro T, Masada Y, Jiang J and Kuribayashi K 1997 Trans. JSCE, 582/III-**41** pp 265-274

Effects of Lime on the Compaction Characteristics of Lateritic Soil in UTM, Johor

N H A Rizal[1], M A Hezmi[1*], R Razali[1,2], N A Wahab[1], M J Roshan[1], A S A Rashid[1], and D Z A Hasbollah[1]

[1] School of Civil Engineering, Universiti Teknologi Malaysia Skudai, Malaysia
[2] Faculty of Civil Engineering, Universiti Teknologi MARA (UiTM) Pasir Gudang, Malaysia
*Corresponding author: azril@utm.my

Abstract. Lime has been investigated as a ground improvement material for soils, particularly for road construction, as an alternative to soil cement stabilisation. However, the effect of lime on lateritic soil compaction characteristics is not completely defined. In this article, soil samples from Universiti Teknologi Malaysia (UTM), Skudai, Johor were subjected to a variety of engineering tests, including specific gravity, sieve analysis, Atterberg's limit, and compaction. According to the results of Atterberg's Limit, lateritic soil has a plasticity index (PI), a plastic limit (PL), and a liquid limit (LL) of 19, 46, and 65, respectively. The obtained specific gravity value is 2.79. Compaction tests were conducted on untreated and lime-treated soils with lime additions of 3%, 5%, 7%, and 9%, respectively, using an automatic compactor. The OMC and MDD values for untreated laterite are 28% and 1.46 g/cm³, respectively. However, as the lime content increases, the OMC increases slightly while the MDD decreases. The dry density of lime-laterite soil decreases from 1.47 to 1.38 g/cm³, but the moisture content increases slightly from 27% to 31%. The MDD of soil treated with 5% lime is lower than that of lateritic soil in its natural state.

1. Introduction

In Malaysia, laterite soil is a common form of residual soil as a result of the hot and humid climatic changes. Due to their broad availability, laterite soils are commonly employed as fill materials for roads, buildings, and slope foundations [1]. Since laterite soils existed through intense weathering, they exhibit clay particles such as montmorillonite, illite, and kaolinite that cause soil swelling and reduce soil strength [2]. Certain soils are loose and imprecisely compacted, or they are formed of small and unreliable particles. These compositions are susceptible to shifting or degrading under pressure or in wet weather conditions. The entire procedure includes utilizing engineering techniques to transform unstable soil into a strong foundation for construction projects. As a result, soil stabilization is necessary to improve engineering properties by limiting the soil's permeability and increasing its total strength, the stabilization process ensures its stability. This is accomplished through controlled compression, proportioning, and the addition of admixtures or stabilizers [3].

Various engineering design, such as railroad subbases, aviation runways, and earth retaining walls, need compaction [4]. The variables that are important in soil compaction consisted of maximum dry density (MDD) and optimum moisture content (OMC) [5]. Soil stabilization with chemical stabilizers (lime and cement) to modify physicochemical characteristics of soil is commonly practiced. Lime treatment was proven to improve soil strength by extensive research on the application of lime-stabilized lateritic soil [1, 3-4, 6-18].

Lime-clay interactions consist of two different processes [19-20]; soil modification, which involves rapid ion-exchange, and soil stabilization, consisting of the time-consuming pozzolanic reactions. Soil modifications occur when there is an interchangeable process between calcium ions in lime and ions in soil, resulting in the rise of exchangeable calcium ions. The increasing calcium ion concentration causes clay particles to flocculate, converting the plastic state soil into a coarser material. The slow pozzolanic reaction produces pozzolanic materials; calcium aluminate hydrate (CAH) and calcium silicate hydrate (CSH) gel that hardens with increasing time [21]. Upon lime addition to the soil, the pore water level of pH increases, producing pozzolanic materials and forming soil-lime mixtures called calcium aluminate silicate hydrate (CASH), pozzolanic materials consist of siliceous and aluminous materials that have few or no cementitious elements, but they can produce a cementitious product in moist condition and ambient temperature [22]. Due to the highly alkaline environment, siliceous and aluminous compounds from clay minerals dissolve. Thus, CSH and CAH minerals are created due to the reaction to existing calcium ions in pore water. These minerals form cementing agent which crystallizes with time and increase the shear strength of stabilized soils [23-25].

Researchers from other countries have reported that increasing lime content decreases MDD and increases the OMC of lateritic soil [1, 3, 6-7]. Lime is a more suitable stabilizer for lateritic soil with high clay and silt content [3]. However, research by Ogundipe and Adekanmi [1] and Amu et al. [8] has shown that maximum MDD is at 6% lime and decreases beyond that percentage. Hezmi et al. [26] studied the compaction properties of tropical soil in Selangor using hydrated lime and found out that the MDD of soil treated with 5% lime content is lower than untreated, although 5% is more than the initial consumption lime. Previous studies showed different regions exhibit different soil compaction properties may be due to the different mineralogical content of lateritic soils. Therefore, this study attempts to establish the compaction characteristics of the untreated and lime-treated soils and to determine impact of various lime concentrations towards the compaction properties of laterite soil at UTM, Johor.

2. Materials and methods

The lateritic residual soil samples were obtained using a backhoe at a borrow pit (approximately 1.0 m below the ground surface) at the coordinate of 1°33'34.4"N 103°38'40.0"E which is within Universiti Teknologi Malaysia (UTM), Skudai, Johor. The soil samples were properly stored in airtight containers at the UTM geotechnical laboratory. After removing stones and plant roots, the soils were air-dried for approximately two days under environmental conditions. The samples were oven dried for three days at 60°C. Pulverized oven-dried soils were sieved through a 2mm sieve for subsequent laboratory testing such as Atterberg's Limit, specific gravity, natural moisture content, sieve analysis, and compaction test.

Quicklime was used in this research and obtained from the Lhoist (Malaysia) Sdn. Bhd. The bag of lime was carefully stored in an airtight container to avoid lime carbonation. The exposure of lime to air causes production of calcium carbonate due to the reaction with carbon dioxide [27], thereby decreasing the calcium content used for soil stabilization, indicating poor lime quality [28]. The results of the quicklime testing provided by Lhoist (Malaysia) Sdn. Bhd is shown in Table 1.

Table 1. Quicklime testing provided by Lhoist (Malaysia) Sdn. Bhd.

Test parameters	Results
CaO Available (%)	88.58
Fineness 200# (%)	91.30
T60 (min)	2.03
Tmax (°C)	72.80

The particle size distribution was determined by combining two methods: wet sieving and a Particle Size Density (PSD) Analyzer. Cohesionless to fine sand-sized soil can be sieved using the wet technique

IOP Conf. Series: Earth and Environmental Science **971** (2022) 012031 doi:10.1088/1755-1315/971/1/012031

based on BS EN ISO 17892-4: 2016 [29]. Particle Size Distribution Analyzer model LA-960V2 is automated equipment that examines the particle size of fine-grained soil or cohesive soil with particle sizes ranging from 10nm to 3mm.

To evaluate the ideal lime dose for soil stabilization, the Initial Consumption Lime (ICL) test, proposed by Eades and Grim [30] was conducted. This test is necessary to establish the optimal lime content needed to achieve a pH level of 12.4 in the lime-soil combination. The lime content was calculated as a percentage by the dry weight of the soil sample. The Modified Proctor Compaction test was based on BS EN ISO 17892-2:2014 [31] to evaluate the maximum dry density (MDD) and optimum moisture content (OMC) of natural soil and soil treated with lime. Both soil conditions were prepared immediately without curing. The quicklime concentrations of 3%, 5%, 7%, and 9% were thoroughly mixed with the soil-water mixture until the lime-soil mixture is uniformly mixed. The graph of moisture content and dry density fluctuation with lime concentration could be plotted after the moisture. Initially, the bulk density, ρ_b, in kg/m³ were computed by the following equations [32]:

$$\rho_b = \frac{M}{V} \quad (1)$$

Where M represents the mass of compacted soil and V represents the volume of the mould used. Next, the dry density, ρ_d, can be obtained from equation 1 [32]:

$$\rho_d = \frac{\rho_b}{1+w} \quad (2)$$

3. Results and discussions

Table 2 shows the engineering properties of natural lateritic soil obtained. The specific gravity indicates that the soil consisted of good lateritic material [33]. Moreover, the laterite soil is categorized as 'high plasticity' because the liquid limit is between 50% and 70% [34-35]. According to the USCS (Unified Soil Classification System) grading curve, the soil is classified as sandy elastic silt. The results of compaction test for natural lateritic soil show MDD of 1.46 g/cm³ and OMC of 28% (Figure 1).

Table 2. Engineering Properties of Lateritic Soil.

Test parameters	Results
Specific gravity	2.79
Natural moisture content (%)	53.19
Liquid limit, LL (%)	65
Plastic limit, PL (%)	46
Plasticity index, PI (%)	19
Maximum dry density (g/cm³)	1.46
Optimum moisture content (%)	28
Soil Classification	MH (sandy elastic silt)
Cu	4.29 (well-graded)
Cc	1.34 (well-graded)

IOP Conf. Series: Earth and Environmental Science **971** (2022) 012031 doi:10.1088/1755-1315/971/1/012031

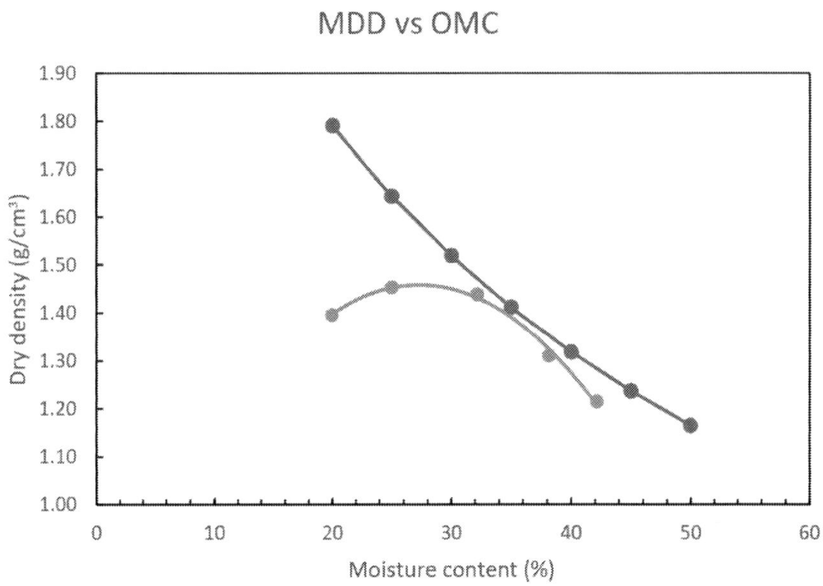

Figure 1. MDD vs OMC of natural lateritic soil.

In this research, ICL test yielded 3.4% as shown in Figure 2. For comparison, ICL value obtained by previous researchers for lateritic soil is shown in Table 3. A quicklime content of 3%, 5%, 7%, and 9% were applied to the laterite soil based on the ICL value procured. For soil modification, 2% to 3% of lime is sufficient. However, for soil stabilization, 5% to 10% of lime is needed for pozzolanic reactions [39-40]. Kassim and Chern [24] determined that the optimal lime concentration for stabilization in typical Malaysian cohesive soils is about 3% to 6%.

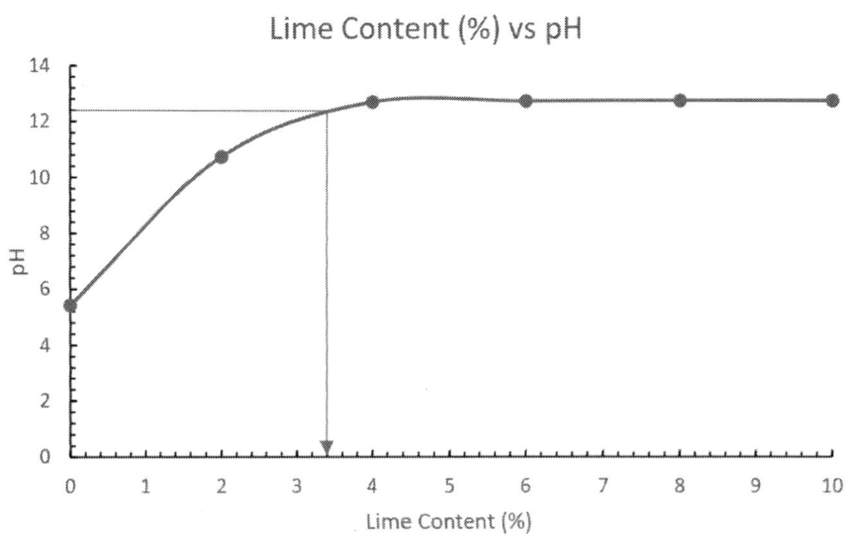

Figure 2. The Initial Consumption Lime (ICL) test result.

IOP Conf. Series: Earth and Environmental Science **971** (2022) 012031 doi:10.1088/1755-1315/971/1/012031

Table 3. Typical ICL values for lateritic soil from previous authors.

ICL value	Author
5	Eisazadeh et al. [36]
5	Saeed et al. [37]
5	Saeed et al. [38]
3.4	Hezmi et al. [26]

The natural and lime-treated lateritic soils' compaction curves are presented in Figure 3. It is shown that the curve shifted slightly to the right with increasing lime concentrations. As the lime content rises, the OMC slightly increases, whereas the MDD decreases. The MDD is lower at 5% lime than in untreated lateritic soil, which is consistent with Hezmi et al. [26]. The decrease in MDD is caused by the cation exchange. When soil is amended with lime, the soil particles flocculate and form large lumps of soil, resulting in larger pore spaces between soil aggregates, hence, leading to reduced dry density [23, 41-42]. Higher lime content decreases the frictional forces between soil particles, preventing proper soil orientation and reduces MDD [26]. For OMC, it shows as the lime content increases, the OMC increases as more water is required for hydration and cation exchange reactions. [3, 43]. Overall, the trend of dry density and moisture content with lime content is illustrated in Figure 4.

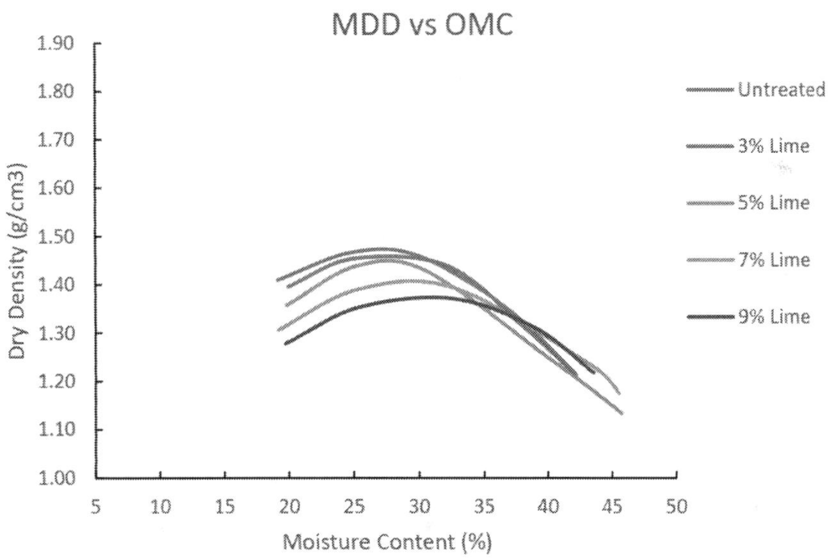

Figure 3. Compaction curve of natural and treated lateritic soil with various concentrations.

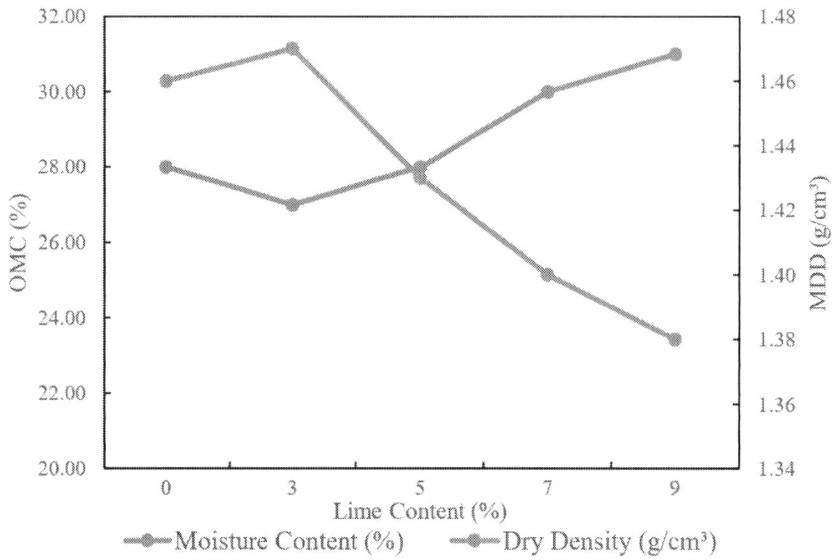

Figure 4. The trend of moisture content and dry density versus various lime concentrations.

The comparison of natural and lime-treated lateritic soil microstructure can be seen by conducting a Scanning Electron Microscopy Test (SEM) as performed by Elhassan et al. [44] in Figure 5. Large pore spaces can be seen in Figure 5 (b) in lime-treated soil whereas for untreated soil, small pores are visible Figure 5(a). Addition of lime to soil creates cementitious gels such as CSH and CASH, thus, high concentration of lime induces formation of these excessive gel materials [45]. As lime does not provide frictional nor cohesion properties, excessive lime serves as a lubricant that reduces soil strength [23]. The existence of cementitious material can be detected by X-ray diffraction (XRD) test as shown in Figure 6.

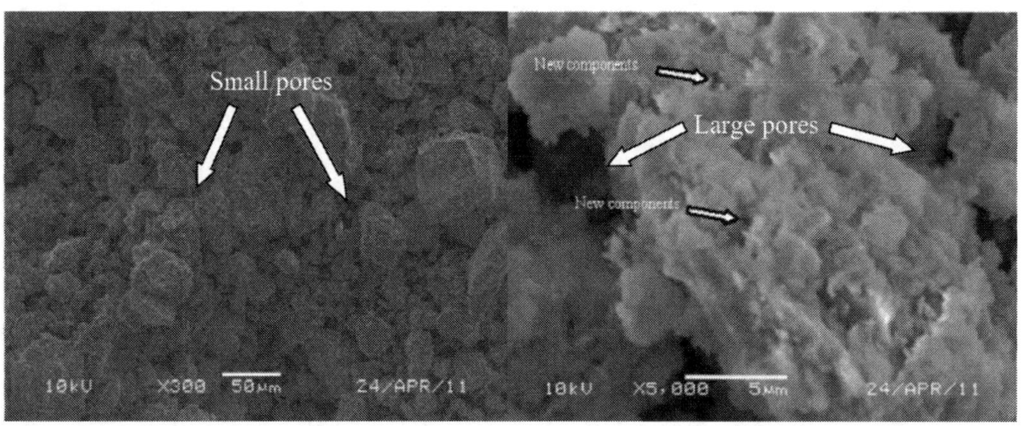

(a) (b)

Figure 5. Microstructure of (a) natural laterite and (b) lime-treated laterite [43].

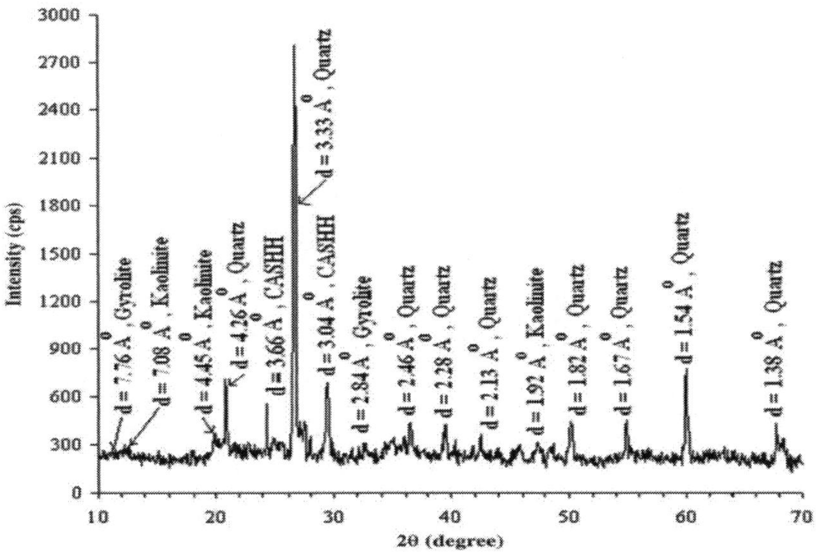

Figure 6. The pattern of X-ray diffraction test for lateritic soil treated with lime [44].

4. Conclusions

The impact of lime on lateritic soil compaction characteristics was studied. The following conclusions can be concluded from the result:

- The maximum dry density (MDD) reduces as the lime concentration rises. The addition of lime to soil forms flocculation and agglomeration of soil particles resulting in large pore spaces and reduces the dry density.
- Excessive lime reducing the frictional forces between soil particles as lime rarely has frictional properties. Lime creates cementitious material (CSH and CASH) when combined with soil. Excessive lime induces an excessive number of cementitious gels.
- The slight increase in OMC with increased lime concentration is due to higher water consumption during the lime-soil reaction to facilitate hydration and cation exchange processes.
- The MDD of soil treated with 5% lime content is lower than the untreated soil, coherent with the result studied by Hezmi et al. [26].

5. References

[1] Ogundipe O M and Adekanmi J S 2019 Assessment of the Effects of Lime and Cement on Geotechnical Properties of Laterites, *FUOYE Journal of Engineering and Technology*. **4** 2.

[2] Dumbleton M 1963 Clay mineralogy of Borneo soils in relation to their origin and their properties as road making materials *Proc. Asian Reg. Conf. Soil Mech. Found. Eng., 2nd, Tokyo* **2** 63

[3] Afolayan O D 2017 Evaluation of the Effect of Lime and Cement on the Engineering Properties of Selected Soil in a University in Southwestern Nigeria. reactions **6** 12.

[4] Horpibulsuk S, Suddeepong A, Chamket P and Chinkulkijniwat A 2013 Compaction behavior of fine-grained soils, lateritic soils and crushed rocks *Soils and foundations* **53** 166-172.

[5] Road Research Laboratory 1952 *Soil mechanics for road engineers (Vol. 1)* (HM Stationery Office)

[6] Ajayi E S 2012 Effect of Lime Variation on the Moisture Content and Dry Density of Lateritic Soil in Ilorin, Nigeria

[7] Amadi A and Okeiyi A 2017 Use of quick and hydrated lime in stabilization of lateritic soil: comparative analysis of laboratory data, *International Journal of Geo-Engineering*, **8** 1-13
[8] Amu O O, Bamisaye O F and Komolafe I A 2011 The suitability and lime stabilization requirement of some lateritic soil samples as pavement, *Int. J. Pure Appl. Sci. Technol*, **2** 29-46.
[9] Bhengu P and Allopi D 2016 Review on the lime-soil stabilization as one of the successful technologies used in the improvement of road-soil-pavement layers strength
[10] Elsharief A M, Elhassan A A and Mohamed A E 2013 Lime stabilization of tropical soils from Sudan for road construction, *International Journal of GEOMATE: Geotechnique, Construction Materials and Environment* **4** 533-538
[11] Far S Z, Kassim K, Eisazadeh A and Khari M 2013 An evaluation of the tropical soils subjected physicochemical stabilization for remote rural roads, *Procedia Engineering* **54** 817-826.
[12] Olinic T and Olinic E 2016 The effect of quicklime stabilization on soil properties, *Agriculture and agricultural science procedia* **10** 444-451
[13] Oyelami C and Van Rooy J L 2016 A review of the use of lateritic soils in the construction/development of sustainable housing in Africa: A geological perspective, *Journal of African Earth Sciences* **119** 226-237
[14] Portelinha F, Lima D, Fontes M and Carvalho C 2012 Modification of a lateritic soil with lime and cement: an economical alternative for flexible pavement layers, *Soils and Rocks, São Paulo* **35** 51-63
[15] Raheem A, Bello O and Makinde O 2010 A comparative study of cement and lime stabilized lateritic interlocking blocks, *The pacific Journal of science and Technology* **11** 27-34
[16] Saing Z, Samang L, Harianto T and Patanduk J 2016 Quick Lime Stabilization of Ferro Laterite Soil to Subgrade Material Implementation, *International Journal of Science and Research* **5** 582-586
[17] Todingrara Y T, Tjaronge M, Harianto T and Ramli M 2017 Performance of Laterite Soil Stabilized with Lime and Cement as a Road Foundation, *International Journal of Applied Engineering Research* **12** 4699-4707
[18] Joel M and Edeh, J E 2015 Comparative analysis of cement and lime modification of Ikpayongo laterite for effective and economic stabilization, *Journal of emerging trends in engineering and applied sciences* **6** 49-56
[19] Boardman D, Glendinning S and Rogers C 2001 Development of stabilisation and solidification in lime–clay mixes, *Géotechnique* **51** 533-543
[20] Rogers C and Glendinning S 1996 Modification of clay soils using lime *Proceedings of the seminar* (Loughborough University Civil & Building Engineering Department)
[21] Kassim and Huey C S 2000 Consolidation characteristics of lime stabilised soil, *Journal of Civil Engineering* **12** 31-42
[22] Sherwood P 1993 *Soil stabilization with cement and lime*
[23] Bell F 1996 Lime stabilization of clay minerals and soils, *Engineering Geology* **42** 223-237
[24] Kassim and Chern K K 2004 Lime stabilized Malaysian cohesive soils, *Malaysian Journal of Civil Engineering* **16** 13-23
[25] Salehi M and Sivakugan N 2009 Effects of Lime-Clay Modification on the Consolidation Behavior of the Dredged Mud, *Journal of Waterway, Port, Coastal, and Ocean Engineering* **135** 251-258
[26] Hezmi M A, Ahmad K, Yunus N Z M, Kassim K A, Rashid A S A and Hassan N A 2019 Compaction characteristics of lime-treated tropical soil, *IOP Conference Series: Materials Science and Engineering*
[27] Cizer Ö, Van Balen K, Van Gemert D and Elsen J 2006 Carbonation reaction of lime hydrate and hydraulic binders at 20 C, *Proc. First International Conference on Accelerated Carbonation for Environmental and Materials Engineering*

[28] Ruskulis O 2005 Methods for Testing Lime in the Field
[29] BS EN ISO 17892-4 2016 Geotechnical investigation and testing *Laboratory testing of soil Part 4: Determination of particle size distribution British Standards Institution London*
[30] Eades J L and Grim R E 1966 A quick test to determine lime requirement for soil stabilization, Highway Research Record **139**
[31] BS EN ISO 17892-2: 2014 Geotechnical Investigation and Testing *Laboratory Testing of Soil Determination of Bulk Density*
[32] Ghosh R 2013 Effect of soil moisture in the analysis of undrained shear strength of compacted clayey soil, *Journal of Civil Engineering and Construction Technology* **4** 23-31
[33] Oluyinka L G and Olubunmi O C 2018 Geotechnical properties of lateritic soil as subgrade and base material for road construction in Abeokuta, Southwest Nigeria, *International Journal of Advanced Geosciences* **6** 78-86
[34] Whitlow R 1995 *Basic Soil Mechanics, 3rd Edition* (Chicago: Longman Scientific and Technical)
[35] Wahab N A, Rashid A S A, Roshan M J, Rizal N H A, Yunus N Z M, Hezmi M A and Tadza M Y M 2021 Effects of Cement on the Compaction Properties of Lateritic Soil, *IOP Conference Series: Materials Science and Engineering* **1153** 012015
[36] Eisazadeh A, Kassim K A and Nur H 2012 Solid-state NMR and FTIR studies of lime stabilized montmorillonitic and lateritic clays, *Applied Clay Science* **67** 5-10.
[37] Saeed K A, Kassim K A, Nur H and Yunus N Z M 2015 Strength of lime-cement stabilized tropical lateritic clay contaminated by heavy metals, *KSCE Journal of Civil Engineering* **19** 887-892
[38] Saeed K A, Kassim K A, Nur H and Yunus N 2019 Comparison of Compressibility Behaviour of Lime-Cement Stabilized Lateritic Clay Soil Contaminated by Heavy Metals *IOP Conference Series: Materials Science and Engineering*
[39] Das B 1990 *Principles of Geotechnical Engineering* (Boston: PWS-KENT puplishing Company)
[40] Maher M, Marshall C, Harrison F and Baumgaertner K 2005 *Context sensitive roadway surfacing selection guide* No. FHWA-CFL/TD-05-004 (United States: Federal Highway Administration, Central Federal Lands Highway Division)
[41] Ochepo J, Ibrahim M and Joseph V 2013 Effect of oil contamination on lime and cement stabilized laterite soil, *Asian Journal of Engineering and Technology* **1**
[42] Osula D O 1991 Lime modification of problem laterite, *Engineering Geology* **30** 141-154
[43] Salahudeen A B, Eberemu A O and Osinubi K J 2014 Assessment of cement kiln dust-treated expansive soil for the construction of flexible pavements, *Geotechnical and geological engineering* **32** 923-931
[44] Elhassan A A, Elsharief A M and Mohamed A E 2013 The Reaction Products of Lime Treated Tropical Clay Soils and Their Impact on Strength Development
[45] Dash S K and Hussain M 2012 Lime stabilization of soils: reappraisal, *Journal of materials in civil engineering* **24** 707-714

Acknowledgments

The author would like to thank Lhoist (M) Sdn. Bhd. for providing lime for this study, as well as the Malaysian Ministry of Education for funding through the Fundamental Research Grant Scheme (FRGS) (CRG 24.3: Compressibility Characteristics of Lateritic Soil Treated with Lime Under Cyclic Wetting and Drying Conditions) (R.J130000.7351.4B505).

IOP Conf. Series: Earth and Environmental Science **971** (2022) 012032 doi:10.1088/1755-1315/971/1/012032

Assessment on Recent Landslide Susceptibility Mapping Methods: A Review

W A A Manan[1*], A S A Rashid[1], M Z A Abdul Rahman[2] and M F A Khanan[2]
[1] Department of Geotechnics and Transportation, School of Civil Engineering, Universiti Teknologi Malaysia, Johor Bahru 81310, Malaysia;
[2] Department of Geoinformatics, Faculty of Built Environment and Surveying, Universiti Teknologi Malaysia, Johor Bahru 81310, Malaysia
*Corresponding author: wardatunahmar@graduate.utm.my

Abstract. Landslide is a destructive natural hazard that causes severe property loss and loss of lives. Numerous researchers have developed landslide susceptibility maps in order to forecast its occurrence, particularly in hill-site development. Various quantitative approaches are used in landslide susceptibility map production, which can be classified into three categories; statistical data mining, machine learning and deterministic approach. In this paper, we choose two regular models in each category, which are Weight of Evidence (WoE) and Frequency Ratio (FR), Artificial Neutral Networks (ANN) and Support Vector Machines (SVM), Shallow Landsliding Stability Model (SHALSTAB) and YonSei-Slope (YS-Slope). Discussion and assessment on these models are based on relevant literature.

1. Introduction

Landslide is a natural disaster that causes extensive property damages and occasionally results in loss of life, especially in a hilly area. World Health Organisation (WHO) reported that landslides affected an estimated 4.8 million people and caused more than 18 000 deaths between 1998-2017. While in the United States, Japan, Italy, and India estimated the annual losses due to landslides disaster, at least USD1 billion for each country [1]. Due to its complexity, uncontrollability, and destructiveness, landslide poses a formidable challenge to us. In view of this, numerous studies have been conducted to forecast the occurrence of landslides in a specific location. Landslide prediction and mitigation techniques are essential for minimising loss of life and property.

Landslide susceptibility mapping (LSM) is an essential and widely used procedure for the spatial prediction of landslides in many countries. Typically, it is used to map a region's landslide hazard, and it can be classified as qualitative or quantitative [2]. [3], propose that a landslide susceptibility map can be performed either qualitatively or quantitatively depending on the availability and accuracy of data, required outcome and scale of analysis. The qualitative method is a relatively subjective approach that demonstrates landslide's susceptibility levels through descriptive expressions based on expert judgement ([4]- [5].

On the other hand, quantitative models quantify the relationship between slope instability and other controlling factors numerically. Generally, the quantitative analyses can be divided into statistical data mining and deterministic approach [4], which are frequently used in previous landslide susceptibility studies [6]-[7]. The deterministic approach mainly measures the factors of safety, while the statistical method mainly focuses on the correlation of data between the historical landslide distribution and the controlling parameters [8]. In the last three decades, machine learning approaches have been applied in various fields and landslides susceptibility analysis due to their superior performance and ability to deal with complex and varying data.

Therefore, this article comprehensively reviews statistical data mining, deterministic and machine learning approaches. Two representative models from each category which Weight of Evidence (WoE), Frequency Ratio (FR), Artificial Neutral Networks (ANN), Support Vector Machines (SVM), Shallow Landsliding Stability Model (SHALSTAB) and Stability Index Mapping (SINMAP) were discussed in details to analyse and compare the predictive power of landslide susceptibility analysis.

2. Landslide susceptibility mapping (LSM)

Susceptibility to landslides is a term that refers to the likelihood or probability of a landslide occurring in a given area. Thus, landslide susceptibility maps have become a popular method of illustrating where future landslides are likely to occur based on landslide factors. Landslide susceptibility can be assessed in various ways, depending on the available data [2]. However, a landslide susceptibility map can't be produced with limited data, but the most necessary depending on the quality of the data [9] [10].

Based on the previous case studies, identified factors that influence the landslide include slope angle, curvature, aspect, soil type, geology, distance to a river, and drainage. Numerous researchers have agreed that the slope angle and aspect are the most critical variables in landslide spatial analysis ([6], [11]-[12]. The possible explanation for this is that the slope angle is the most significant slope stability analysis parameter with direct effects on shear strength [13].

Susceptibility is a term that refers to the degree to which future slope movements are able to influence terrain [14]. [4] have suggested that it is necessary to answer basic questions such as where what, and how the landslide occurred to define the landslide susceptibility in the studied area. The landslide susceptibility map classifies slope conditions as stable or unstable, indicating the likelihood of a landslide. Besides, it can also be considered an expression of relative hazard, total landslide density of likely frequency [15].

3. The statistical data mining approach

Statistical data mining techniques are applied in LSM to reduce subjectivity in using qualitative approaches. The statistical approach uses GIS tools to assess the spatial distribution of existing landslides with the spatial distribution of various causative factors [4]. Generally, it can be classified into two types: bivariate and multivariate. This article reviews the two models in bivariate analysis: weight of evidence (WoE) and Frequency Ratio (FR).

Bivariate analysis analyses two variables that determine the correlation between a dependent variable and an independent variable [1]. Various types of bivariate models are most commonly applied in previous studies to produce landslide susceptibility mappings. They are Frequency Ratio (FR), Fuzzy Logic (FL), Weight of Evidence (WoE), Statistical Index (Si), Weighted Overlay Model (WOM) and Weighting Factor (Wf). [16]-[17]. The area under the curve (AUC) validates a model's accuracy throughout the model development. Devkota [18] have classified AUC value in four conditions; less than 0.5 is considered as no ability to evaluate, between 0.7 and 0.8 indicates acceptable interpretability, between 0.8 and 0.9 is excellent, and outstanding if it is more than 1.0. AUROC curves are among the most widely used tools in models for evaluating and assessing landslide susceptibility according to the works of literature [19]. Based on Chen *et al.* [20], the AUC can be calculated as in the below equation (1).

$$AUC = \frac{(\sum TP + \sum TN)}{(P + N)} \tag{1}$$

Where P is the total number of landslides, N is the total number of non-landslides, TP is the true positive and TN is the true negative.

3.1 *Weight of Evidence (WoE)*

According to Bayes' theorem, the basis of Weight of Evidence (WoE) is conditional probability. WoE method has been applied to identify the spatial association between the location of a landslide and a set of causal factors [21]. Most of the previous studies [16] [17][22] [23] have been conducted using WoE by focussing on the theoretical background and application. Weight of Evidence (WoE) is derived from the density of observed landslide locations in each class of parameters, referred to as "evidence," such as slope, lithology, and aspect [24].

The spatial correlation between the causal factor's class and the occurrence of a landslide can be determined using the magnitude of the weight contrast. The landslide susceptibility index (LSI) is calculated by adding the contrasts of all the parameters factors. The greater the landslide susceptibility index values are, the higher the probability of landslide to occur.

3.2 *Frequency Ratio Model*

The frequency ratio (FR) approach has been utilised to identify the correlation between landslide distribution prone areas and each causal factor that induces landslides [22]. According to Pradhan *et al.* [25], the following formula in equation (2) has been used to compute the FR value based on the correlation between the variables equation.

$$FR = P_{LO}/P_{LF} \tag{2}$$

P_{LO} is the percentage of landslide-prone area in each factor subcategory and P_{LF} is the percentage of each factor subcategory that influences the landslide.

The landslide susceptibility index (LSI) has been calculated using the following equation (3) by adding all of the FR values. According to [23], the higher the LSI value is, produced, the greater the tendency will be for a landslide to occur.

$$LSI = \sum_{i=1}^{n} FR_i \tag{3}$$

Where n value is refer to the number of influencing factors.

4. Machine learning approach

Machine learning (ML) is the computer programming that used past experience or samples to optimise the performance criterion by learning from the sample [26]. Numerous studies in the different geotechnical fields had used machine learning as an effectively geotechnical application tool in their study [27]–[37]. Machine learning has also gained popularity in producing landslide susceptibility mapping (LSM) as it has evolved, and various types of ML algorithms have been used to produce LSM [6] [50]-[75].

LSM that produced using ML used the main concept: the correlation between a set of landslide conditioning factors and a past data set of the landslide event were evaluated based on the ML algorithm. This was done to determine their spatial connection to predict the potential landslide event in the study area. B. Pradhan *et al.* [65] has clarified that spatial prediction of a landslide is more efficient to be conducted using ML approach than other methods, for instance, analytic method or expert's opinion method. This article then presents further discussions on two types of ML frequently used in producing LSM. They are Artificial Neural Network (ANN) and Support Vector Machines (SVM).

4.1 *Artificial Neuro Network*

The artificial neuron network (ANN) is a well-organised mathematical model that is widely used in many fields and this model was created based on the mechanism of the human brain to solve the problem or replicate the human learning process represented by computerised programming [66]–[68]. The ANN has been used as modelling of prediction method in the geotechnical area, particularly in landslide study, to develop good quality LSM or landslide hazard maps (LHM) [80][81]. The ANN has been effectively used to assess landslide susceptibly in many studies. The selection of the ANN as the assessment tool for landslide susceptibility is due to the ability of this model to learn and analyse a set of landslide causal factors based on a non-linear numerical model [77] [81] [82].

Furthermore, ANN can work on the problem that the statical method is unable to resolve due to the limitation of the theory or data [68]. Besides, the ANN model does not require a specific parameter to develop and assess the landslide susceptibility since it is independent and able to manage vague and uncertain data, making this model suitable to be performed using continuous, categorical and binary data without breaching any assumptions [79] [82].

There are various types of ANN, and this article only focuses on the feed-forward multilayer perceptron's (MLP) with backpropagation learning algorithms as most previous literature used this type of ANN modelling. The structure of the ANN-MLP model consists of three main components: the input, hidden, and output layer. The neuron in the input layer is only connected to the hidden layer as well as the neuron in the hidden layer is connected to the output layer, which means that interconnection of the neuron to others layer does not exist [72] [73] [74].

Figure 1 shows an example of a simple ANN-MLP architecture design that is commonly used in LSM. The development of ANN model is commonly assessed with the area under the ROC curve to determine its performance and efficiency. Hosmer and Lemeshow [75] describe the criteria for rejecting or accepting the area under the curve value (AUC) for the assessed model. If the AUC of the model assessment is less than 0.50, it is considered as fail, between 0.70 and 0.80 is acceptable, between 0.80 and 0.90 is excellent, and between 0.90 and 1.00 is outstanding.

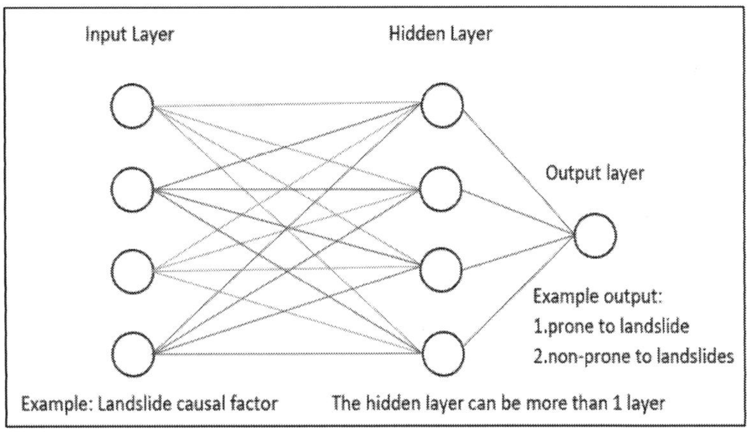

Figure 1. ANN-MLP architecture design.

4.2 *Support Vector Machines (SVM)*

Many studies have been conducted to examine the performance of SVM to produce LSMs either as a single method, hybrid or to compare it with other approaches [66]–[69] [76-89]. Saro *et al.* [90] claim that SVM produces high performance with a minimum setting requirement, making it one of the common ML approaches used to produce LSM. While SVM is categorised as a supervised machine learning method that complements the structural risk minimisation principle [82] [90].

Nuriah *et al.* [78] highlight that SVM can search for an optimal separating hyperplane to map the dimensional future space based on original input space. In other words, the goal of the SVM model during developing LSM is to pursue the optimal separating hyperplane that can differentiate between landslide susceptible area and landslide non-susceptible area. Pradhan and Lee [46] remark that the SVM model is developed based on two key principles: the optimal classification hyperplane and the use of a kernel function.

The basic principle of SVM can be described in figure 2. Dots and squares represent two classes of different data. H1 and H2 are located parallel to the H line, which is defined as their classification line. Those lines move over the data points closest to the classification line, which is the support vectors that create a gap between that two-line known as classification margin [80]. For a given hyperplane, the margin is considered to have the shortest distance from each class's data vector, indicating that the accuracy of classification by the SVM model increases toward the increase of the classification margin [41]. The kernel function role in SVM is to facilitate the input data transformation to the higher dimensional space for it to be linearly classified, as being illustrated in figure 2(b). Thus, the type of kernel function used in SVM has a major influence on the SVM performance, and its selection is crucial [81].

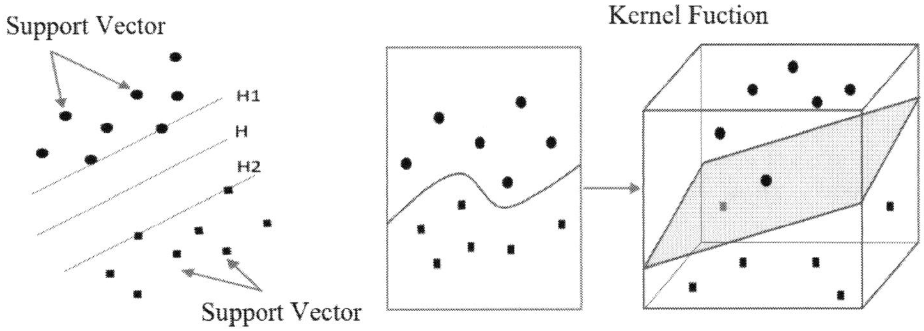

(a) (b)

Figure 2. The principle of SVM (a) Dimensional hyperplane differentiating the two classes by the maximum gap; (b) non-separable case and the slack variables [76]

4.3 *Performance of ANN and SVM*

Many previous studies have used ANN as the sole model to produce LSMs. Yao *et al.* [65] conducted a study in Vaz Watershed, Iran, to evaluate the performance of the ANN model with the two separate algorithms, which was in Multi-Layered Perceptron (MLP) and Radial Base Function ((RBF) in producing LSMs. This study was conducted based on nine causal factors and 136 landslide inventory data. The study showed that both MLP and RBF produced an excellent performance, which was more than 80 per cent, but MLP performance was slightly higher in training and validating datasets. Another study by Lo and Leung [39] used only 34 landslides inventory data with 5 causal factors to produce LSMs using MLP and Probabilistic Neural Network (PNN). Again, in this study, the MLP model showed a higher performance compare to PNN. However, the model's performance in this study was slightly poor but still accepted as the MLP and PNN produce 73 per cent and 68 per cent validation performance, respectively. The reduction in the performance of the ANN-MLP model by Lo and Leung [39] compared to Yao *et al.* [65] may reflect the influence landslide causative factor that was used. Moreover, the study by Lo and Luang [39] validated only with past landslide inventory data, while the study by Zare et al. (2013) used the area under curve (UAC) method to validate the model.

Besides, Ortiz and Martínez-Graña (2018) conducted a study at Capitanejo, Colombia, choosing 14 landslide causal factors and 54 landslide inventory data using MLP-ANN model to develop LSMs. This study found that the MLP-ANN model produced an outstanding LSMs performance as the AUC value for tanning and the validation data set were 98.8 per cent and 92.86 per cent. Other researchers [58] [50] [71] also used MLP-ANN model to develop LSMs in the various study area and found that the model produced an excellent performance as the assessment of the LSMs using the AUC method produced a result of more than 80 per cent both for training and validation data set. However, the study by Lee *et al.* [82] and Hong *et al.* [70] used the same approach to produce a slightly lower performance, which was between 71 and 76 per cent. This may be due to the different validation methods used in this study, which may affect their performance. The LSMs provided by Sharir *et al.*[82] only validated the data set through ANN modelling, while Hong *et al.* [70] validated using the scoring wizard method.

Su *et al.* [83] used SVM as the only model in their study at Wencheng, China. In his study, 354 landslide data was used, with 7 landslide causative and 2 triggering factors being considered. The uniqueness of this study was that the data set of this model did not divide training and validation set as in the usual practice in the ML method. As an alternative, fivefold cross-validation was used to remove the impact of the production on the training dataset. This study found that the LSMs produced an excellent performance with an AUC value of 96 per cent.

Another study by Saro *et al.* [90] also used SVM only to produce LSM but in two different study areas: Pyeong Chang and Inje areas of Gangwon Province, Korea. The selection of landslide causative factor remained the same for both areas except slope length only considered at Inje while land used only considered at Pyeong Chang. The result showed that the LSMs for the Peong Chang area had slightly higher performance than Inje, with the AUC value of 81.36 per cent and 77.39 per cent respectively. The study area may be subject to the change in a spatial distribution corresponding to an event in that area, leading to slight differences in the model performance. The sensitivity analysis was conducted in this study to determine the high impact causative factor to the LSMs. The study found that factors that related to the topography of the area such as SPI, TWI, slope, aspect, and slope length had a strong influence on the LSMs performance.

W. Chen *et al.* [89] and Feizizadeh *et al.* [95] conducted a study to compare the performance of SVM with the four different kernel functions which were LN,PL, RBF and SIG to produce LSMs. The result of both studies showed that the SVM-SIG model had the lowest performance and SVM-PL had the second-highest performance for both studies. The main difference between those two studies was the SVM-RBF produced the most outstanding performance of AUC value in a study by W. Chen *et al.* [89] while in the study Feizizadeh *et al.* [95] the SVM-LN had the highest performance. The difference in the model performance in those studies may be due to the difference in a causative factor employed in the study, which may affect the effectiveness of four different kernel functions of SVM in producing LSMs. Almost all the SVM models that are discussed above have produced an AUC value of more than 80 per cent which indicates that the SVM model has a good performance in producing LSMs.

5. Deterministic approach

A deterministic approach to landslide susceptibility mapping can be identified in geotechnical fields. The deterministic approach includes the interaction between hydrology, topography, soil properties, and in some cases, vegetation to understand and predict the location and timing of occurrences of landslides. Mostly, models generated from the deterministic approach generally consist of a methodology that combines the hydrological model for analysing the pore-water pressure and the infinite slope-stability model for computing its safety factors such as Stability Index Mapping, SINMAP [85], Transient Rainfall Infiltration and Grid-based Regional Slope-stability Model, TRIGRS [86], Shallow Slope Stability Model, SHALSTAB [87] and YonSei-Slope, YS-Slope Model [88].

The fundamental of these models is the combination of hydrological and geotechnical approaches based on slope stability analysis. The discrepancies between models are mostly due to the different models used in failure criterion interpretation and hydro-geotechnical consideration. The YS-Slope and TRIGRS models, for example, estimated the Factor of Safety (FS) in unsaturated soil conditions, but SHALSTAB and SINMAP used saturated soil conditions. Although both the YS-Slope and the TRIGRS incorporate water infiltration into the subsurface, the YS-Slope Model is more advanced in addressing groundwater recharge conditions, which can result in deep seated failure in geotechnical assessments.

5.1 *Stability index mapping (SINMAP)*

SINMAP was established by Pack *et al.* [85] to provide terrain stability mapping tools that could complement stability mapping methods currently that were practised in Columbia at the time. Theoretically, this method is applicable for a shallow translational landslide. This model improved the stability mapping methods using Grid-based DEMs configured free extension to a GIS platform so that analysis is easy to use and widely available. In this model, some parameters are derived from the Digital

Elevation Model (DEM). These can be slope angle, flow direction, and catchment area. In contrast, the remaining parameter, especially geotechnical data, needs to be set manually or it can straight away use the default values proposed by the authors. This model is decent in which the cohesion value either for soil or root strength is considered in the infinite slope stability model and can be set to zero for the cohesionless situation. Simultaneously, this model incorporates parameter uncertainty through uniform probability distributions with lower and upper bounds on uncertain parameters.

The SINMAP model calculates the stability index (SI), which refer to the probability of a specific location in terms of stability, assuming uniform distributions of the parameters over these uncertainty ranges in each grid cell of the map. The value ranges between 0 (most unstable) and 1 (least unstable). Table 1 and 2 shows the range of parameters and SI based on stability classes. The SI is a dimensionless form of the infinite slope stability model, which calculates the factor of safety assuming the wet and dry density is equally based on Hammond *et al.* [89]. :

$$\mathrm{SI} = \frac{C_r + C_s + \cos^2\theta[\rho_s g(D - D_w) + (\rho_s g - \rho_w g)D_w]\tan\phi}{D\rho_s g \sin\theta\cos\theta}, \quad (4)$$

Where Cr is root cohesion, Cs is soil cohesion, h is slope angle, qs is the wet soil density, qw is the density of water, D is the vertical soil depth, Dw is the vertical height of the water table within the soil layer, and / is the internal friction angle of the soil.

Table 1. Parameters and ranges.

Parameters	Range
Transmissivity/effective recharge	2000 to 3000 m
Dimensionless cohesion	0 to 0.25
Internal friction angle	30 to 45°
Soil density.	2000 kg/m^3

Table 2. Stability Index (SI) based on stability classes.

Condition	Class	Prediction slope Zone
SI > 1.5	1	Stable
1.5 > SI > 1.25	2	Moderate
1.25 > SI > 1.0	3	Quasy-stable
1.0 > SI > 0.5	4	Lower threshold
0.5 > SI > 0.0	5	Upper threshold
0.0 > SI	6	Defended

5.2 *Yonsei Slope Model, YS-Slope*

YS-Slope model is a GIS-based physical landslide prediction model developed by Kim *et. al.* [88]. This model considers rainfall infiltration depth, recharge, and groundwater flows, whereas the SINMAP model did not include this component in their model. This model was developed using GIS-based raster data and input parameters. It consists of internal and external parameters related to the slope's stability, as shown in Figure 3. This model improved on the hydro-geotechnical approach by incorporating unsaturated soil parameters such as the soil-water characteristic curve and matric suction, as well as internal friction angle and cohesion value is obtained from soil strength properties. The infinite slope failure model was improved in order to compute the factor of safety (FS) of rainfall-induced landslides by taking into account the other influential component in increasing soil strength and interception loss due to plant growth, as shown in the equation below:

$$FS = \frac{(c_s' + c_r') + [(\gamma_{sat} - \gamma_w)(Z_w + D_w) + \gamma_t \cdot (D - Z_w - D_w) + q_0]cos^2\beta \cdot \tan\emptyset'}{[\gamma_{sat} \cdot (Z_w + D_w) + \gamma_t \cdot (D - Z_w - D_w) + q_0] \cdot \sin\beta \cdot \cos\beta} \quad (5)$$

Where c_s' is soil cohesion, c_r' is the shear strength increase by root reinforcement, γ_{sat} is the saturated unit weight of soils, γ_w is the unit weight of water, γ_t is the total unit weight of soils, q_0 is the forest tree load, Z_w is wetting front depth, D_w is groundwater table from the bedrock, D is the thickness of dry soil, $\emptyset'$ is internal friction angle of soil and β is slope inclination.

The modified Green-Ampt model by Mein and Larson [91] was used in hydrological evaluation to analyse rainfall infiltration, and the wetting front depth was estimated based on this analysis. The cumulative infiltration and runoff are computed in transient rainfall-infiltration analysis by comparing the infiltration capacity with the rainfall intensity.

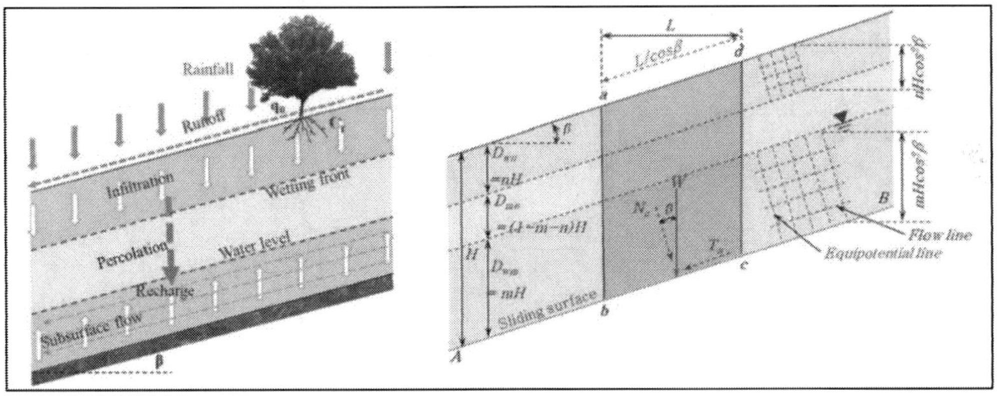

Figure 3. Infinite slope plane for slope stability analysis in YS-Slope model.

5.3 *Performance of SINMAP and YS Slope*

The YS-Slope model has purposed to analyse the rainfall-induced landslides in the spatial based predicting model. Jeong *et al.* [92] used this model to analyse rainfall-induced landslides and found that these procedures can be applied to predicting landslides and debris flows. furthermore, Hong *et al.* [93] propose a numerical method to simulate landslides and debris flows by combining rainfall infiltration analysis and the YS-Slope model in their study. While Samseong *et al.* [92] employed the YS-slope Model and prediction approach to developing a unified prediction method to examine the susceptibility map of landslides at Umyeonsan Mountain in Korea.

Similar to the SINMAP model, the YS-Slope model also can be used for shallow landslide studies. However, YS-Slope can be further used for deep-seated landslides prediction. Jeong *et al.* [92] prove that YS-slope can predict shallow landslides and deep-seated landslides and reasonably agree with historical landslides. Besides, the researcher confirmed with ROC analysis and showed that relatively high accuracy compared to other predictions.

Duc Long and Dung [94] used the YS-Slope model by incorporating the unsaturated soil properties to develop the LSM. The assessment of landslide susceptibility is based on the contributions related to spatial data such as elevation, slope angle and groundwater table spatial distribution map and input data corresponding to unsaturated permeability function, rainfall intensity, soil-water characteristic curve (SWCC). They found that incorporating the unsaturated soil characteristic becomes more realistic in predicting the landslide susceptibility area in terms of factor of safety.

Furthermore, Kim *et al.* [95] investigate the effectiveness of YS-Slope by coupling the effects of hydro-mechanical processes on the infiltration behaviour of unsaturated soils. A few sets of infiltration analyses for a soil column were carried out under varied soil conditions, and the results were compared using the software GEO-SLOPE [96]. The research reveals that the volume change of the soil influences the transient seepage analyses in deformable soils. It is relevant that whenever the matric suction on a slope changes due to rainfall infiltration, consequently changes the effective stress that directly relates to seepage processes. Therefore, the infiltration rate of the YS-Slope is slower than the Geo-slope due to permeability function based on the porosity that altered the matric suction distribution. Thus, the coupled effects of hydro-mechanical behaviour of soils considered in the YS-slope model has a positive effect on unsaturated soil stability, especially during rainfall.

6. Conclusion

Statistical data mining, machine learning, and deterministic approach are widely used in LSM. The statistical approach is the simplest way to compute LSM compared to machine learning and the deterministic approach. Machine learning needs knowledge of computer programming, while deterministic approaches are more toward needing geotechnical input. The advantages and disadvantages of each approach were listed in Table 3 and summaries of statistical, machine learning, and deterministic approaches were discussed in Table 4. Based on previous research works, each approach can yield accurate results in determining the landslide-prone area based on qualified data representing the condition of the study area, scale, and method of data acquisition. In other words, descriptive, reliable and easily accessible input parameters should be selected as the parameter used in a study. Due to this condition, it is necessary to map the landslide-prone areas by forecasting and specifying the region where reliable and accurate landslide susceptibility maps can be helpful for land planners, decision-makers, and risk assessment personnel.

Table 3. The advantages and disadvantages of each approach.

Type of Approach	Advantages	Disadvantages
Statistical Approach	• Simple and easily implemented in GIS using remote sensing inputs. • Minimum data required	• loss of data quality and accuracy with the oversimplification of thematic input data, and loss of data sensitivity in forced individual analysis of causative factors • basically in linear based, so they were inefficient in capturing the nonlinearity and complexity of landslide phenomena.
Machine Learning Approach	• Can map out the detailed occurrence of past landslides • To collect sufficient information on the variables that are the occurrence of landslides	• Difficult to prepare data and need a longer time. • Just susceptibility assessment • Not readily be extrapolated to the neighbouring area
Deterministic Approach	• Able to quantitatively calculate the safety factor. • Externally available models can be used instead of spending time implementing model algorithms in GIS. • Encourage the detailed investigation and analysis of geotechnical parameters.	• The data requirements can be prohibitive, and it is usually difficult to procure the necessary input data to be used in the models effectively.

Table 4. Summaries of statistical, machine learning, and deterministic approaches.

IOP Conf. Series: Earth and Environmental Science **971** (2022) 012032 doi:10.1088/1755-1315/971/1/012032

Weight of Evident (WoE) and Frequency Ratio (FR)	Artificial Neural Network (ANN) and Support Vector Machine (SVM)	SINMAP and YS-Slope
• Statistical approaches generally require the collection of a large amount of data. They are particularly useful for medium-scale (regional) mapping. • is used to validate a model's accuracy throughout the model development • WoE appears to be efficient but has limitations - WoE model assessment had a lower performance validation result.	• ANN and SVM can yield accurate results in determining landslide-prone • SVM model outperformed other models of ML in most of the study • unique features in SVM - more efficient such as robustness to noise, non-linear decision boundaries, easily implementable probabilistic outcome, and an inherent ability to deal with high dimensional classification problems.	• these models have covered a shallow translational landslide controlled by groundwater convergence • SINMAP model greatly limits the results, which should only be treated as a general susceptibility indicator, not a true safety factor (FS) value. • The failure criterion used to derive the FS equation in YS-Slope consider unsaturation condition, while SINMAP focus on the saturated condition

Acknowledgments
This research proposal was sponsored by the UTM High Impact Research Grant awarded by UTM R&D FUND Research (Development of UTM Landslide Susceptibility Map -Q.J130000.3651.03M35)

7. References

[1] H Gao, P S Fam, H C Low, L T Tay, and H Lateh 2019 An overview and comparison on recent landslide susceptibility mapping methods Disaster Adv. **12** 46–64

[2] F Guzzetti, A Carrara, M Cardinali and P Reichenbach 1999 Landslide hazard evaluation: a review of current techniques and their application in a multi-scale study, Central Italy *Geomorphology* **31** 181-216

[3] M Ciurleo, L Cascini and M Calvello 2017 A comparison of statistical and deterministic methods for shallow landslide susceptibility zoning in clayey soils *Eng. Geol.* **223** 71–81

[4] P Aleotti and R Chowdhury 1999 Landslide hazard assessment: Summary review and new perspectives *Bull. Eng. Geol. Environ.* **58** 21–44

[5] M S Sulaiman, A Nazaruddin, N M Salleh, R Z Abidin, N D Miniandi and A H Yusoff 2019 Landslide occurrences in Malaysia based on soil series and lithology factors *Int. J. Adv. Sci. Technol.* **28** 1–26.

[6] A D Regmi *et al.* 2014 Application of frequency ratio, statistical index, and weights-of-evidence models and their comparison in landslide susceptibility mapping in Central Nepal Himalaya *Arab. J. Geosci.* **7** 725–742

[7] L Borrelli, M Ciurleo and G Gullà 2018 Shallow landslide susceptibility assessment in granitic rocks using GIS-based statistical methods: the contribution of the weathering grade map *Landslides* **15** 1127–1142

[8] N Saadatkhah, A Kassim and L M. Lee 2014 Qualitative and quantitative landslide susceptibility assessments in Hulu Kelang area, Malaysia *Electron. J. Geotech. Eng.* **19** 545–563

[9] R J Fabbri, A G Chung and C F Cendrero A 2003 Is Prediction of Future Landslides Possible with a GIS? *Nat Hazards* **30** 487–499

[10] J Coe, J Godt, R Baum, R Bucknam and J Michael 2004 Landslide susceptibility from topography in Guatemala *Landslides Eval. Stab. Terrain Eval. Stabilisation* **1** 69–78

[11] M Teixeira, C Bateira, F Marques and B Vieira 2015 Physically based shallow translational landslide susceptibility analysis in Tibo catchment, NW of Portugal *Landslides*

[12] B C Vieira, N F Fernandes, O Augusto Filho, T D Martins and D R Montgomery 2018 Assessing shallow landslide hazards using the TRIGRS and SHALSTAB models, Serra do Mar, Brazil *Environ. Earth Sci.* **77** 260-274

[13] Q Q Wang, Y Guo, W Li, J He and Z Wu 2019 Predictive modeling of landslide hazards in Wen County, northwestern China based on information value, weights-of-evidence, and certainty factor *Geomatics, Nat. Hazards Risk* **10** 820–835

[14] P Reichenbach, M Rossi, B D Malamud, M Mihir, and F Guzzetti 2016 A review of statistically-based landslide susceptibility models *Earth-Science Rev.* **180** 60–91[15] J Chacón, C Irigaray, T Fernández and R El Hamdouni 2006 Engineering geology maps: Landslides and geographical information systems *Bull. Eng. Geol. Environ.* **65** 341–411

[16] V Vakhshoori and M Zare 2016 Landslide susceptibility mapping by comparing weight of evidence, fuzzy logic, and frequency ratio methods *Geomatics, Nat. Hazards Risk* **7** 1731–1752

[17] S Pasang and P Kubíček 2020 Landslide susceptibility mapping using statistical methods along the asian highway, bhutan *Geosci.* **10** 1–26

[18] K C Devkota *2013* Landslide susceptibility mapping using certainty factor, index of entropy and logistic regression models in GIS and their comparison at Mugling-Narayanghat road section in Nepal Himalaya *Nat. Hazards* **65** 135–165

[19] J L Zêzere, S Pereira, R Melo, S C Oliveira and R A C Garcia2017 Mapping landslide susceptibility using data-driven methods *Sci. Total Environ.* **589** 250–267

[20] W Chen 2018 Landslide susceptibility modeling based on GIS and Novel Bagging-based Kernel logistic regression *Appl. Sci.* **8** 12

[21] T Kavzoglu, E K Sahin, and I Colkesen 2014 Landslide susceptibility mapping using GIS-based multi-criteria decision analysis, support vector machines, and logistic regression *Landslides* **11** 425–439

[22] E Nohani 2019 Landslide susceptibility mapping using different GIS-Based bivariate models *Water (Switzerland)* **11** 1–22
[23] A Ozdemir and T Altural 2013 A comparative study of frequency ratio, weights of evidence and logistic regression methods for landslide susceptibility mapping: Sultan mountains, SW Turkey *J. Asian Earth Sci.* **64** 180–197
[24] I Armaş 2012 Weights of evidence method for landslide susceptibility mapping. Prahova Subcarpathians, Romania *Nat. Hazards* **60** 937–950
[25] B Pradhan, L Saro, and M F Buchroithner 2010 A GIS-based back-propagation neural network model and its cross-application and validation for landslide susceptibility analyses *Comput. Environ. Urban Syst.* **34** 216–235
[26] R. T. Sataloff, M. M. Johns, and K. M. Kost 2010 Introduction To Machine learning, 2nd Edition *The MIT Press*
[27] M K Lo and Y F Leung 2019 Bayesian updating of subsurface spatial variability for improved prediction of braced excavation response *Can. Geotech. J.* **56** 1169–1183
[28] I. Papaioannou and D Straub 2017 Learning soil parameters and updating geotechnical reliability estimates under spatial variability – theory and application to shallow foundations *Georisk Assess. Manag. Risk Eng. Syst. Geohazards* **11** 116–128
[29] P Xiong 2021 Towards advancing the earthquake forecasting by machine learning of satellite data *Sci. Total Environ.* **771** 145256
[30] D J Lary, A H Alavi, A H Gandomi and A L Walker 2016 Machine learning in geosciences and remote sensing *Geosci. Front.* **7** 3–10
[31] W W Zhang, A T C Goh, Y Zhang, Y Chen, and Y Xiao 2015 Assessment of soil liquefaction based on capacity energy concept and multivariate adaptive regression splines *Eng. Geol.* **188** 29– 37
[32] X Liang, H He and Y Zhang 2019 Optimization design of micro-piles in landslide safety protection based on machine learning *Saf. Sci.* **118** 861–867
[33] M Zhang, M Li, J Zhang, L Liu, and H Li 2019 Onset detection of ultrasonic signals for the testing of concrete foundation piles by coupled continuous wavelet transform and machine learning algorithms *Adv. Eng. Informatics* **43** 10-15
[34] M Mittal, S C Satapathy, V Pal, B Agarwal, L M Goyal and P Parwekar 2021 Prediction of coefficient of consolidation in soil using machine learning techniques *Microprocess. Microsyst.* **82** 103830
[35] Y Bouslihim, A Rochdi, N El, and A Paaza 2021 Machine learning approaches for the prediction of soil aggregate stability *Heliyon* 7
[36] L Zhang, L Yang, T Ma, F Shen, Y Cai and C Zhou 2020 A self-training semi-supervised machine learning method for predictive mapping of soil classes with limited sample data *Geoderma*, **384** 114809
[37] R Jena 2020 Earthquake hazard and risk assessment using machine learning approaches at Palu, Indonesia *Sci. Total Environ.* **749** 141582
[38] S Lee, J H Ryu, J S Won and H J Park 2004 Determination and application of the weights for landslide susceptibility mapping using an artificial neural network *Eng. Geol.* **71** 289–302
[39] L Ermini, F Catani and N Casagli 2005 Artificial Neural Networks applied to landslide susceptibility assessment *Geomorphology* **66** 327-343
[40] A M S Pradhan and Y T Kim 2016 Landslide susceptibility mapping of Phewa catchment using multilayer perceptron artificial neural network *Nepal J. Environ. Sci.* **4** 1–9
[41] S Lee, M J Lee and H S Jung 2017 Data mining approaches for landslide susceptibility mapping in Umyeonsan, Seoul, South Korea *Appl. Sci.* 7
[42] S A Salleh, A S A Abd Rahman, A N Othman and W M N Wan Mohd 2018 Comparative study of landslides susceptibility mapping methods: Multi-Criteria Decision Making (MCDM) and Artificial Neural Network (ANN) *IOP Conf. Ser. Earth Environ. Sci.* **117** 1–6
[43] L Yu, Y Cao, C Zhou, Y Wang and Z Huo 2019 Landslide susceptibility mapping combining information gain ratio and support vector machines: A case study fromWushan Segment in the Three Gorges Reservoir Area, China *Appl. Sci.* 9

[44] D H Lee, Y T Kim and S R Lee 2020 Shallow landslide susceptibility models based on artificial neural networks considering the factor selection method and various non-linear activation functions *Remote Sens.* 12

[45] W Chen, X Chen, J Peng, M Panahi and S Lee 2021 Landslide susceptibility modeling based on ANFIS with teaching-learning-based optimization and Satin bowerbird optimizer *Geosci. Front.* **12** 93–107

[46] X Yao, L G Tham and F C Dai 2008 Landslide susceptibility mapping based on Support Vector Machine: A case study on natural slopes of Hong Kong, China *Geomorphology* **101** 572–582

[47] M Marjanović, M Kovačević, B Bajat and V Voženílek 2011 Landslide susceptibility assessment using SVM machine learning algorithm *Eng. Geol.* **123** 225–234

[48] D Tien Bui, B Pradhan, O Lofman and I Revhaug 2012 Landslide susceptibility assessment in vietnam using support vector machines, decision tree, and nave bayes models *Math. Probl. Eng.*

[49] B Pradhan 2013 A comparative study on the predictive ability of the decision tree, support vector machine and neuro-fuzzy models in landslide susceptibility mapping using GIS *Comput. Geosci.* **51** 350–365

[50] S. Lee 2007 Landslide susceptibility mapping using an artificial neural network in the Gangneung are, Korea *Int. J. Remote Sens.* **21** 4763–4783

[51] N Micheletti 2014 Machine Learning Feature Selection Methods for Landslide Susceptibility Mapping *Math. Geosci.* **46** 33–57

[52] H Hong, B Pradhan, C Xu and D Tien Bui 2015 Spatial prediction of landslide hazard at the Yihuang area (China) using two-class kernel logistic regression, alternating decision tree and support vector machines *Catena* **133** 266–281

[53] B T Pham, B Pradhan, D Tien Bui, I Prakash and M B Dholakia 2016 A comparative study of different machine learning methods for landslide susceptibility assessment: A case study of Uttarakhand area (India) *Environ. Model. Softw.* **84** 240–250

[54] A Shirzadi 2017 A comparative study between popular statistical and machine learning methods for simulating volume of landslides *Catena* **157** 213–226

[55] H J Oh, P R Kadavi, C W Lee, and S Lee 2018 Evaluation of landslide susceptibility mapping by evidential belief function, logistic regression and support vector machine models *Geomatics, Nat. Hazards Risk* **9** 1053–1070

[56] O Ghorbanzadeh, T Blaschke, K Gholamnia, S R Meena, D Tiede and J Aryal 2019 Evaluation of different machine learning methods and deep-learning convolutional neural networks for landslide detection *Remote Sens.* 11

[57] B Peethambaran, R Anbalagan, D P Kanungo, A Goswami and K V Shihabudheen 2020 A comparative evaluation of supervised machine learning algorithms for township level landslide susceptibility zonation in parts of Indian Himalayas *Catena* **195** 104751

[58] B Pradhan and S Lee 2009 Landslide risk analysis using artificial neural network model focussing on different training sites *Int. J. Phys. Sci.* **4** 1–15

[59] B Pradhan and S Lee 2010 Regional landslide susceptibility analysis using back-propagation neural network model at Cameron Highland, Malaysia *Landslides* **7** 13–30

[60] M. Khamehchiyana, P. Abdolmaleki, and B. Rakei 2011 Landslide susceptibility mapping using backpropagation neural networks and logistic regression: The Sephidargole case study, Semnan, Iran*Geomech. Geoengin* **3** 237–250

[61] H. C. Quan and B. G. Lee 2012 GIS-based landslide susceptibility mapping using analytic hierarchy process and artificial neural network in Jeju (Korea) *KSCE J. Civ. Eng.* **7** 1258–1266

[62] S Pascale 2013 Landslide susceptibility mapping using artificial neural network in the urban area of Senise and San Costantino Albanese (Basilicata, Southern Italy) *Lect. Notes Comput. Sci. (including Subser. Lect. Notes Artif. Intell. Lect. Notes Bioinformatics)* **7974** 473–488

[63] M S Alkhasawneh, U K Ngah, L T Tay, N A Mat Isa and M S Al-Batah 2014 Modeling and testing landslide hazard using decision tree *J. Appl. Math.*

[64] M S Al-Batah, M S Alkhasawneh, L T Tay, U K Ngah, H Hj Lateh and N A Mat Isa 2015 Landslide Occurrence Prediction Using Trainable Cascade Forward Network and Multilayer

Perceptron *Math. Probl. Eng.*
[65] M Zare, H R Pourghasemi, M Vafakhah and B Pradhan 2013 Landslide susceptibility mapping at Vaz Watershed (Iran) using an artificial neural network model: A comparison between multilayer perceptron (MLP) and radial basic function (RBF) algorithms *Arab. J. Geosci.* **6** 2873–2888
[66] A Abbaszadeh Shahri, J Spross, F Johansson and S Larsson 2019 Landslide susceptibility hazard map in southwest Sweden using artificial neural network *Catena* 183
[67] H J Oh and S Lee 2017 Shallow landslide susceptibility modeling using the data mining models artificial neural network and boosted tree *Appl. Sci.* **7** 1–14
[68] M Conforti, S Pascale, G Robustelli and F Sdao 2014 Evaluation of prediction capability of the artificial neural networks for mapping landslide susceptibility in the Turbolo River catchment (northern Calabria, Italy) *Catena* **113** 236–250
[69] X Hu, H Zhang, H Mei, D Xiao, Y Li and M Li 2020 Landslide susceptibility mapping using the stacking ensemble machine learning method in lushui, southwest China *Appl. Sci.* 10
[70] A M Nuriah, R Ruslan and W I Wan Mohd Muhiyuddin 2018 Spatial Modeling Various Types Of Slope Failure Using Artificial Neural Network (ANN) In Pulau Pinang, Malaysia *J. Teknol. Sciences Eng. 804* **1** 135–146
[71] A R Rasyid, U Hasanuddin, N P Bhandary, and R Yatabe 2018 GIS based Landslide Susceptibility Mapping using Artificial Neural Network (ANN) Model in South Sulawesi Province, Indonesia *Int. J. Earth Sci. Eng.* **11** 107–111
[72] H Moayedi, M Mehrabi, M Mosallanezhad, A S A Rashid and B Pradhan 2019 Modification of landslide susceptibility mapping using optimized PSO-ANN technique *Eng. Comput.* **35** 967–984
[73] S Benchelha 2019 Landslide susceptibility mapping in the municipality of Oudka, northern Morocco: A comparison between logistic regression and artificial neural networks models *Int. Arch. Photogramm. Remote Sens. Spat. Inf. Sci. - ISPRS Arch.* **42** 41–49
[74] B Peethambaran, R Anbalagan, D P Kanungo, A Goswami and K V Shihabudheen 2020 A comparative evaluation of supervised machine learning algorithms for township level landslide susceptibility zonation in parts of Indian Himalayas *Catena* 195
[75] D. W. Hosmer and S. Lemeshow 2000 *Applied logistic Resgression*, vol. 2nd edition
[76] F Ren, X Wu, K Zhang and R Niu 2015 Application of wavelet analysis and a particle swarm-optimized support vector machine to predict the displacement of the Shuping landslide in the Three Gorges, China *Environ. Earth Sci.* **73** 4791–4804
[77] D Tien Bui, T A Tuan, H Klempe, B Pradhan and I Revhaug 2016 Spatial prediction models for shallow landslide hazards: a comparative assessment of the efficacy of support vector machines, artificial neural networks, kernel logistic regression, and logistic model tree *Landslides* **13** 361–378
[78] W Chen, H Chai, Z Zhao, Q Wang and H. Hong 2016 Landslide susceptibility mapping based on GIS and support vector machine models for the Qianyang County, China *Environ. Earth Sci.* **75** 1–13
[79] M Abedini, B Ghasemian, A Shirzadi and D T Bui 2019 A comparative study of support vector machine and logistic model tree classifiers for shallow landslide susceptibility modeling *Environ. Earth Sci.* **78** 1–15
[80] Y Huang and L Zhao 2018 Review on landslide susceptibility mapping using support vector machines *Catena* **165** 520–529
[81] C Xu, F Dai, X Xu and Y H. Lee 2012 GIS-based support vector machine modeling of earthquake-triggered landslide susceptibility in the Jianjiang River watershed, China *Geomorphology* **145–146** 70–80
[82] K Sharir, R Roslee, L K Ern, and N Simon 2017 Landslide Factors and susceptibility mapping on natural and artificial slopes in Kundasang, Sabah *Sains Malaysiana* **46** 1531–1540
[83] C Su, L Wang, X Wang, Z Huang, and X Zhang 2015 Mapping of rainfall-induced landslide susceptibility in Wencheng, China, using support vector machine *Nat. Hazards* **76** 1759–1779
[84] B Feizizadeh, M S Roodposhti, T Blaschke and J Aryal 2017 Comparing GIS-based support

vector machine kernel functions for landslide susceptibility mapping *Arab. J. Geosci.* 10
[85] R. T. Pack, D. G. Tarboton, and C. N. Goodwin 2016 Terrain stability mapping with SINMAP, technical description and users guide for version 1.0
[86] R. L. Baum, W. Z. Savage, and J. W. Godt 2008 TRIGRS — A Fortran Program for Transient Rainfall Infiltration and Grid-Based Regional Slope-Stability Analysis, Version 2.0 *U.S. Geol. Surv.* **78** 2008–1159
[87] D. R. Montgomery and W. E. Dietrich 1994 A physically based model for the topographic control on shallow landsliding *Water Resour. Res.* **4** 1153–1171
[88] J. Kim, K. Lee, S. Jeong, and G. Kim 2014 GIS-based prediction method of landslide susceptibility using a rainfall infiltration-groundwater flow model *Eng. Geol.* 63–78
[89] C Hammond, D Hall, S Miller and P Swetik 1992 Level I Stability Analysis (LISA) documentation for version 2.0 *Gen. Tech. Rep. - US Dep. Agric. For. Serv.*
[90] S Lee, S M Hong and H S Jung 2017 A support vector machine for landslide susceptibility mapping in Gangwon Province, Korea *Sustain.* **9** 15–19
[91] Mein R.G and C. Larson 1973 Modeling Infiltration during a Steady Rain *Water Resour. Res.* 384–394
[92] S Jeong, J Park, M Hong, J Kim and K. Lee 2019 Unified Prediction Method for the Rainfall-Induced Landslides and Debris Flows *Springer* 28.
[93] M Hong, S Jeong, and J Kim 2020 A combined method for modeling the triggering and propagation of debris flows *Landslides* **17** 805–824
[94] P Duc Long and N T Dung 2020 Susceptibility assessment of slope failures in Singapore using GIS- based prediction models
[95] Y Kim, S Jeong and J Kim 2016 Coupled infiltration model of unsaturated porous media for steady rainfall *Soils Found.* **56** 1071–1081
[96] GEO-SLOPE 2012 User's Manual. Geo-Slope International Ltd, Canada

IOP Conf. Series: Earth and Environmental Science **971** (2022) 012033 doi:10.1088/1755-1315/971/1/012033

Seismic Damage-Level Prediction Model of Cable-Stayed Bridge Using Nonlinear Time History Analysis for Intelligent Structural Health Monitoring under Earthquake Events

N H Aizon[1], Adnan[2*], and N Arjuna[3]

[1,2,3] School of Civil Engineering, Faculty of Engineering UTM, Skudai, Johor, Malaysia
[2] Engineering Seismology and Earthquake Engineering Research (eSEER), Institute of Noise and Vibration, UTM, Kuala Lumpur, Malaysia
*Corresponding author e-mail: azlanadnan@utm.my

Abstract. The current condition of a bridge should be given the highest priority to ensure its safety and its serviceability to the bridge users. Therefore, this paper explained the development of an intelligent decision-support tool in Bridge Health Monitoring system using a Neural Network model as a prediction of seismic damage-level for cable-stayed bridge. A total of eight earthquake loads scaled to various Peak Ground Acceleration (PGA) values. The input and output data which were fed into Feed Forward Artificial Neural Network (ANN) for damage level prediction model were developed based on acceleration responses from the Non-Linear Time History (NLTH) analysis of the cable-stayed bridge. A total of 16620 data were used as the input data. The damage-level categorization is based on FEMA 356. Data used for the ANN training are 70% for training, 15% for validation, and 15% for testing. The damage level prediction can greatly help bridge authority in order to maintain their bridge structure integrity by identifying and predicting the probability of damage occurring under earthquake loads.

1. Introduction

Recently, bridge structural health monitoring has been developed for condition assessment and operational management of the structures. The possible structural damage and probability of collapse under unexpected events such as earthquake is the important part in Bridge Health Monitoring. Neves *et al.* [1] presented a new approach for damage detection in bridges Gaussian Process and came up with a damage index. Meanwhile, Cheng and Dong [2] investigated the ANN method to predict the natural frequency of cable-stayed bridge that is important for dynamic behaviour of the bridge. Unfortunately, the empirical method was not enough to provide a mass training data input due to subjectivity in assigning each building to a damage state or a lack of precision in determining the ground motion affecting the region thus it is suggested to apply the results of a finite element method to predict better. Arangio and Bontempi [3] applied Bayesian neural network model to predict damage based on the dynamic behaviour of vibration, and did not focus on the seismic loading. Suryanita [4] has adopted Genetic Algorithm in the ANN model to predict the damage level due to earthquakes loading for a simply supported span of bridge. The study was able to predict the damage level with an accuracy of over 80%. However, the prediction of damage level due to earthquakes is not applicable for cable-stayed bridges due to the complexity of the structures. Therefore, Neves *et al.* [5] suggested that a wider range of possible locations and multiple damage scenarios for damage should be considered to provide a better prediction for cable-stayed bridge.

Artificial neural network (ANN) can be used as the intelligent decision maker and damage detector during bridge monitoring. Back-Propagation Neural Network is one of the Artificial Neural Network (ANN) widely used for its simple learning method and high accuracy. It is also a systematic method for training multilayer artificial neural networks. It has a solid mathematical foundation and is therefore extremely practical for most engineering problems. Therefore, ANN is expected to serve as a powerful decision support tool for problems requiring parallelism and optimization under the complex and many variables of sensor data processing. In addition, there are four possible ANN solutions for different problems, namely auto-association, pattern recognition, classification, and unique detection. There are many ANN applications worldwide in various fields of knowledge that were found in the literature [1,4,7–10]. Therefore, in this paper, the ANN model was developed to predict the damage level of the cable-stayed bridge.

2. Application of Neural Network Model of Cable-stayed Bridge Damage Assessment

The feedforward network is the network structure used in this study. The multilayer feedforward networks use the gradient descent extension based on the delta learning rule, usually known as the backpropagation rule, to provide a computationally efficient method for changing the weights in a feedforward network with differentiable activation function units to learn a training set from input-output data [6]. Various nets come under the feedforward type of nets. One of the most popular types of feedforward networks is the back-propagation network that is used to train the data input. It is crucial to select appropriate training parameters, namely, training function, learning rate, the momentum of coefficient, and stopping criteria (e.g., minimum gradient magnitude, maximum number of validation increases, maximum training time, minimum performance value, and maximum number of training epochs) for the efficiency of the network algorithm in operating the training session as shown in Table 1. The selection of parameters for the network algorithm is based on trial and error, such as whether the parameter should be changed or whether they should be gradually increased or decreased. However, some studies have specified parameters to solve their problems [5,7,11–13]. Therefore, every specific problem requires a unique algorithm that fits its needs. Steps for the neural network design process in this research include, data collection (input and output data), create the network, configure the network, initialized random weights and biases, train the network, test the network, and validate the network. Table 1 shows the Parameter of ANN training used in the training process.

Table 1. Parameter of ANN Training

Parameter	Value
Training Function	LM(Levenberg-Marquardt)
Transfer function	Log Sigmoid
Training Epoch Trial	10,000 to 50,000
Learning rate	0.1
Momentum of Coefficient	0.9
Performance Indicator	MSE (Mean Squared Error) and Regression (R)
Performance Goal	0.05^2
Minimum Performance Gradient	$1e^{-15}$
Maximum Performance Fail	10,000 – 50,000
Hidden Layer Neurons	2n+1
Data division Percentage (train, test, validate)	70, 15, 15

This study used the Neural Network Back Propagation for the prediction of the optimum weight and damage levels. This method depends solely on the selection of suitable initial weight, learning rate, momentum coefficient, networks architecture model, and activation function. The weight defines the acceleration or discontinuation of the input signals. The architecture model for this study has 2n+1 number of input neurons, one hidden layer, and an output layer. There are two models developed to investigate the best damage level prediction model for the cable-stayed bridge. The first model used 2 inputs (acceleration and time), while the second model used 11 inputs (acceleration response and time). The variation between 2 input data and 11 input data is to investigate whether one sensor and time series is enough to get an accurate damage level prediction or not. The performance of this model is presented in the next section. Figure 1 shows the architecture of the design models.

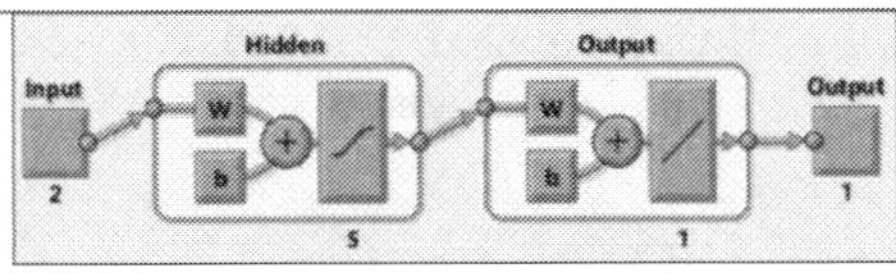

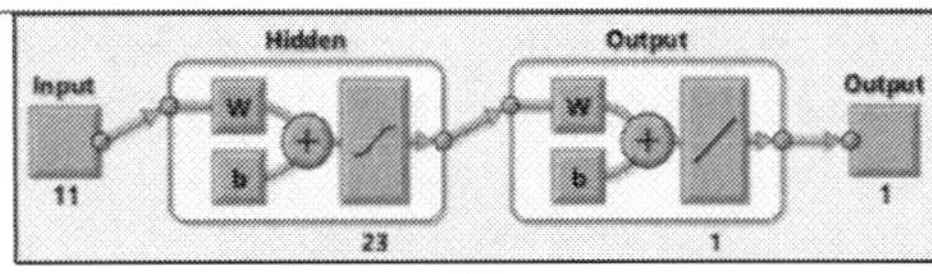

Figure 1. (a) The architecture of Model testing 1; 2 input with 1 hidden layer, (b) The architecture of Model testing 2; 11 input with 1 hidden layer

3. Input and Output Data from Finite element modelling

The bridge has two pylons (east pylon and west pylon) with 143 m height from the pile-cap, two rows (left and right) of semi-fan cable system on each pylon (84 cables in total), and 22 concrete piers along the bridge. Deck consists of a composite (in the middle span) and concrete box girder with 3.2 m in depth and 25 m in width for four lanes; the box girder rigidly connects with the two pylons and spans to each side. The bridge has a total span length of 1708 m. The cable is stayed at a distance of 12 m from each other along the bridge span with a carriage width of 3.65 m at each of the 4 lanes. The cable-stayed bridge finite element model was analysed under eight earthquake ground motions (New Zealand earthquake, Ranau Earthquake, Loma Prieta Earthquake, Northridge Earthquake, Kobe Earthquake, Chi Chi Earthquake, Padang Earthquake, and Aceh Earthquake) with PGA varied between 0.75 g, 1.0 g, and 1.5 g. The acceleration response at ten(10) locations of the main component of the bridge was observed and recorded as shown in Figure 2.

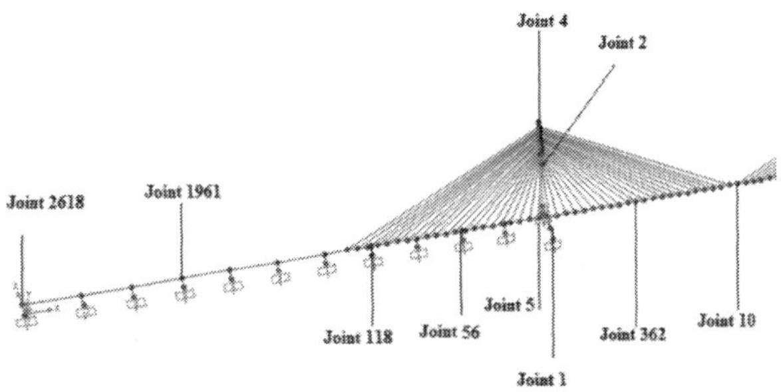

Figure 2. Location observed on cable-stayed bridge

Bridge damage level is based on FEMA 356 [14], which consists of Immediate-Occupancy (IO), Life-Safety (LS), and Collapse-Prevention (CP) from the Pacific Earthquake Engineering Research (PEER) database website. In this study, the acceleration response from the nonlinear analysis (direct integration) was used as the input to feed the network algorithm. The damage level of the hinge formation of the cable-stayed bridge was taken as the target output for the network training process. Table 2 also shows the ending time of processing finite element analysis. The ending time indicated the duration of the structure receiving the earthquake load in finite element analysis. As can be seen from Table 2, not all ground motion developed every damage level from IO to CP; the damage level that did not occur is notified as NA (not available). Out of 24 cases, five of them developed all four types of damage levels (IO to CP): three cases from Loma Prieta of 0.75 g, 1.0 g and 1.5 g and two cases from Chi-Chi earthquake of 0.75 g and 1.5 g. However, all the collected data were used to obtain the ANN prediction model for the cable-stayed bridge. In the real situation, the ground motion that excites the bridge does not necessarily cause all damage levels. Therefore, it is more appropriate to include all the acceleration responses from the finite element modelling simulation.

IOP Conf. Series: Earth and Environmental Science **971** (2022) 012033 doi:10.1088/1755-1315/971/1/012033

The inputs were the time-acceleration domain in training and testing neural network while the output layer was the damage level of the bridge categorized 0 for the safety level (S), 1 for the IO level, 2 for the LS level, and 3 for the CP level. Level B in the table indicates a safe condition before the damage starts to emerge. The input and output for the training was mounted to 16620 data, resulting from the finite element analysis of eight earthquake ground motions. The damage detection time was the first seconds of the damage levels that occurred.

Table 2. Damage Level of Bridge for ANN Training Data

NO.	TIME HISTORY	NO. OF DATA	DAMAGE DETECTION TIME (sec) B	IO	LS	CP	END
1.	Northridge 0.75g	683	2.55	25.7	NA	NA	34.1
2.	Northridge 1.0g	733	2.00	22.3	NA	NA	36.6
3.	Northridge 1.5g	538	1.95	22.3	NA	NA	26.85
4.	Kobe 0.75g	837	7.0	29.3	NA	NA	41.8
5.	Kobe 1.0g	647	6.9	17.5	NA	NA	32.3
6.	Kobe 1.5g	674	6.65	13.8	31.2	NA	33.65
7.	Loma Prieta 0.75g	46	0.15	0.7	1.65	1.95	2.25
8.	Loma Prieta 1.0g	44	0.15	0.65	1.15	1.65	2.15
9.	Loma Prieta 1.5g	44	0.15	0.60	1.15	1.65	2.15
10.	Chi-chi, Taiwan 0.75g	848	2.6	30.95	41.05	42.35	42.35
11.	Chi-chi, Taiwan 1.0g	505	1.65	17.25	NA	NA	25.2
12.	Chi-chi, Taiwan 1.5g	437	1.15	13.5	17.8	19.85	21.8
13.	Aceh, Indonesia 0.75g	1601	46.25	NA	NA	NA	80
14.	Aceh, Indonesia 1.0g	1601	9.6	NA	NA	NA	80
15.	Aceh, Indonesia 1.5g	1601	1.9	NA	NA	NA	80
16.	Padang,Indonesia 0.75g	1601	0.5	NA	NA	NA	80
17.	Padang,Indonesia 1.0g	1601	0.45	54.8	NA	NA	80
18.	Padang,Indonesia 1.5g	1061	0.2	12.55	NA	NA	80
19.	New Zealand 0.75g	278	3.4	7.85	NA	NA	13.85
20.	New Zealand 1.0g	293	3.4	7.55	NA	NA	14.6
21.	New Zealand 1.5g	222	3.25	7.5	7.9	NA	11.05
22.	Ranau, Malaysia 0.75g	287	4.35	10.6	NA	NA	14.3
23.	Ranau, Malaysia 1.0g	227	4.35	9.85	NA	NA	11.3
24.	Ranau, Malaysia 1.5g	211	4.25	9.65	NA	NA	10.5
	Total data	16,620					

Note: NA=Not Available

4. Result

The mean squared error (MSE) is the parameter used to assess the performance of the ANN model. Typically, gradient descent is used as an optimization method in the back-propagation. Therefore, to minimize the sum squared error or MSE, the training function (Levenberg Marquardt) used the gradient descent backpropagation between the output and the specified target values. With respect to the weight in the network, the method calculated the gradient loss function. Thus, the iterations during training and testing were repeated until the MSE of the output reached less than 1% of error (MSE < 0.01). Another performance parameter used was Regression (R), which is a prediction performance measured to indicate how well the model could predict the data. The regression indicator is from 0 to 1. Therefore, the higher value of R indicates a better prediction performance of the model. Both of these parameter are in unitless value.

Table 3 and Table 4 show the regression of model testing 1 and model testing 2. Meanwhile, Figure 3 shows the MSE of model testing 1 and model testing 2. For model 1, it can be concluded that the lowest MSE value achieved by 10,000 to 50,000 of the testing epoch was 0.2300, with the regression value of 0.6765. The 11 input data achieved lower error (MSE) and higher regression better than 2 input data models. The average of Regressions (R-mean) for acceleration data 11 input was recorded above 80%. The R-mean value that signified the damage values from the acceleration data was predicted to be 80% closer to the actual damage values. Therefore, the mean squared error (MSE) was more accurate in model testing 2, which had the lowest MSE value compared with the other models. This model obtained the MSE value of 0.1549 and the regression value of 0.8354.

Table 3. Regression for Model testing 1; 2 input with 1 hidden layer

epochs	Reg training	Reg testing	Reg validation	Reg mean
50000	0.67915	0.69167	0.66914	0.67965

Table 4. Regression for Model testing 2; 11 input with 1 hidden layer

epochs	Reg training	Reg testing	Reg validate	Reg mean
50,000	0.85748	0.77954	0.78947	0.83535

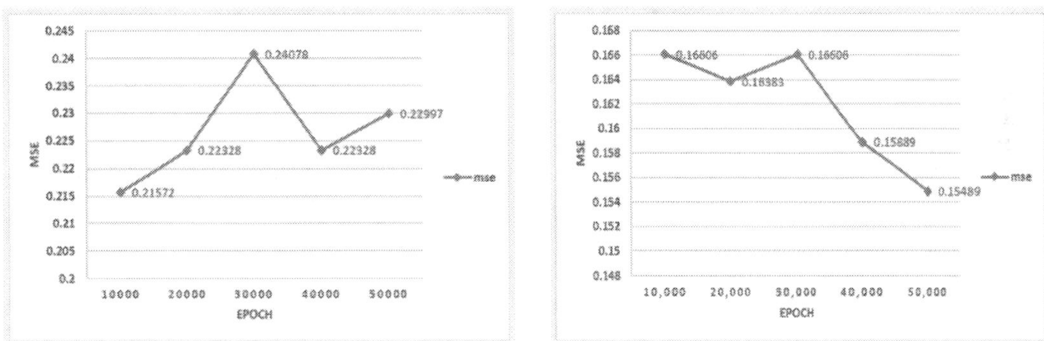

(a) (b)

Figure 3. (a) MSE for Model testing 1; 2 input with 1 hidden layer (b) MSE for Model testing 2; 11 input with 1 hidden layer

5. Discussion

The model adopted in this study could be beneficial for monitoring the complex cable-stayed structure with the aid of an earthquake sensor such as accelerometer. This sensors are installed on the cable-stayed bridge to observe the behaviour of the bridge, including deformations and damages. These sensors is also capable of feeding the local server with the essential data, which includes the acceleration via the data logger with the help of internet data transmission such as 3G antenna. More than one location data are required to catch the damaging trend of the bridge. The new data via the data logger could enable the monitoring systems to learn and improve the prediction from the previous simulation data. The simulation data for the ANN models were obtained earlier in this study from the finite element analysis in the bridge analysis software (CSI Bridge).

In applications, the acceleration data from the sensor is needed as input neurons to predict the four possible outputs of the damage level in codes. The damage level codes are 0 (zero) for the safety level (S), 1 (one) for the IO level, 2 (two) for the LS level, and 3 (three) for the CP level.

6. Conclusion

In conclusion, for model 1, the lowest MSE value achieved by 10,000 to 50,000 of the testing epoch was 0.2300, with the regression value of 0.6765. The 11 input data achieved lower error (MSE) and higher regression than the two-input data model. The model with 11 inputs was chosen as the best ANN architecture model in predicting damage to the cable-stayed structures whenever the structure was subjected to earthquake loadings. The average of regressions (R-mean) for acceleration data of 11 inputs was recorded above 80%. The R-mean value that signified the damage values from the acceleration data was predicted to be 80% closer to the actual damage values. Therefore, the mean squared error (MSE) was more accurate in model testing 2, which had the lowest MSE value compared to the other models. This model obtained the MSE value of 0.15489 and regression value of 0.8354.

7. References

[1] Neves A C, Gonz I and Leander J 2018 A New Approach to Damage Detection in Bridges Using Machine Learning *Exper. Vibra. Analy. for Civ. Struct.* Springer International Publishing AG **5** 73-84
[2] Cheng J and Dong F 2016 A simplified method for free vibration analysis of cable-stayed bridges *Int. J. Steel Struct.* **16** 151–62
[3] Arangio S and Bontempi F 2015 Structural health monitoring of a cable-stayed bridge with Bayesian neural networks *Struct. Infrastruct. Eng.* **11** 575–87
[4] Suryanita R 2014 Integrated bridge health monitoring, evaluation and alert system using neuro-genetic hybrids *Doctoral Thesis:Universiti Teknologi Malaysia* 250
[5] Neves A C, González I, Leander J and Karoumi R 2017 Structural health monitoring of bridges: a model-free ANN-based approach to damage detection *J. Civ. Struct. Heal. Monit.* **7** 689–702
[6] Rumelhart D E, Hinton G E and Williams R J 1986 Learning representations by back-propagating errors *Nature* **323** 533–6
[7] Elbayoumi M, Ramli N A and Yusof N F F M 2015 Development and comparison of regression models and feedforward backpropagation neural network models to predict seasonal indoor PM2.5–10and PM2.5concentrations in naturally ventilated schools *Atmos. Pollut. Res.* **6** 1013–23
[8] Tohidi S and Sharifi Y 2016 A new predictive model for restrained distortional buckling strength of half-through bridge girders using artificial neural network *KSCE J. Civ. Eng.* **20** 1392–403
[9] Sharifi A, Mazaheri T and Emad K Management System Designing of Bridge's Maintenance Costs *2nd International Conference on Engineering System Management and Applications, ICESMA 2010.*
[10] Suryanita R 2016 The Application of Artificial Neural Networks in Predicting Structural Response of Multistory Building in The Region of Sumatra Island *KnE Eng.* **1** 1–6
[11] Shamisi M Al, Assi A and Hejase H 2009 Using MATLAB to Develop Artificial Neural Network Models for Predicting Global Solar Radiation in Al Ain City–UAE *Intechopen.Com* 219–38
[12] Yin C, Rosendahl L and Luo Z 2003 Methods to improve prediction performance of ANN models *Simulation Modelling Practice and Theory* vol **11** pp 211–22
[13] Premalatha N and Valan Arasu A 2016 Prediction of solar radiation for solar systems by using ANN models with different back propagation algorithms *J. Appl. Res. Technol.* **14** 206–14
[14] FEMA356 2000 Prestandard and Commentary for the Seismic Rehabilitation of Buildings *Federal Emergency Management Agency, Washington, DC, Developed by the Applied Technology Council.* **1** 1–518

Acknowledgments
Financial support for this study was provided by a grant from Ministry of Higher Education of Malaysia (MOHE) for Universiti Teknologi Malaysia (UTM) under Research Grant Votes no. 14H69 and R.J130000.4J223.

IOP Conf. Series: Earth and Environmental Science **971** (2022) 012034 doi:10.1088/1755-1315/971/1/012034

Characterization of Subgrade Soil of Lowland Areas of Semarang City

A Rochim[1*] and L Fitriyana[1]

[1] Civil Engineering Department, Universitas Islam Sultan Agung, Semarang 50112, Indonesia
*Corresponding author: abdulrochim@unissula.ac.id

Abstract. The northern part of Semarang city has been suffering a lot from settlement and inundation caused by rainfall and tidal flood. The settlement is predicted to emerge from the subgrade soil containing soft soil with low bearing capacity and high compressibility. This study aims to characterize subgrade soil of lowland areas of Semarang city. In order to define the depth of soft soil, site class, soil classification, and the settlement, soil data was taken from twelve locations distributed in five sub districts of Semarang lowland areas. The minimum depth of subgrade soil is 30 m below the ground surface. To predict the settlement, subgrade soil was loaded by a surcharge load of 80 kPa, and a safety factor of 2 was taken into account. The analysis was based on N-SPT, cone resistance, index properties, and engineering properties. The result shows that the subgrade soil of Semarang lowland areas is dominated by soft clay with a depth of 5 m to 20 m with the site class from SE to SD. Soil is classified as CL and CH. The predicted settlement is from 0.321 m (lasts one year) to 1.395 m (lasts 13 years).

1. Background

From the perspective of civil engineering, there are five kinds of problematic soils: soft clay, peat soil, swelling soil, collapsible soil, and liquefied soil. It is called problematic soil since if it is used directly as a subgrade for a foundation or as a building material, it will cause problems. One of these problematic soils is soft clay with a very low bearing capacity and a high compressibility. In order to have sufficient bearing capacity, soil improvement on soft clay should be conducted.

Around 20 million ha or 10 percent of the total land area of Indonesia are estimated to be soft soil (soft clay and peat). The largest distribution of soft soil is on the islands of Kalimantan, Sumatra, Papua, and Java [1]. The distribution of soft soil on the island of Java is in the northern part of the island of Java which is close to the coastal area. Moreover, according to Satibi [2], the soft soil area in Indonesia is around 60 million ha or about 30 percent of Indonesia's land area. Some building failures resting on soft soil can be found in some cases as follows:

1) Settlement of the approaching bridge [3]. In the main bridge, there is no issue on the bridge's foundation since it is laid on hard soil, whereas the approaching bridge that does not use the same foundation system as the bridge approaching settlement. The settlement of the approaching bridge is caused by soil consolidation which results in a fracture between the border of the bridge road and the approaching bridge,
2) The 30 meter-long settlement of the toll road of Palembang - Indralaya (Palindra) [4]. This toll road was built on a saturated soft soil subgrade in 2015 and completed in October 2017. It is reported that the settlement of this road was due to the imperfect soil consolidation process.

Several studies that present soft soil improvement can be group into two kind methods:

- Using additional materials: grout modular column [5], controlled modulus column [6], platform column [7]. These methods are without changing the properties of soft soil, the foundation is rested on the soft soil while the upper structure is made as light as possible to have a small settlement,
- Using non-additional material: prefabricated vertical drains (PVD) [8], [9], [10], PVD with vacuum consolidation [11], [12]. In these methods, the soft soil is consolidated with a combination of preloading and PVD to increase the bearing capacity of the soil so that heavy structures do not cause a large settlement. Data on soft soil depth, bearing capacity, and soil settlement are required from the previous studies.

Current conditions in the field use both construction technologies, so that database of soft soil is very useful for foundation design purposes and estimated construction cost. Unfortunately, mapping soft soil in Semarang containing data based on soft soil depth, soil bearing capacity, and possible settlement has not been widely conducted. This study was conducted to propose a data database on soft soil in Semarang that can be a secondary source.

2. Research method

This method describes in detail the materials and equipment used to achieve the database of soft soil in Semarang. The laboratory data from the soil data of the city of Semarang was collected in order to analyse the depth of the soft soil, the class of the soil site, and the soil classification. Furthermore, the soil data was used to calculate the magnitude of settlement and the time required to reach the possible settlement.

Soil data used for this research was a soil data of the city of Semarang, especially in the lowland part of Semarang, which contains a thick layer of soft soil. Soft soil with a sufficiently thick depth (minimum 8 meters) is desirable for analysis. Soil data obtained for analysis were taken from field testing using CPT or SPT. This soil data were obtained from the Laboratory of Geotechnics of Engineering Faculty, UNISSULA. Several activities were conducted: 1) observation of the location where soil data were collected, 2) selection of soil parameters to be used in the analysis, 3) determination of soft soil depth, calculation of soil site class [13] and classification of the soil, 4) analysis of the magnitude and duration of soil settlement.

2.1 Research Location

The existence of soft soil on Java island is often discovered in areas along the northern coast of Java [1]. Topographically, the city of Semarang consists of coastal areas, lowlands and hills, which is grouped as the lower city and the upper city of Semarang [14]. From this definition, coastal areas and lowlands can be categorized as lower cities, whereas hills are categorized as upper cities. Based on the definition of [14], the lower and upper cities refer to their location altitude from sea level (masl). In this research, the writer defines Lowland of Semarang and Hill of Semarang based on their altitude from sea level as well. Administratively, the city of Semarang is divided into 16 subdistricts (Figure 1) and the altitude of the locations in Semarang measured from the sea level based on the sub districts is presented in Table 1[15]. Considering from the height from sea level, 11 areas of Semarang city are below 11 meters while 5 others are above 100 meters and even 3 areas are 300 meters high.

It can be seen from the map of subdistrict of Semarang and referring to the information presented in Table 1 that subdistricts whose masl below 11 meters in the northern part of Semarang are close to the North Coast of Java. Each year, these locations experience flooding, either flooding due to rain or flooding due to high tide. Based on Bappeda Semarang [14], some of these locations: Candi Baru, Jatingaleh, Gombel, Mijen, and Gunungpati are categorized as hills or upper Semarang which has heights above sea level between 90.56 meters and 348 meters. Based on this data, in this study the authors define Lowland of Semarang, as the research location, is an area has a height above sea level less than 10 meters which is potential to flooding every year, on the other hand, those with a height more than 10 meters are defined as Hilly Semarang.

IOP Conf. Series: Earth and Environmental Science **971** (2022) 012034 doi:10.1088/1755-1315/971/1/012034

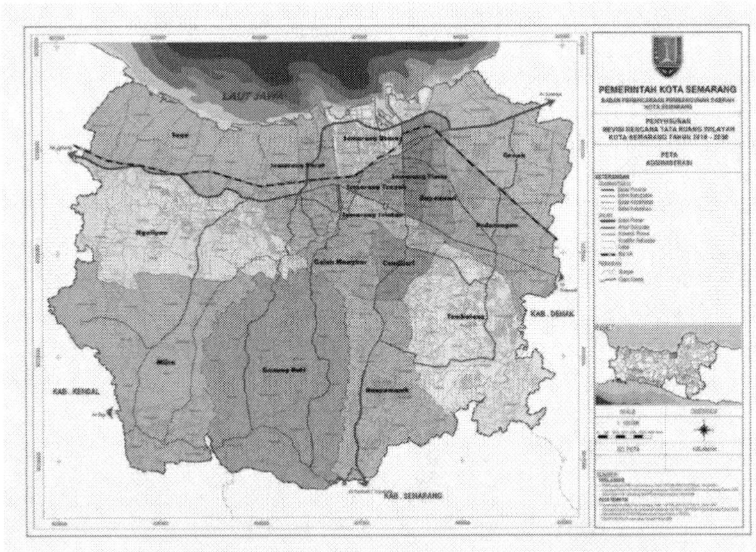

Figure 1. Map of subdistricts in Semarang [16].

Table 1. Area height above sea level (masl) corresponding to subdistricts in Semarang [15].

No.	Subdistrict	Capital of subdistrict	masl (m)	No.	Subdistrict	Capital of subdistrict	masl (m)
1	Mijen	Mijen	311	9	Genuk	Gebangsari	2
2	Gunungpati	Gunungpati	300	10	Gayamsari	Gayamsari	3.5
3	Banyumanik	Banyumanik	300	11	Semarang Timur	Rejosari	2
4	Gajah Mungkur	Gajah Mungkur	150	12	Semarang Utara	Panggung Lor	1
5	Semarang Selatan	Lamper Kidul	6	13	Semarang Tengah	Miroto	2
6	Candisari	Jatingaleh	2.5	14	Semarang Barat	Karangayu	3
7	Tembalang	Tembalang	125	15	Tugu	Tugurejo	1
8	Pedurungan	Gemah	6	16	Ngaliyan	Ngaliyan	11

masl : meter above sea level

From the data of the height as presented in Table 1, of 12 subdistricts, 11 subdistricts are defined as the Lowland of Semarang having masl lower than 11 meters. Considering the number of districts having masl lower than 11 meters, it can be concluded that the city of Semarang is dominated by Lowland of Semarang, however according to the area as shown in Table 2, the total area of Lowland of Semarang is only 145.1 km^2 amounting to 38.83 %. In terms of population density, Lowland of Semarang has a greater density as its consequence as being the centre of government, trade, industry, educational and cultural activities.

Table 2. Area and population density corresponding to subdistricts in Semarang [14].

Location	Subdistrict	Area (square km)	Population density (/square km)	Density classification [17]
1	Mijen	57.55	0.76	low
2	Gunungpati	54.11	1.148	low
3	Banyumanik	25.69	4.349	very dense
4	Gajahmungkur	9.07	6.662	very dense
5	Semarang Selatan	5.93	14.453	very dense
6	Candisari	6.54	12.317	very dense
7	Tembalang	44.2	2.62	dense
8	Pedurungan	20.72	7.453	very dense
9	Genuk	27.39	2.636	dense
10	Gayamsari	6.18	10.794	very dense
11	Semarang Timur	7.7	10.865	very dense
12	Semarang Utara	10.97	11.371	very dense
13	Semarang Tengah	6.14	12.581	very dense
14	Semarang Barat	21.74	7.146	very dense
15	Tugu	31.78	0.804	low
16	Ngaliyan	37.99	2.619	dense
Total area (square km)		373.7		

This research took place both in Lowland of Semarang and Upper Semarang. Soil data at these locations came from a secondary data. Some of the soils in the Upper Semarang locations were analysed to compare with soils in Lowland of Semarang. The results of this comparison can be used to clarify the soil characteristics in the two areas. In the Lowland of Semarang area, 12 locations were sampled to be analysed, while 7 locations were taken from the Upper Semarang area. The sample locations in the Lowland of Semarang and Upper Semarang areas are presented in Table 3 and Table 4, respectively. To ease the reading of data, each location is coded 1 to 12 for Lowland of Semarang and *a* to *g* for Upper Semarang. In the next tables, this code was used to replace the sample locations.

According to Bappeda Semarang [14], most of the soil consists of sand and clay with the land used mostly for roads, housing, buildings, yards, industrial areas, ponds, and rice fields. Unlike Lowland of Semarang, Upper Semarang consists mostly of igneous rock. Clay soil is considered as problematic soil because it has low soil bearing capacity and high compressibility. The shear strength parameter of clay is generally characterized by the small value of the shear angle in the soil but the dominance of the cohesion value. These two parameters greatly affect the soil shear strength or soil bearing capacity. Therefore, there are many studies on clay soil stabilization with the aim of obtaining higher soil shear strength parameters [18] [19]. In Mohd Yusof *etc.* [18] using Electro Osmotic Consolidation can increase soil shear strength and decrease the plasticity index and soil water content. With the addition of 0.5% nano-SiO2 and 0.3% fiber into the natural clay material of Iran, Changizi *etc.* [19] show the value of cohesion and shear angle in soil increased from 38 kPa to 73 kPa (cohesion) and 13.5° to 27° (deep soil shear angle). The high compressibility of clay soil also depends on the external load. With a very dense population density, external loads also increase, not only the burden from the housing and the people themselves but also the burden due to trading activities, transportation etc. This condition makes Lowland of Semarang soil potential to decline throughout the year. The characterization of subgrade in the Lowland of Semarang is explained in the following subsections with the aim of recognizing more detail the character of the soil containing clay as an addition to the database.

Table 3. Research locations in the Lowland of Semarang.

Location	Research locations	Subdistricts
1	Retention Ponds & Pump Houses Kampung Bahari, Tambak Lorok	Semarang Utara
2	RSIGM Sultan Agung, Kaligawe Km 4 Semarang	Gayamsari
3	Polairud office, Kalibaru Barat Pelabuhan Tanjung Semarang	Semarang Utara
4	Pasar Kubro bridge, Arteri Yos Sudarso Semarang	Gayamsari
5	Shared lecture building (GKB) UNISSULA, Kaligawe	Gayamsari
6	Multi Function Tower, Terboyo Raya area Blok ABC	Semarang Barat
7	DEPO 2, Jl. Arteri Yos Sudarso Semarang	Gayamsari
8	Tanjung Laut Puri Anjasmoro, Bandara Internasional Ahmad Yani	Genuk
9	Madukoro Raya Semarang, Fly over Ahmad Yani #1	Semarang Barat
10	BPBD office building, Jl. Imam Bonjol Semarang	Semarang Utara
11	Customs Directorate Building, Ahmad Yani Semarang	Semarang Tengah
12	Kalibanger, Kemijen Semarang Timur	Semarang Barat

Table 4. Research locations in the Upper Semarang.

Location	Research locations in Upper Semarang	Subdistricts
a	New building of Faculty of Social, 3rd Campus UIN Walisongo	Ngaliyan
b	New building of Faculty of Science, 3rd Campus UIN Walisongo	Ngaliyan
c	New building of Faculty of Tarbiyah, 3rd Campus UIN Walisongo	Ngaliyan
d	New building of Library, 3rd Campus UIN Walisongo	Ngaliyan
e	Office building of DISHUBKOMINFO, Tambak Aji Raya No. 5	Ngaliyan
f	Sport center of Jatidiri, Gajahmungkur Semarang	Gajahmungkur
g	Gandhi Memorial Intercontinental School Grha Candi Golf	Candisari

3.Results and discussion

This section explains locations in the Lowland of Semarang where the soil data was taken in this study. Several locations from the hilly area of Semarang were also taken for comparison. Furthermore, from each location, the depth of soft soil with the *N-SPT* equals 4, which was the author's problem limitation. Then the analysis of the soil site class in a location was based on the N-SPT value referring to SNI [13]. The analysis results were used to determine the consistency of the soil, whether it is classified as soft soil or not. The results of this analysis are also useful for determining the magnitude of the amplification of earthquake acceleration at ground level. Subsequent technical studies were conducted on several soil parameters such as soil classification, soil shear strength parameters c and ϕ, and moisture content and plasticity values.

3.1 Soft Soil Depth

The data analysed was obtained from secondary data from boring, SPT and / or CPT testing in the area of Semarang, both Lowland of Semarang and Upper Semarang. In determining the depth of soft soil, the consistency of the soil used from very soft clay to soft clay with an N-SPT equals to 4. For site class, soft soil is defined if the average *N-SPT* is less than 15 for 30 m top soil [13]. For the purposes of determining the depth of soft soil, data from the CPT test cone resistance value (q_c) must be converted first to *N-SPT* using the empirical formula q_c is 4 *N-SPT* [20]. Table 5 and Table 6 present the results of the soft soil depth analysis for the Lowland of Semarang and Upper Semarang areas respectively. Data for 4 location points in the Lowland of Semarang area used cone resistance data, while the others used *N-SPT* data. The depth of soft soil in the Lowland of Semarang area based on the *N-SPT* value or q_c

ranges from 2 meters to 20 meters (Table 5), with the greatest depths being in the North Semarang and Genuk districts. When viewed on the map in Figure 1, these two sub-districts are directly adjacent to the North Coast of Java and have a masl of 1-2 meters (Table 1). This data confirms that the areas adjacent to the North Coast of Java have thicker soft soil compared to other areas. The existence of soft soil becomes a problem when it becomes the foundation for building construction. Inaccuracy in characterizing soft soil may result in the construction not functioning properly or further resulting in building failure. In contrast to the condition of the Lowland of Semarang, the subgrade conditions of Upper Semarang as presented in Table 6 at a depth of 5 meters has reached soil with a very stiff consistency even some of the consistency is hard soil.

Based on this analysis, it can be seen that soft soil is not found in the Upper Semarang area, even if there is this layer only a few meters (1 to 2 m) below the soil surface and it could be caused by weather factors such as the presence of a layer of humus on the soil surface. The existence of a hard soil layer that is not far from the ground level makes the design and cost of foundation construction are less complicated and less expensive when compared to the foundation construction in the Lowland of Semarang which has a big problem due to the presence of a very thick water-saturated soft soil layer.

Table 5. Depth of soft soil and site class in the Lowland of Semarang.

Location	Depth (m)	*N-SPT*	q_c (kg/cm^2)	Average *N-SPT* up to the depth of 30 m	Site class
1	20	4		4	Soft soil
2	15	4		9	Soft soil
3	15	4		9	Soft soil
4	15	3		8	Soft soil
5	5	5		10	Soft soil
6	13		10	*	-
7	10	5		12	Soft soil
8	10		7	*	-
9	10	4		5	Soft soil
10	5	3		19	Stiff soil
11	2	7		15	Stiff soil
12	10	4		24	Stiff soil

Table 6. Depth of soil and site class in the Upper Semarang.

Location	Depth (m)	*N-SPT*	q_c (kg/cm^2)	Average *N-SPT* up to the depth of 30 m	Site class
a	5	29		52	Hard soil
b	5	25		46	Stiff soil
c	5	31		46	Stiff soil
d	5	41		55	Hard soil
e	2		75		-
f	5	35		89	Hard soil
g	5	21		43	Stiff soil

3.2 Site Class

In a building construction plan it is necessary to know the classification or site class of the subgrade where the construction rests. Determining the class of site is important in addition to knowing the thickness of the compressible subgrade as well as determining the proper soil parameters as input to a design. To determine the class of the soil site, based on SNI-1726-2019 [13], a minimum soil depth of 30 meters measured from the ground surface should be tested by SPT or CPT tests. Likewise, boring tests are required as supporting data to provide a more definite picture of the properties of the soil. According to SNI-1726-2019 [13], this site class is divided into five types of soil, namely SA (hard rock), SB (rock), SC (very dense soil and soft rock), SD (soil moderate), and SE (soft soil), where the determination of the class of the soil site is based on the N-SPT value or the value of the shear wave propagation velocity (V_s). The value of the shear wave propagation velocity (V_s) or the maximum shear modulus (G_{max}) is usually correlated with the shear strength obtained from laboratory tests or from field tests such as the *N-SPT* and q_c values. Based on the data presented in Table 5 and Table 6 the average value of *N-SPT* is calculated from the *N-SPT* value up to a depth of 30 meters. Some location points cannot determine the class of the soil site because the field test does not reach a depth of 30 meters. Table 5 presents the data from the calculation of soil site classes in the Lowland of Semarang which results in the classification of soft soil (SE, average *N-SPT* less than 15) and medium soil (SD, $15 \leq N\text{-}SPT \leq 50$).

The subdistricts of Genuk, Gayamsari, and North Semarang are found to have a lot of subgrade classified as soft soil, although there are also stiff soils in these subdistricts. Table 5 presents site class corresponding to *N-SPT* value, with stiff soil has an average *N-SPT* value ranges from 15 to 50 so that the lowest soil site class will be used. In the Upper Semarang city area as shown in Table 6, it is not found soft soils (SE) but stiff soils (SD) and very dense soils and soft rocks (SC, N-SPT average higher than 15). Stiff soil site class here *N-SPT* averages close to 50 or close to the SC site class. This analysis result is consistent with data from Badan Perencanaan Pembangunan Daerah [14] which presents the geological structure of Upper Semarang is mostly composed of igneous rock.

3.3 Soil characteristics based on undrained soil parameters c and ϕ and soil gradation

The results of laboratory tests of the parameter values of the shear strength of soil: cohesion (c) and the angle of shear in the soil (ϕ) are presented in Table 7. Some locations do not have soil property data since CPT tests were not equipped with boring tests. In order to focus the study on soft soil only, the values of c and ϕ are presented only in the location points in the Lowland of Semarang. Soil property data in Table 7 is data on depth according to Table 5. Based on the analysis, it is known that the soil in the study location is dominated by fine grained soils (clay and silt soils) with a percentage higher than 50 %. Table 7 presents the value of c with almost all of the c values are less than 25 kPa, some are slightly higher than 25 kPa. Based on this c value data, it can be concluded that the soil in the study location at that depth (depth in Table 5) is categorized as soft soil (c < 25 kPa). Corresponding to the existing ϕ value (derived from sand), almost all of the soil is categorized as loose soil ($\phi < 30°$), only a few have values slightly greater than 30°. From the study of the shear strength parameters c and ϕ, the results are consistent with the technical study of the *N-SPT* value that the soil in the Lowland of Semarang is dominated by soft clay soil.

3.4 Soil characteristics based on water content and soil plasticity

The results of laboratory tests for moisture content and soil plasticity are presented in Table 7. Several locations where CPT tests were not equipped with boring tests so that there was no data on moisture content and soil plasticity. Almost all the water content (w) in the study location is greater than 30 percent (w > 30 percent) which indicates that the soil is soft clay soil, there are even soils that have moisture content higher than 100 percent (it may a peat). Only two location points have moisture content less than 30 percent (16% to 23%). From the review of the plasticity values of *LL* (liquid limit), *PL* (plastic limit), and *PI* (plasticity index), the soil in Lowland of Semarang is dominated by high plasticity ($IP > 17$) and medium ($7 < IP < 17$) soils, only a few are low plasticity ($IP < 7$). High soil plasticity can also be seen from its *LL* value which *LL* > 50% is categorized as high plasticity soil. From the available plasticity data, it can be concluded that the Lowland of Semarang soil is dominated by soft clay soil. Plasticity only exists in fine-grained soils (clay and silt), not coarse-grained soils (sand and gravel, *IP* = 0). From the study of the value of water content and soil plasticity, the results are consistent with the technical study of N-SPT and shear strength parameters *c* and ϕ, that the soil in the Lowland of Semarang is dominated by water-saturated soft clay.

Table 7. Soil properties and soil classification in the Lowland of Semarang.

Location	*c* (kPa)	φ (°)	Water content (%)	*LL*	*PL*	*PI*	USCS Soil classification
1	12	22.86	78.06	38	26.32	11.68	CL
2	5	23.99	71.51	73.0	41.57	31.43	MH
3	10	21.72	132.95	51	43.04	7.96	MH
4	7	29.2	71.27	25.77	20	5.77	CL-ML
5	0	25.08	23	39	38.37	0.63	SM
6	0	0	-	-	-	-	-
7	10	32.31	58.81	57.0	25.23	31.77	CH
8	0	0	-	-	-	-	-
9	24.1	11.91	56.39	60	39.25	20.75	MH
10	31.3	11.46	62.24	70.64	36.69	33.95	MH
11	27.8	31.34	16.76	non plastic soil			-
12	3.2	27.27	69.96	27	18.68	8.32	CL-ML

3.5 Soil characteristics based on soil classification

From the soil properties of the laboratory test results presented in Table 7, the soil classification is listed in the last column. Soil in the research location in the Lowland of Semarang is covered by clay, silt, and sand with classification as follows: high plasticity (CH) clays, low or medium plasticity (CL) clays, high plasticity silts (MH), and low or medium plasticity silts (ML). Only two locations are dominated by sand, namely silty sand (SM). It can be seen from the results of soil classification that the soil in Lowland of Semarang is dominated by fine grained soils with plasticity from low to high. The higher the plasticity, soil can have low soil bearing capacity and causes land settlement. From the study of soil classification, the results are consistent with the technical study of N-SPT, shear strength parameters *c* and ϕ, moisture content, and plasticity, that the soil in Lowland of Semarang is dominated by saturated soft fine grained soils.

3.6 Determination of preloading load

The magnitude of soil settlement in subgrade was estimated using preloading. The determination of the preloading load takes into account: a) the traffic load, b) the safety factor taken is equal to 2. The bearing capacity of water-saturated soft subgrade is defined as the traffic load multiplied by the safety factor. After the soil is given a preloading load, excess pore water pressure (ΔU) emerges. By the time after the ΔU dissipates, the soil experiences a change in soil volume or soil settlement. This whole process starts from ΔU dissipation until soil settlement is called the consolidation process which results in an increase in soil bearing capacity. In terms of soil bearing capacity, this increases due to the increase in soil shear strength parameter (gain strength). This new value of c is used to calculate the bearing capacity of the soil (ultimate stress), where the ultimate stress is equal to 2 times the traffic load. The traffic load used here is the "T" truck load, the load that is applied to road bridges which refers to [21]. This "T" truck load uses the most critical condition modelling with the largest axle load equal to 112.5 kN. The formula for calculating the ultimate stress is as q_{ult} equals to c (N_c). From the calculation, total load is 215.53 kPa.

In the process, preloading is not applied all at once but is carried out gradually (stage construction) considering that if it is applied in one time, the preloading soil will sink into the subgrade. Table 8 presents the undrained parameter values of soil shear strength c and ϕ average at each location point. Some locations do not have soil shear strength parameters since the test is not conducted. For the calculation of the preloading load, the c value of the initial existing soil (called c_0) is used which is taken from the average value of all points of the study location. To avoid the average value that does not represent the most values, the value of c_0 at location points 10 is omitted because the values are too low from the values at other locations. From the calculation results, the average value of c_0 and ϕ is 14.60 kPa and 25.20° respectively. By providing a preloading load ($\Delta\sigma$) of 80 kPa (equivalent to 5 meters of embankment), the c_1 (gain strength value after consolidation) is 33.11 kPa. The gain strength of c is calculated from the empirical formula $c_1 = c_0 + 0.22\ \Delta\sigma$. According to Begemann [22] the c value of 33.11 kPa is categorized as medium stiff soil, and its shear failure can be categorized as local shear failure. Furthermore, the average value of c_1 and ϕ is used to calculate the ultimate shear value of the subgrade. Corresponding to $\phi = 25.20°$, Terzaghi's bearing capacity factor N_c equals to 15 and then the magnitude of the ultimate shear is 496.65 kPa. The calculated total traffic load T is 215.53 kPa, and with a used safety factor of 2, then the calculated load is 431.06 kPa. Since the ultimate shear is 496.65 kPa more than load 431.06 kPa, the preloading load of 80 kPa has been satisfied.

Table 8. Average undrained shear strength parameter c and ϕ

Location	Average c_0 (kPa)	Average φ (°)
1	13.74	28.53
2	14	18
3	13.92	21.34
4	7.5	26.26
5	18.62	20.05
6	-	-
7	11.84	28.19
8	-	-
9	17.43	25.06
10	0,00	12.17
11	25.5	31.66
12	9.05	27.71

3.7 Estimation of Soil Settlement

From the preloading load calculation, it is determined that the load applied to the subgrade is 80 kPa. In the calculation of soil settlement, with soft soil conditions, the soil is assumed to be normally consolidated clay (OCR = 1). The settlement analysed here is only a primary consolidation settlement using 1D-Terzhagi's formula, as in equation (1). Soil volume weight (γ) used is γ average and the soil compression index value C_c is used is the average value of soil C_c with soft soil consistency. Since the compression index value of C_c is not available for all locations, for this the C_c used is the mean C_c value from Sophian [23] which ranges from 0.166 to 0.772, which is 0.444. In Sophian [23] also gave the C_v value in the coastal city of Semarang in the range of 0.00022 to 0.002 cm^2/second. Table 9 presents the results of the analysis of settlement at 14 locations in the Lowland of Semarang which have soft clay depth from 5 m to 20 m, where the magnitude of settlement ranges from 0.321 to 1.395 m.

$$S_c = \frac{H.c_c}{1+e_0} log \frac{p_0' + \Delta p}{p_0'} \quad (1)$$

Table 9. Soil settlement magnitude in the Lowland of Semarang

Location	Research locations	Settlement* (m)
1	Retention Ponds & Pump Houses Kampung Bahari, Tambak Lorok	1.395
2	RSIGM Sultan Agung, Kaligawe Km 4 Semarang	1.174
3	Polairud office, Kalibaru Barat Pelabuhan Tanjung Semarang	0.833
4	Pasar Kubro bridge, Arteri Yos Sudarso Semarang	1.097
5	Shared lecture building (GKB) UNISSULA, Kaligawe	0.321
6	Multi Function Tower, Terboyo Raya area Blok ABC	-
7	DEPO 2, Arteri Yos Sudarso Semarang	0.914
8	Tanjung Laut Puri Anjasmoro, Bandara Internasional Ahmad Yani	-
9	Madukoro Raya Semarang, Fly over Ahmad Yani #1	1.319
10	BPBD office building, Imam Bonjol Semarang	0.881
11	Customs Directorate Building, Ahmad Yani Semarang	0.582
12	Kalibanger, Kemijen Semarang Timur	0.430

* The settlement is referred to primary consolidation settlement and under the loading of 80 kN

4.Conclusion

The Lowland of Semarang city has the potential to have large soil settlement due to subgrade soil containing saturated soft soil. The subgrade soil of the Lowland of Semarang is dominated by very soft clay and soft clay with the depth of 5 m to 20 m with the site class from SE to SD. The clay soils are defined as CL and CH (with $OCR = 1$ and $C_v = 0.002$ cm^2/second). The predicted soil settlement ranges from 0.321 m (lasts one year) to 1.395 m (lasts 13 years).

5. References

[1] Departemen Permukimam dan Prasarana Wilayah 2002 Panduan Geoteknik 1: Timbunan Jalan pada Tanah Lunak – Proses Pembentukan dan Sifat-sifat Dasar Tanah Lunak WSP International Jakarta

[2] Satibi S 2009 *Numerical Analysis and Design Criteria of Embankments on Floating Piles* PhD Thesis (Institut fur Geotechnik der Universitat Stuttgart)

[3] Azis A A 2018 *Ketika Tanah Lunak Tak Ada Flyover yang Tak Retak* Retrieved on 3 April 2020 from https://kaltimkece.id/warta/terkini/ketika-tanah-lunak-tak-ada-flyover-yang-tak-retak

[4] Ariyanti F 2017 *Tanah Tol Palembang-Indralaya Ambles, Ini Penjelasan Menteri PUPR* Retrieved on 3 April 2020 from https://www.liputan6.com/bisnis/read/2996348/tanah-tol-palembang-indralaya-ambles-ini-penjelasan-menteri-pupr

[5] Afif A K, Ahmad M R and Rekardi H 2019 Studi Kasus Kolom Grout Modular (KGM) terhadap Penurunan dan Stabilitas Timbunan Oprit Jembatan Di Atas Tanah Lunak *Proc. 23rd Annual Indonesian National Conference on Geotechnical Engineering* (Jakarta)

[6] Menard 2019 Karakteristik Transfer Beban Dan Pola Penurunan Terhadap Waktu Pada Controlled Modulus Column Menggunakan Analisis Numerik Tiga Dimensi *Proc. 23rd Annual Indonesian National Conference on Geotechnical Engineering* (Jakarta)

[7] Pratama A W S, Rekardi H, and Karlinasari R 2019 Studi Kasus Soil Treatment Minipile Dengan Platform Sebagai Zona Transisi Vacuum Consolidation Method Dengan Area Box Jalan Pada Jalan Tol Pemalang Batang *Proc. 23rd Annual Indonesian National Conference on Geotechnical Engineering* (Jakarta)
[8] Mission J L, Kim H J and Won M S 2012 Ground Improvement Optimization with Prefabricated Vertical Drains (PVD) and Surcharge Preloading *Proc. The 2012 World Congress on Advances in Civil, Environmental, and Materials Research (ACEM '12)* (Seoul)
[9] Long P V, Bergado D T, Nguyen L V and Balasubramaniam A 2013 Design and Perfomance of Soft Ground Improvement Using PVD with and without Vacuum Consolidation. *Geotechnical Engineering Journal of the SEAGS and AGSSEA* **44** 36-51
[10] Dong P H 2018 Influence of Vertical Drains on Improving Dredged Mud by Vacuum Consolidation Method *Journal of Science and Technology in Civil Engineering* **12** pp. 63-72
[11] Long P. V, Nguyen L V, Bergado D T and Balasubramaniam A 2015 Performance of PVD Improved Soft Ground Using Vacuum Consolidation Methods with and without airtight Membrane *Journal of Geotectiles and Geomembranes* **43** 473-483
[12] Seah T H, Kim T B and Nguyen T D 2016 Ground Improvement Via Vacuum Consolidation Method in Vietnam *Geotechnical Engineering Journal of the SEAGS and AGSSEA* **47** 1-9
[13] SNI-1726-2019 Tata Cara Perencanaan Ketahanan Gempa untuk Struktur Bangunan Gedung dan Non Gedung Badan Standarisasi Nasional (BSN)
[14] Badan Perencanaan Pembangunan Daerah (Bappeda) Kota Semarang 2005 *Rencana Pembangunan Jangka Panjang Daerah (RPJPD) Kota Semarang Tahun 2005-2025* Bappeda Kota Semarang
[15] Open Data Provinsi Jawa Tengah 2018 *Tinggi Wilayah di Atas Permukaan Laut (DPL) Menurut Kecamatan di Semarang 2016* Retrieved on May 5, 2021 from http://data.jatengprov.go.id/id/dataset/tinggi-wilayah-di-atas-permukaan-laut-dpl-menurut-kecamatan-di-semarang-2016/resource/50b40914-b58f-4a40-b1e7-5fbb58d5125b
[16] Badan Perencanaan Pembangunan Daerah (Bappeda) Kota Semarang 2008 *Citraudara Quickbird Kota Semarang Kedalaman 100 meter* Bappeda Kota Semarang
[17] SNI 03-1733-2004. *Tata Cara Perencanaan Lingkungan Perumahan di Perkotaan* Badan Standarisasi Nasional (BSN)
[18] Mohd Yusof K N, Ahmad F, Abdullah M M A B and Mohd Tahir M F 2014 Effects of Electro Osmotic Consolidation in South West of Johor: Small Laboratory Scale *Materials Science Forum* **803** 255–264
[19] Changizi F and Haddad A 2015 Strength properties of soft clay treated with mixture of nano-SiO2 and recycled polyester fiber *Journal of Rock Mechanics and Geotechnical Engineering* **7** 367-378
[20] Tanuwijaya E, Kawanda A and Hendy Wijaya H 2019 Studi korelasi nilai tahanan konus sondir terhadap parameter tanah pada proyek di Jakarta Barat *Jurnal Mitra Teknik Sipil* **2** 169-176
[21] RSNI T-02-2005 *Standar Pembebanan untuk Jembatan* Badan Standarisasi Nasional (BSN)
[22] Begemann HKSPh 1965 The maximum pulling force on a single tension pile calculated on the basis of results of the adhesion jacket cone *Proceedings of the 6th international conference on soil mechanics and foundation engineering* (Montreal), 229-233
[23] Sophian R I 2010 Penurunan muka tanah di kota-kota besar pesisir pantai utara Jawa (Studi kasus kota Semarang) *Bulletin of Scientific Contribution* **8** 41-60

Acknowledgments
We would like to thank Faculty of Engineering and Institute Research and Community Service Universitas Islam Sultan Agung for funding this research through the internal research funding 2020-2021.

AUTHOR INDEX